タンクヘッド

TANKHEAD®

MECHANICAL ENCYCLOPEDIA ARTBOOK

タンクヘッド

TANKHEAD®

MECHANICAL ENCYCLOPEDIA ARTBOOK

CREATED BY

EMERSON TUNG

STORY BY

TIM POPELIER
EMERSON TUNG

MECHANICAL DESIGN AND ART BY

EMERSON TUNG

WRITTEN BY

TIM POPELIER

COPY EDITORS

MAX GABRIELE
M. CHANDLER

ENVIRONMENT LINE ART BY

RYAN BARRY/9MONKEYS

ENVIRONMENT COLORS BY

EMERSON TUNG

CULTURAL CONSULTANT

ANDY CHAISIRI

BOOK LAYOUT & DESIGN

MATT MOYLAN

SPECIAL THANKS:

JANDRUFF
FELIX LEYENDECKER
GIACOMO "LAZY FORGER" PANTALONE
ALBA "DASBILLIGEALIEN" VOSS
IRYS CHING

CONTENTS

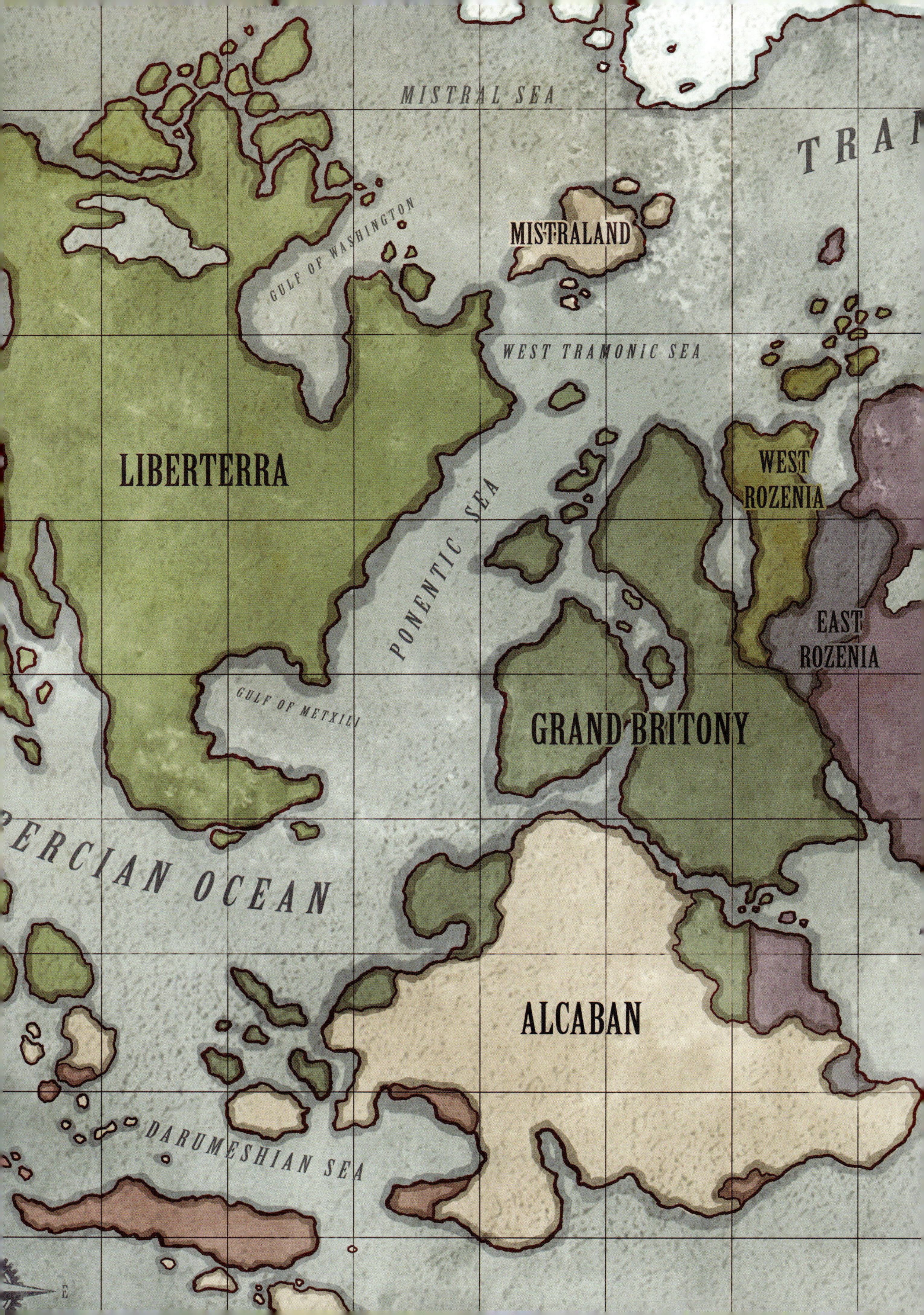
MISTRAL SEA
TRA
MISTRALAND
GULF OF WASHINGTON
WEST TRAMONIC SEA
LIBERTERRA
WEST
ROZENIA
PONENTIC SEA
EAST
ROZENIA
GULF OF METXILI
GRAND BRITONY
ERCIAN OCEAN
ALCABAN
DARUMESHIAN SEA
E

INTRODUCTION

The story of TANKHEAD centers around a global conflict called the Neverending War, also known as the Neverwar, waged between three great powers on the world of Earde: the Britonica Alliance, the Greater Kyzerian Empire, and the Yamaichi Dominion. In the early years of the conflict, the Alliance and Empire fought until arriving at a stalemate with the introduction of trench warfare.

All that changed with the secretive Zeehond Expedition, when the Kyzerians unearthed ancient artifacts on the Britonican island of Isle à Von. They used the knowledge gained from studying these artifacts to create devastating weapons known as tankheads, walking armored vehicles that were able to easily overrun trenches and break through the enemy's defenses on the frontlines.

The waning fires of the Neverwar were rekindled. They would burn on for nearly 50 years.

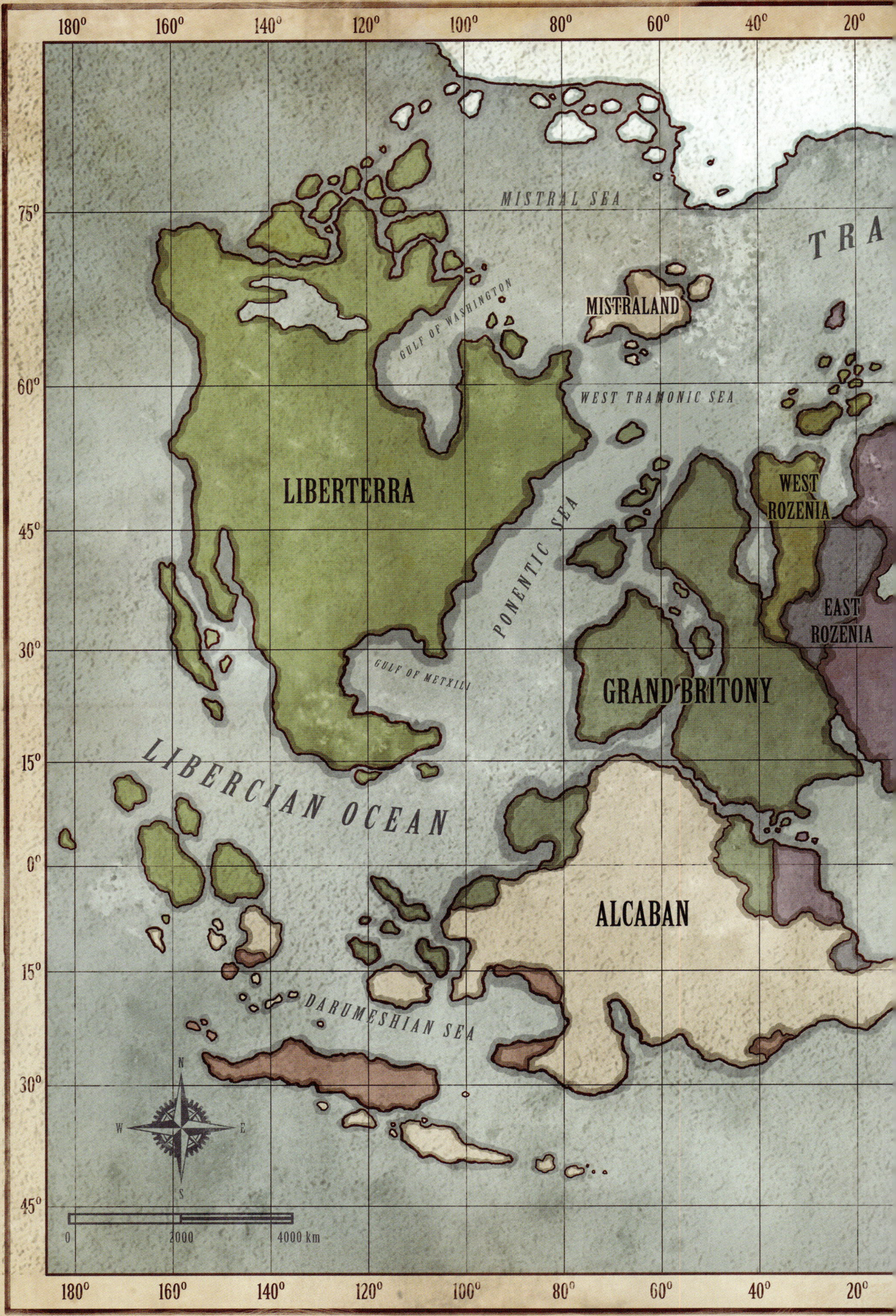

180° 160° 140° 120° 100° 80° 60° 40° 20°
MISTRAL SEA
TRA
MISTRALAND
GULF OF WASHINGTON
WEST TRAMONIC SEA
LIBERTERRA
PONENTIC SEA
WEST ROZENIA
EAST ROZENIA
GULF OF METXILI
GRAND BRITONY
LIBERCIAN OCEAN
ALCABAN
DARUMESHIAN SEA
N
W
E
S
0
2000
4000 km
75°
60°
45°
30°
15°
0°
15°
30°
45°
180° 160° 140° 120° 100° 80° 60° 40° 20°

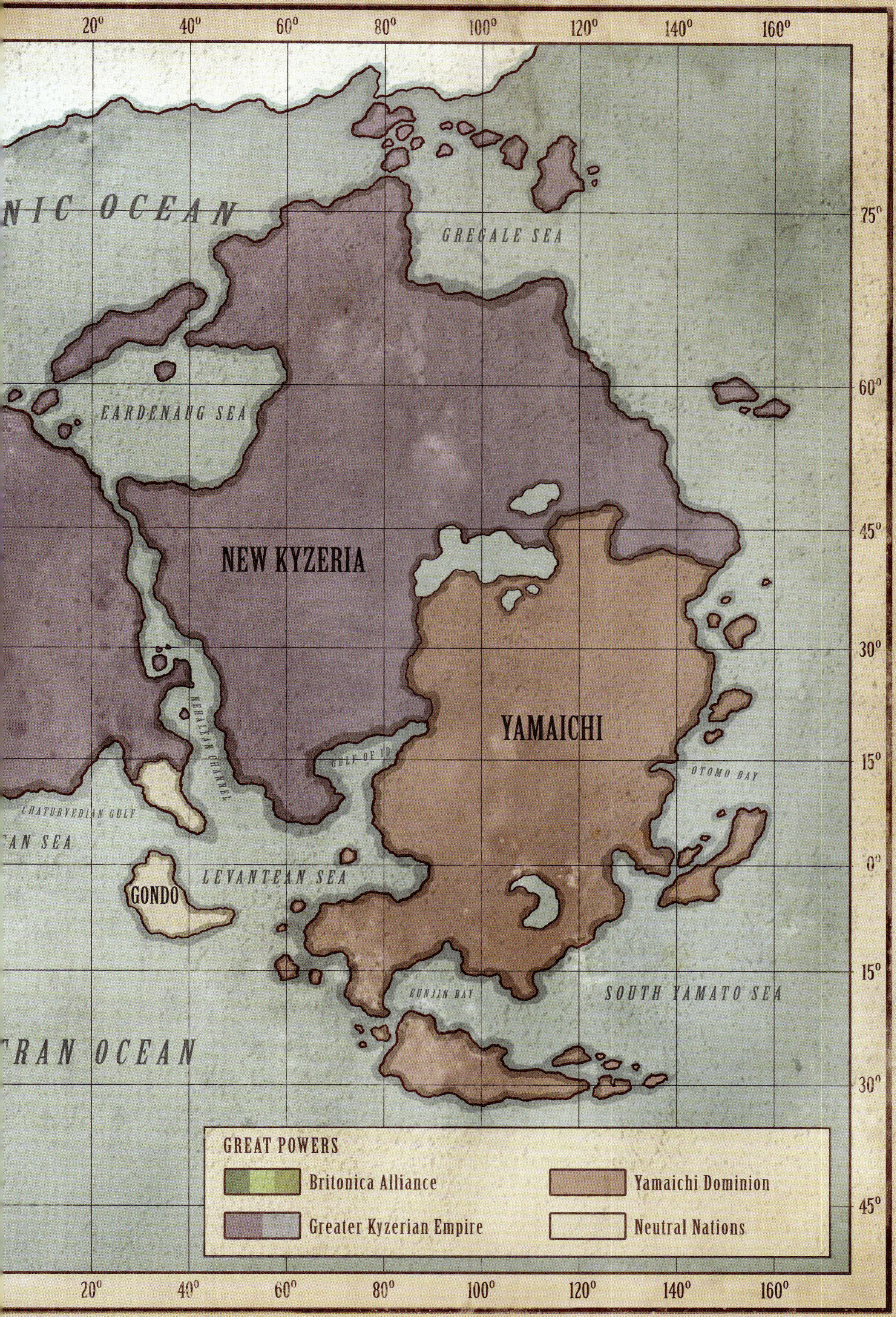

The world map of Earde, displaying the Great Powers and their spheres of influence circa WC025.

BRITONICA ALLIANCE

AFFILIATED NATIONS

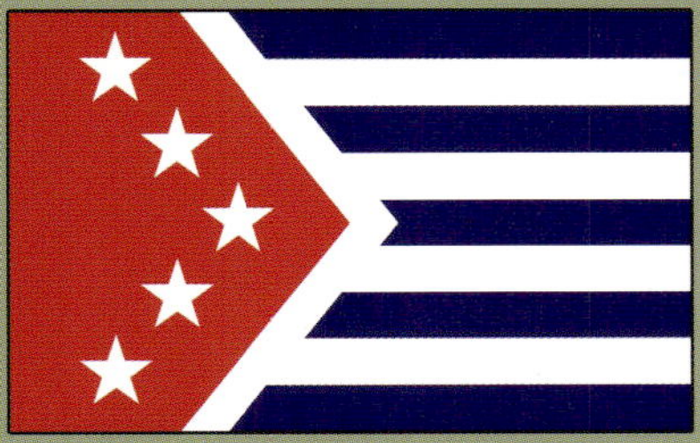

Liberterra

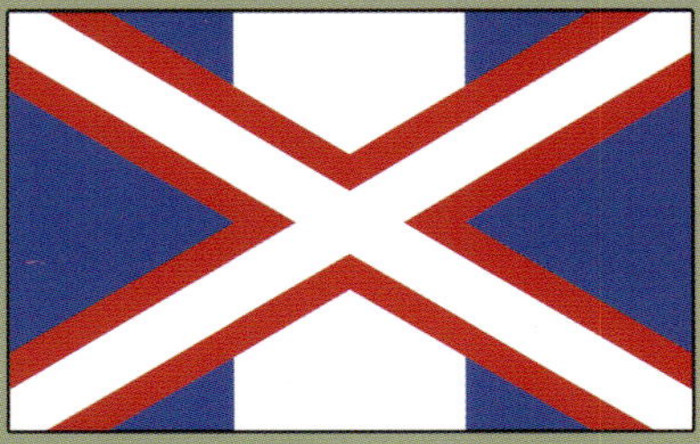

Grand Britony

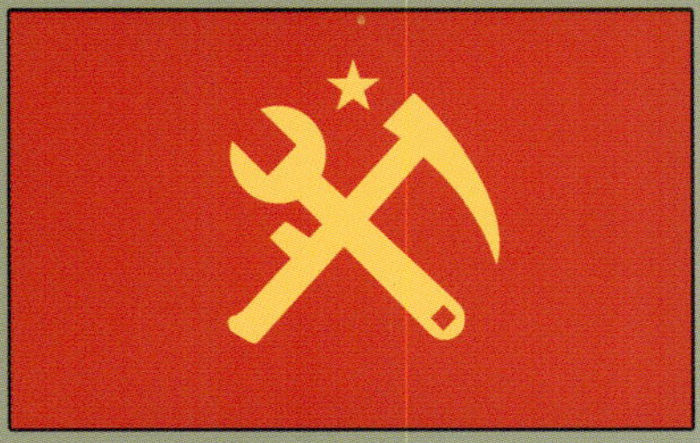

West Rozenia

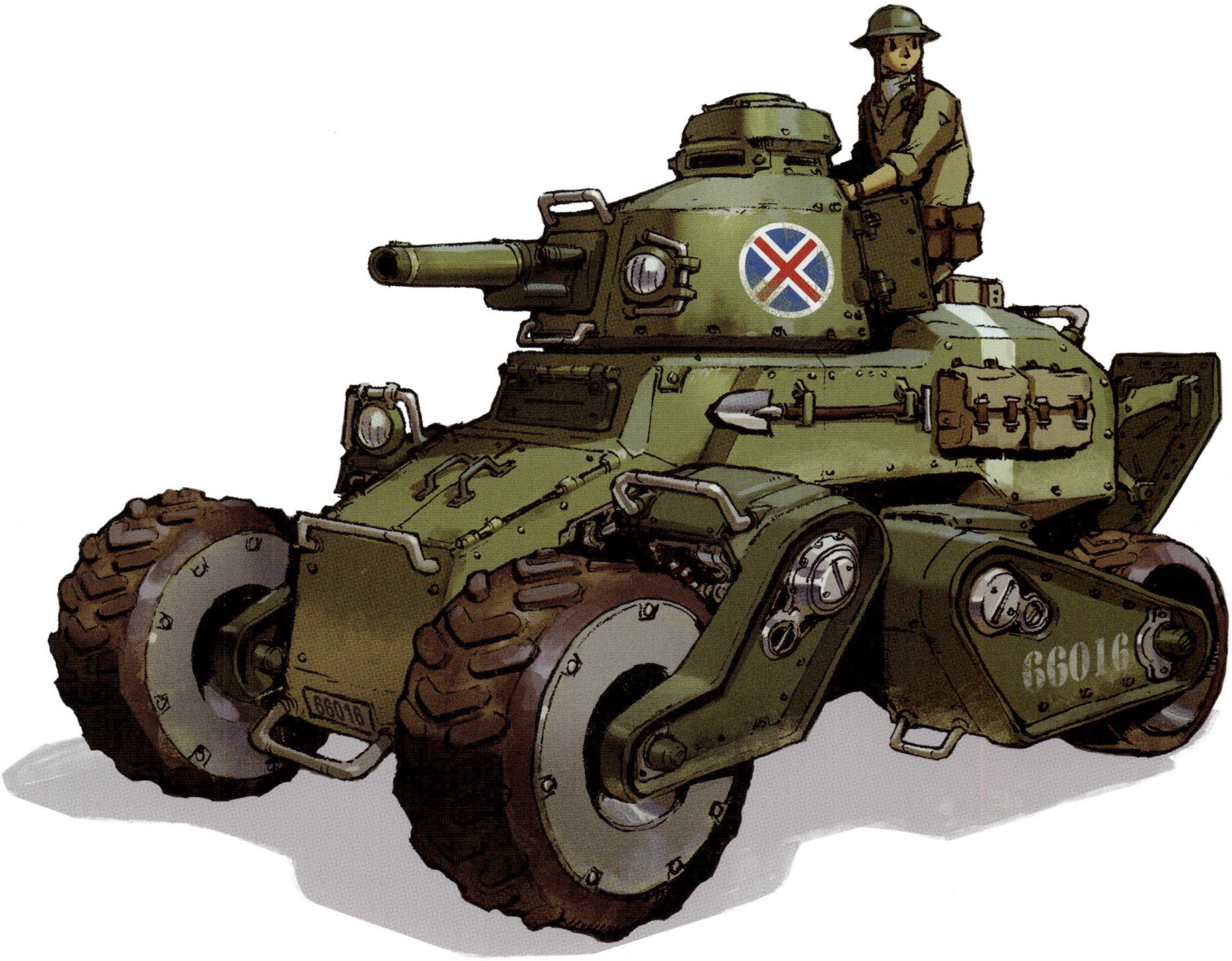

"An engineer is an artist who thinks in math."
- Sir George Reginwald

REGINWALD LT17

Modern historians point to Reginwald's LT17 as the first truly modern tank and the forefather of tankheads. Many believe that Sir George Reginwald had conceived of his so-called "tilt lever armored cart" many years before he presented the initial concepts to the Britonica Defense Ministry. Reginwald was heavily involved in the budding automobile industry during the time of the Trout Incident (BWC 10) and the Hilland revolts (BWC 8 – BWC 3). He observed the difficulty early armored vehicles faced in traversing rough terrain, often being blocked by even light rubble or debris. This flaw, along with the fixed weapons—which required the entire vehicle to turn in order to aim—made for a bad combination. Likely as a response to these limitations, Reginwald secretly developed the "swing arm suspension vehicle with gyrating weaponry" concept.

The iconic suspension arms were already in use in agricultural machinery and some military scout vehicles but had not yet been implemented on much heavier armored cars. Far more significant, however, was the introduction of the turret or, as Reginwald called it, the "gyrating weapon system." Any weaponry larger than machine guns had been in fixed mounts up to this point, and the notion of putting the entire weapon and firing mechanism inside a rotating armored cupola was, objectively, revolutionary.

When the war ministry came to Reginwald's automobile factory, looking to extend production of military units already in use, they were instead swept off their feet by the avid designer who presented them with radical ideas. While a number of the liaisons considered Reginwald's designs too outlandish, enough of them were convinced to at least let him produce a prototype. The speed at which Reginwald was able to present a working machine indicates how much thought he had already given to it in his private time. While the prototype suffered some setbacks, it still easily passed the trials set for it, even managing to traverse the infamous Swidford obstacle course, which had broken all previous machines attempting it.

The evolution of the LT17 and the many conflicts in which it took part are too extensive to summarize here; suffice it to say that, even prior to the Neverwar, it had already seen combat on all continents. While many considered it obsolete as a mainline battle unit at the outbreak of the Neverwar, it still saw service in many rearguard units and was employed by smaller nations on both sides of the conflict. Even towards the end of the conflict, the LT17s saw use as fast action or scouting units, demonstrating again how ahead of its time the design had been.

The influence of the LT17 is apparent in all turreted vehicles—a type which quickly became nearly ubiquitous once it had been introduced. The turret introduced in the LT17 led to the turret "heads" of tankheads, inspiring their namesake. Less well known is the fact that the pioneering engineering concepts of moving limb articulation essential to their development were also first conceived in the LT17.

Most people today point to the Kyzerian Stormwalker as the first significant step in the birth of tankheads. While it is true that the Kyzerians reverse-engineered the technology needed for tankhead development from the Isle à Von artifacts they stole from Grand Britony shores, their engineers still drew heavily on conventional tank technology to bring those early Stormwalkers to life. The years of applied engineering knowhow derived from the LT17 swing arm were essential in turning the unprecedented notion of a walking tank to reality. While the first tankheads were built by Kyzerian engineers, it was a marriage of two technologies of Brit origin: the mysterious Isle à Von artifacts and the humble LT17, making the tankhead as much a Britonican legacy as it was a Kyzerian innovation. The artifacts were responsible for the intuitive leap from conventional carriage design to a bipedal one, but that leap would have not have been possible without Brit resources and proven Brit engineering.

SPECIFICATIONS

COUNTRY: Grand Britony

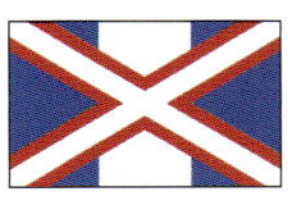

UNITS BUILT: 1,950
NO. OF CREW: 2
MAIN ARMAMENT:
1x75 mm smoothbore gun
ARMOR: 8–23 mm
MAX SPEED: 70 km/h

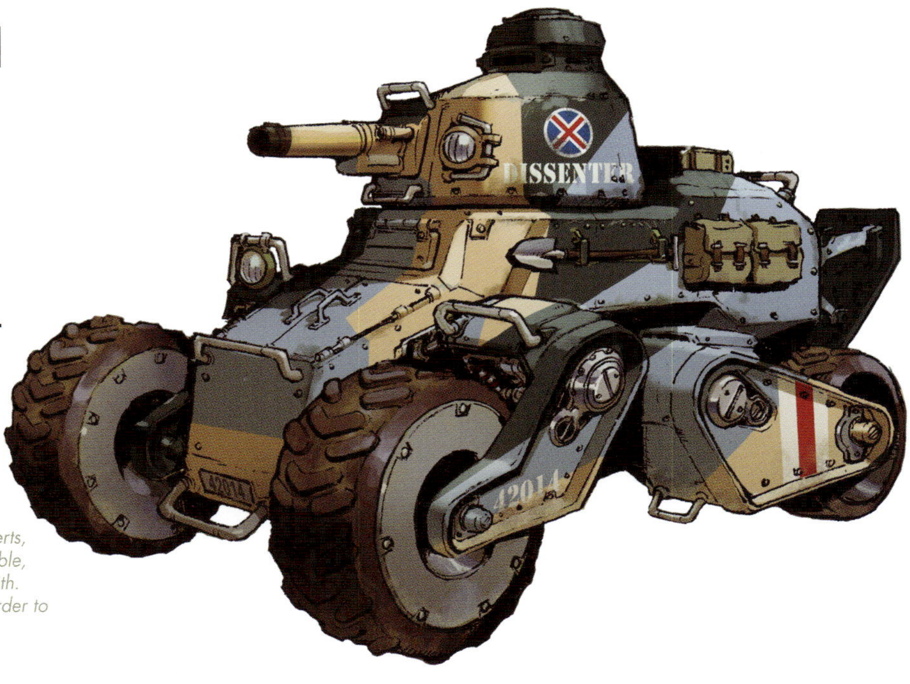

A LT17 in distinctive angular disruptive camouflage. This pattern was used in deserts, where remaining hidden is nearly impossible, as there is little to no terrain to blend in with. Disruptive camouflage instead made it harder to identify the type of vehicle.

SCAMMELL
EXPLORER RECOVERY VEHICLE

An often-overlooked aspect of military loss is recovery. A vehicle can be disabled by any number of defects, whether through damage or general wear and tear, many of which can make a vehicle entirely inoperable until repaired. Jammed transmissions, broken axles, or simply getting bogged down in deep mud will put any tank out of commission, and sometimes there is no way for the crew to fix the problem. This is where recovery vehicles like the Scammell Explorer come in. During the Neverwar, these unsung guardian angels would draw broken tanks and other disabled vehicles to safe positions or engineering bays where they could be restored to working order. It is easy to think of an army only in abstract numerical representations—ammo reserves, weapon stockpiles, armor tonnage—but in any conflict, the quality of recovery and retrieval is as important as armor thickness or enemy firepower.

The Explorer incorporates a simplified and strengthened swing arm suspension first pioneered in the Reginwald LT17. The heavy beams on the wheels enable the Explorer to reliably carry and tow extreme loads while doubling as armor for the wheel and axle. The rest of the vehicle, however, is not armored or armed, as it's purely meant to be a rear area support unit.

SPECIFICATIONS

COUNTRY: Grand Britony
UNITS BUILT: 6,525
NO. OF CREW: 2
TOW LIMIT: 20 ton
MAX SPEED: 40 km/h

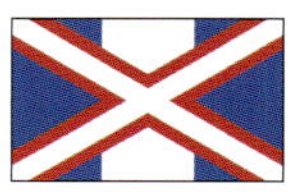

The Explorer served throughout the war in multiple roles from tank recovery to artillery tractor, in towing field guns from one theater to another, or even into effective positions during battle. Many of these early model Explorers were lost in WC010 when the first Kyzerian Stormwalkers broke through the Southern Front and the Alliance was forced to make an unexpected retreat. The Explorers, ironically, were ungratefully abandoned and destroyed by their crews or captured as the steadfast support vehicles failed to keep up with the retreat from the Kyzerian monsters.

In WC012, the first versions that could transport tankheads were delivered to Alliance forces. The Explorer could reliably move these tremendously heavy machines reliably over long distances, including the muddy dirt roads of the countryside.

Many of these later models, including newer versions of the recovery and tractor designs, would survive the war and continue service. Uniquely, the Scammell Explorer enjoyed an active and productive retirement, as they proved very popular among civilians after they had been rotated out of military use. Heavy industrial transport companies were willing to pay good money for the rugged trucks, and emergency service providers—firefighting forces, for instance—converted surplus Explorers to fit their specific needs, particularly in less developed areas or those areas which had been hit especially hard by the war.

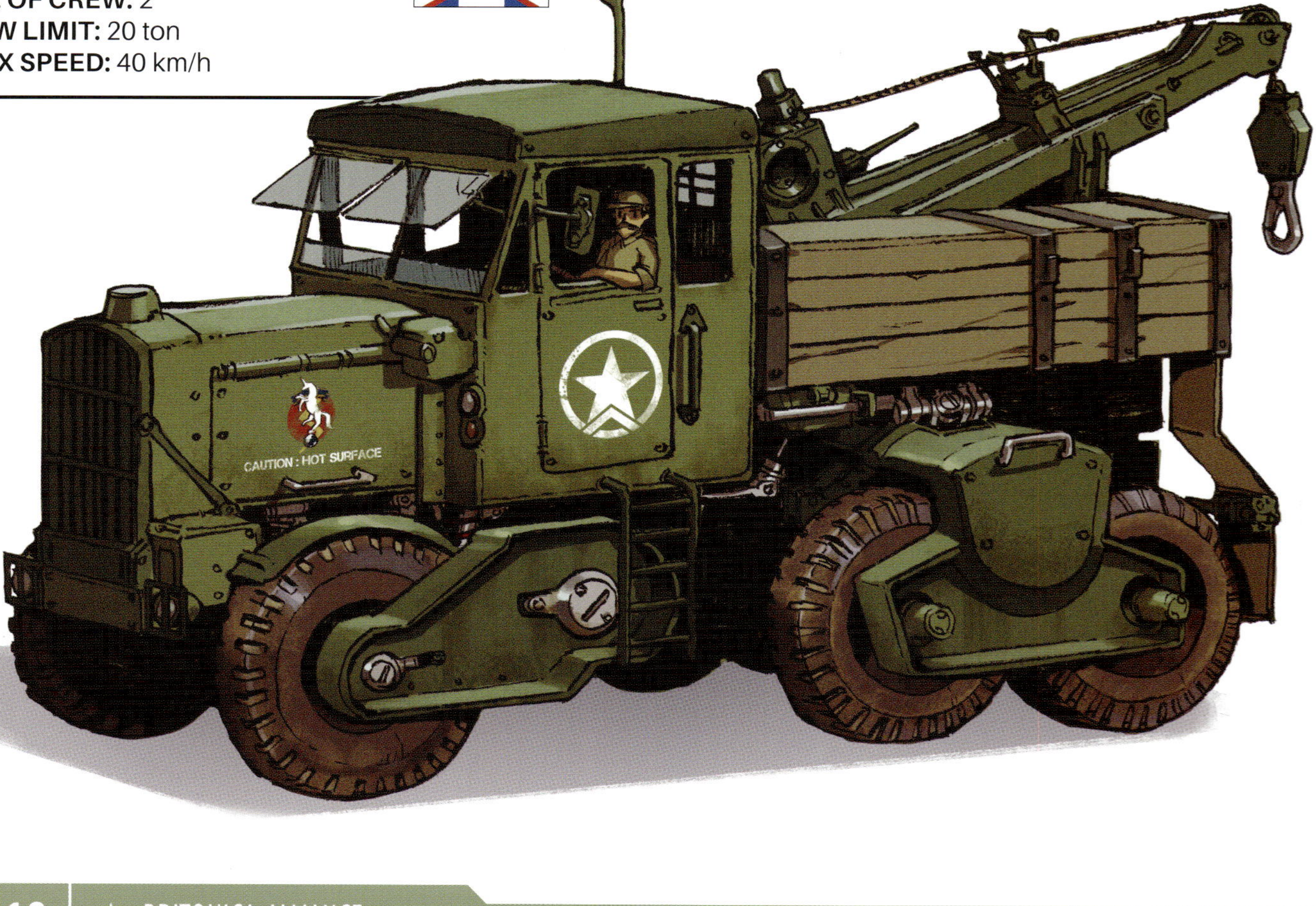

T17E1 STAGHOUND

Although a Liberterran design, the Staghound only ever saw use by Britonican forces. A reliable machine with decent speed and light offensive capabilities, the Staghound was often in the vanguard during the early battles of the Neverwar. That changed when the front lines became expansive trenchworks. A dedicated command version without a turret was produced later in the war, but the "Nimbus" antenna array field modification shown here was quite popular among daring regimental commanders who preferred to lead from the front.

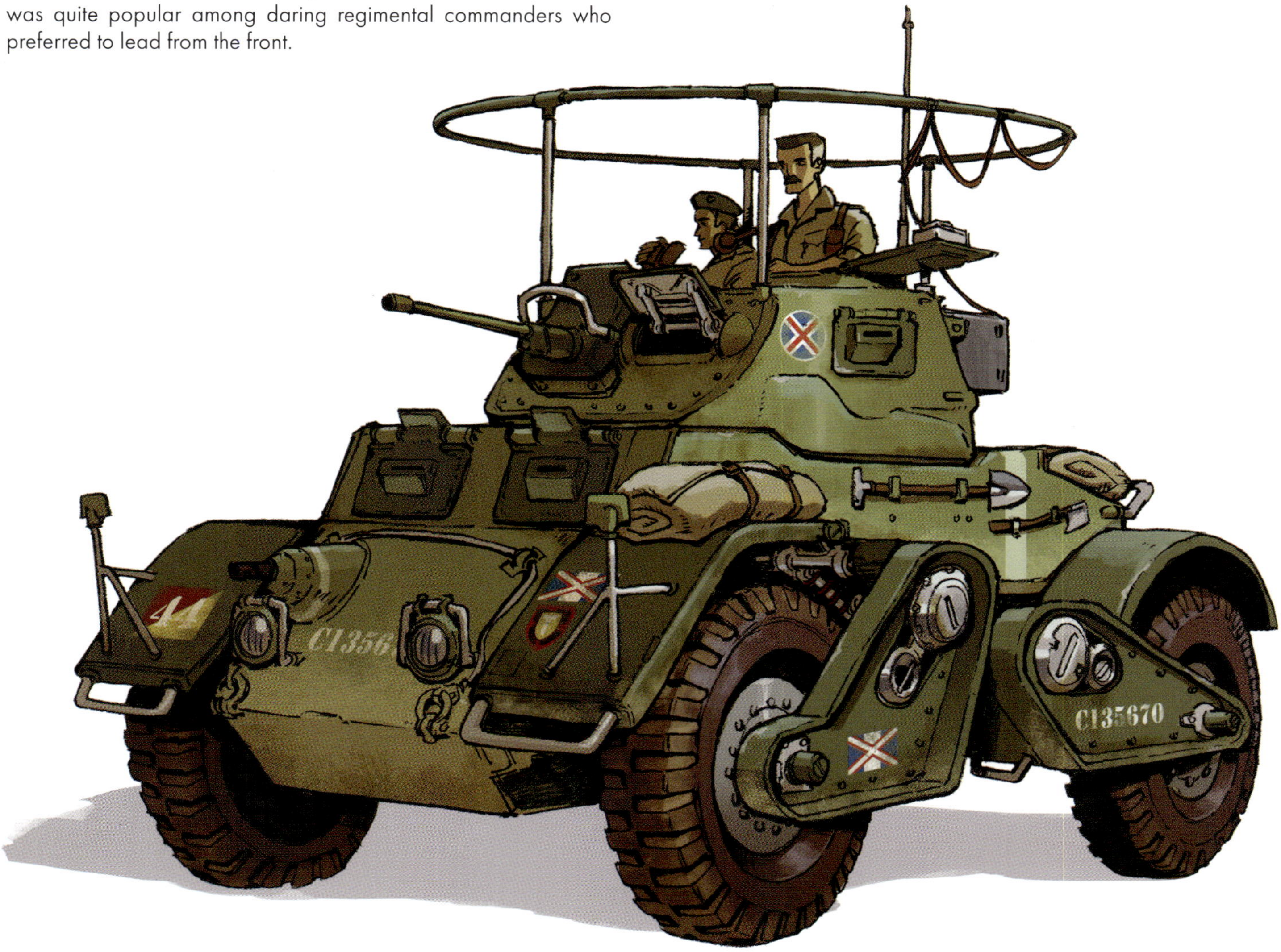

The Staghound might not be as flashy as a tankhead, but in a way it perfectly embodies the ideal Brit mentality. A vehicle that emphasizes mobility and maneuvering, used by daring officers to lead their units from the front lines.

SPECIFICATIONS

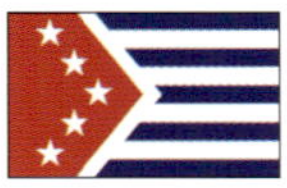

COUNTRY: Liberterra
UNITS BUILT: 7,635
NO. OF CREW: 4–5
MAIN ARMAMENT: 1 x 37 mm gun
SECONDARY ARMAMENT: 1 x 30-cal machine gun
ARMOR: 8–43 mm
MAX SPEED: 88 km/h

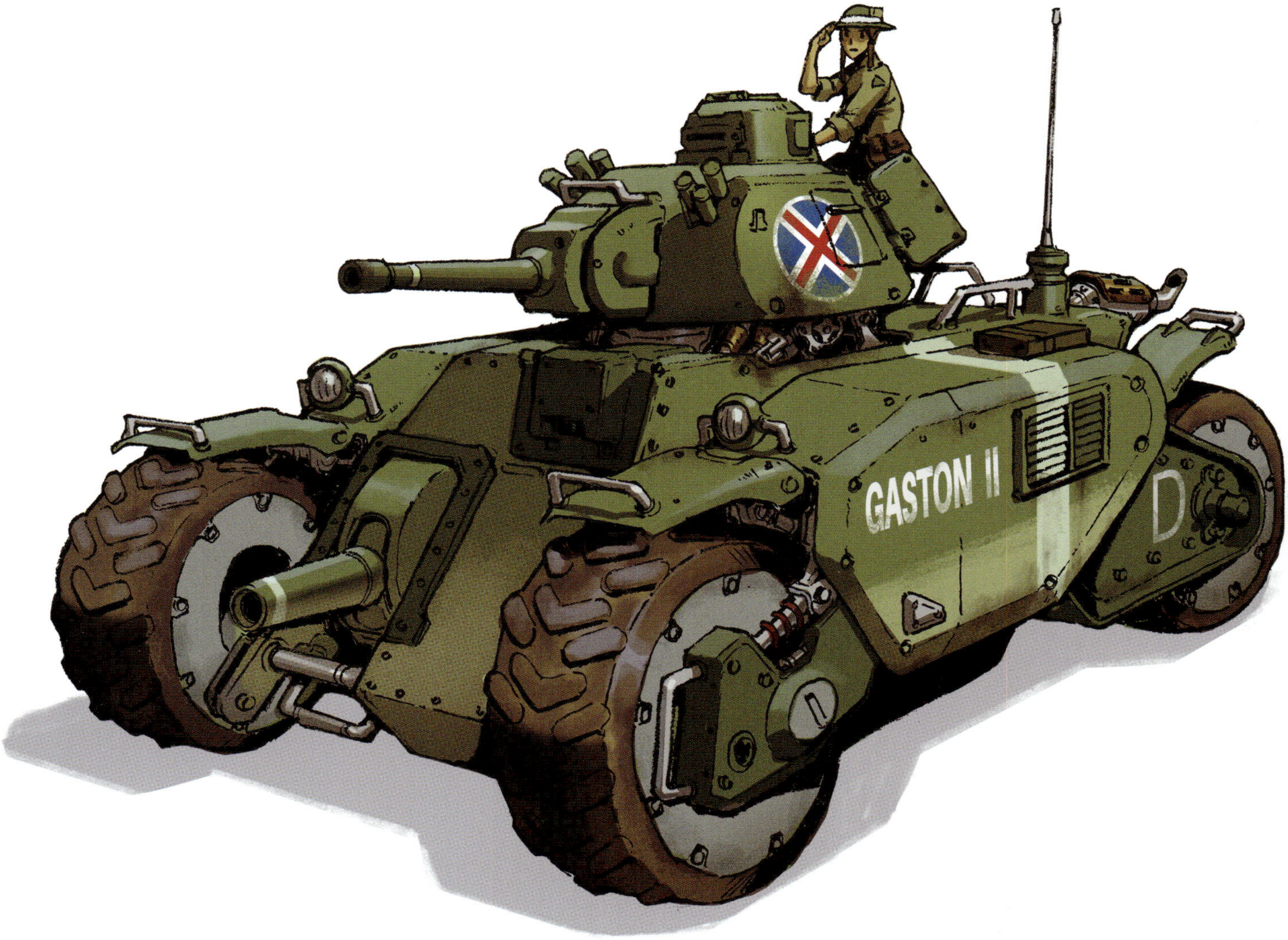

"It's absurd that such a formidable tank should be wasted on a hull-mounted cannon that never fires during combat! I'd call it a waste if they had been on our side, but instead I call it lucky."

- Otis Theoson, Kyzerian Tank Platoon Commander

REGINWALD B1

The instantly recognizable Reginwald B1 is a study in contradiction. It was deemed a failure by most of Alliance High Command due to logistic issues stemming from its long-suffering design process. It managed to perform exceptionally in the field, yet failed to have any real impact in the grand scheme of things; a machine designed for a role that proved to be not nearly as important as it was expected to be.

It is important to realize that while the B1 was first deployed well after the start of the Neverwar, its development started much earlier than that. While there were small-scale skirmishes and some noteworthy battles in the years immediately preceding the War Century, such as the Trout Incident and the Hilland revolts between BWC 10 and BWC 2, it was a period generally lacking full-scale armored conflict. Rather, most conflicts were localized rebellions conducted from central, fortified positions, and the B1, intended to act as a self-propelled assault gun which could get into range of armored fortifications to create breaches for attacking infantry, was very much a product of its time. However, such actions were rare during the stormy early years of the Neverwar, when battlelines were fluid and field engagements were the norm.

The design process of the B1 was plagued with setbacks and delayed by political infighting within the Alliance Armored Corps and Infantry Command. Expectations regarding the vehicle's speed, armament, weight, and armor thickness varied wildly and were entirely unreasonable when compared to the project's meager funding. Successive prototypes differed vastly from one iteration to another. For instance, while the iconic offset hull-mounted howitzer was a consistent feature from the earliest prototype to the final production model, the wheel configuration underwent many changes; one version had no less than ten wheels, four pairs of small ones and one pair of enormous ones. Despite a high degree of industrial competitiveness between the workshops involved in the project, many trial versions proved disappointingly unreliable on anything but tarred roads. The eight-wheeled design suffered significant breakdowns during the first trials and some sources claim it was entirely destroyed by its own power tearing the machine apart. (It has been suggested, however, that this is an exaggerated rumor circulated by the engineers working on the competing design.)

In the meantime, the Reginwald LT17 was developed and, while still flawed (chief among those flaws being lackluster armament), its turret and swing arm suspension performed well. Subsequently, the B1 was updated with a small turret incorporating LT17 technology, though only armed with twin anti-infantry machine guns. With other vehicles performing so much more promisingly, the B1 saw its funding cut time and time again. Along with interminable internal strife and bitterness between developers, the project seemed destined for termination, only its promising hull-mounted howitzer justifying continued investment.

When the Neverwar began in WC001 and the border between the Alliance and the Empire south of Rozenia saw serious conflict for the first time in a decade, the B1 project was reinvigorated. Its faltering wheel arrangements were replaced by something very close to the LT17, though heavier and better armored. The turret was also upgraded with a heavier 50mm short-barreled cannon, bringing its firepower to a level actually surpassing the LT17.

However, problems continued to plague the B1. Its limited operational range and demanding upkeep led to the decision to move production and repair facilities closer to the frontlines.

Still, many of its features were dated and near-obsolete by the time it went into full production. For instance, its riveted armor was less effective than more modern cast methods already in use in many contemporary vehicles. As a result, coupled with its short range which often made it an easy target for artillery, the B1s suffered heavy losses in engagements with Kyzerian anti-tank guns.

When the tank was employed in its intended role, assaulting fortifications and bunkers, it performed remarkably well. Its main gun had a limited range but against small arms and even other tanks, the design proved itself supremely resilient. Many early-war assaults were spearheaded by the B1, creating and often forcing their way into the breach, wreaking havoc on defenses and opposing tanks alike. In one remarkable action in WC007, a B1 commanded by then Captain (later General) Peter Planke single-handedly destroyed thirteen Kyzerian tanks that had tried to ambush it.

SPECIFICATIONS

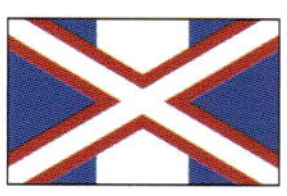

COUNTRY: Grand Britony
UNITS BUILT: 1,810
NO. OF CREW: 4
MAIN ARMAMENT: 1 x 75 mm howitzer
SECONDARY ARMAMENT:
1 x 47 mm anti-tank gun
6 x smoke grenade launchers
ARMOR: 38 mm
MAX SPEED: 56 km/h

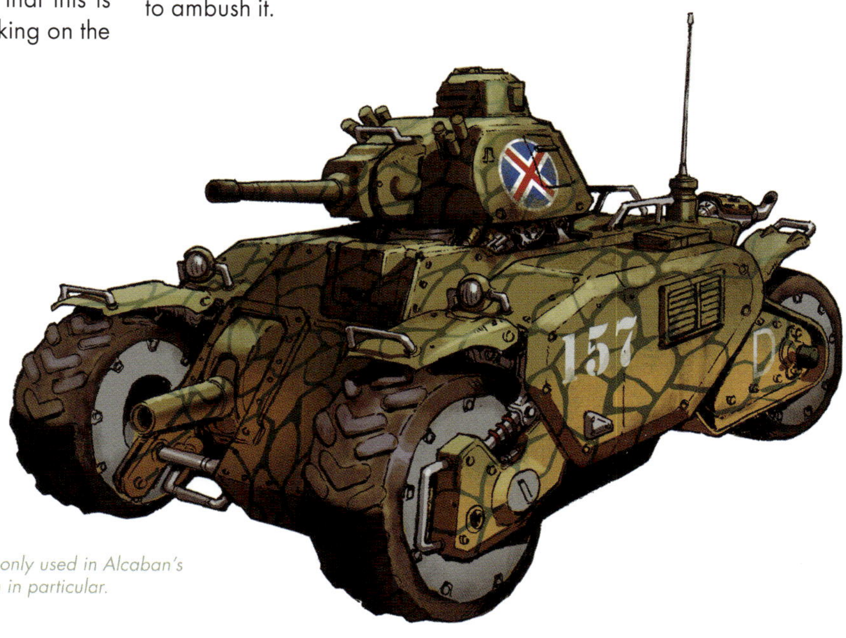

B1 in the camouflage pattern commonly used in Alcaban's dusty grasslands, the Hilland region in particular.

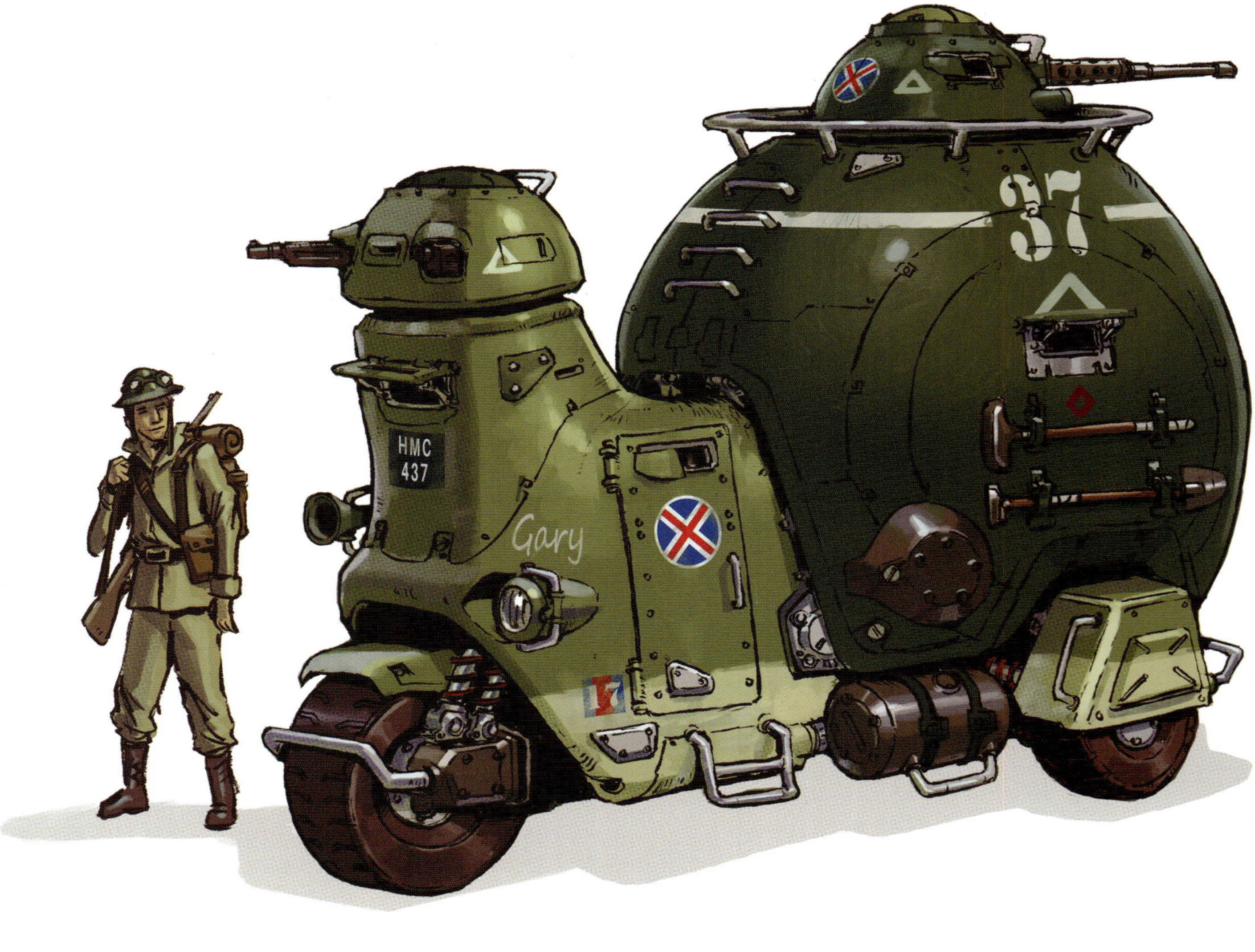

"The Snail rolls remarkably well on her shell, sir!"
- Test Driver JS Smithe

ES CARGO TRANSPORT

When Alliance Army Command first commissioned the Equipment Supply Cargo Transport unit, the specifications called for a light one-man vehicle capable of quickly transporting ammunition from ammo dumps to artillery or tankheads. Early prototypes took the shape of a short, powerful, moped-like vehicle with a towed trailer. These included features such as a quick-release towing hook that enabled rapid switching of trailers to optimize delivery. The driver would deliver a full trailer and take back an empty trailer to the ammo dump, where it in turn would be switched out for another full one.

When the design team demonstrated the prototype for the combined Supply and Armored Division staff, they were met with a few "simple" requests for revisions. This snowballed into a practically unending list of modifications that would transform this modest unit into the machine now fondly known as "the Snail."

One of the earliest changes requested was that the cargo should be protected; the machine would be transporting live ammo for tankheads or artillery, and stray light arms fire was a concern. Even a particularly ill-placed rifle round might set off the cargo. This led to fabrication of an armored trailer. During trials, the new trailer overturned repeatedly, so it was given a round shape that would roll rather than crash. This iconic look would survive throughout the rest of the design changes.

When the Snail proved slower than hoped, there was a call for armaments to be added so drivers could defend themselves in the event of an attack. Twin machine guns mounted on the moped's windshield were tried but, while this looked intimidating, it proved to be impractical as it only allowed the machine to fire in one direction. Simultaneously, Command demanded a redesign that would adequately protect the driver from incoming fire along with the cargo. Thus, the trailer concept was abandoned for a larger armored cargo hold with a cramped driver compartment.

The now dangerously conspicuous profile this machine presented on the battlefield would draw more dangerous threats so, it was decided, heavier firepower than light machine guns was necessary. The design team tested a number of driver-operated weapon systems, but they concluded that it was unreasonable to demand a single crewman to do all that was now being required of him. In the end, the designers settled for a small manned turret on top of the cargo compartment, significantly reducing its cargo capacity and doubling its crew.

Among many other regrettable decisions, the final design still maintained the tri-wheel arrangement from early in the design process. The gearbox used was changed multiple times, but the actual engine never received much attention throughout the vehicle's haphazard evolution. The production model Snail ended up being slow to accelerate, but could in fact build up a decent speed, so long as it had a straight road and firm, even ground. Overturning remained an issue, however, especially when fully loaded. What's more, this tendency to topple was increased with the addition of the high gun turret. Despite all this, the ES Cargo was easy to produce and functional in the field, which resulted in large numbers seeing use throughout the Neverwar. Even today, it is not uncommon to see surplus Snails converted into quirky campervans or food trucks.

SPECIFICATIONS

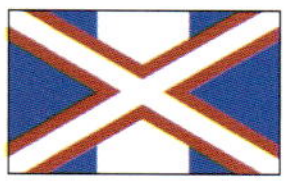

COUNTRY: Grand Britony
UNITS BUILT: 243,000
NO. OF CREW: 2
MAIN ARMAMENT: 1 x 50-cal heavy machine gun
SECONDARY ARMAMENT:
2 x 30-cal light machine guns
ARMOR: 16 mm
MAX SPEED: 40 km/h

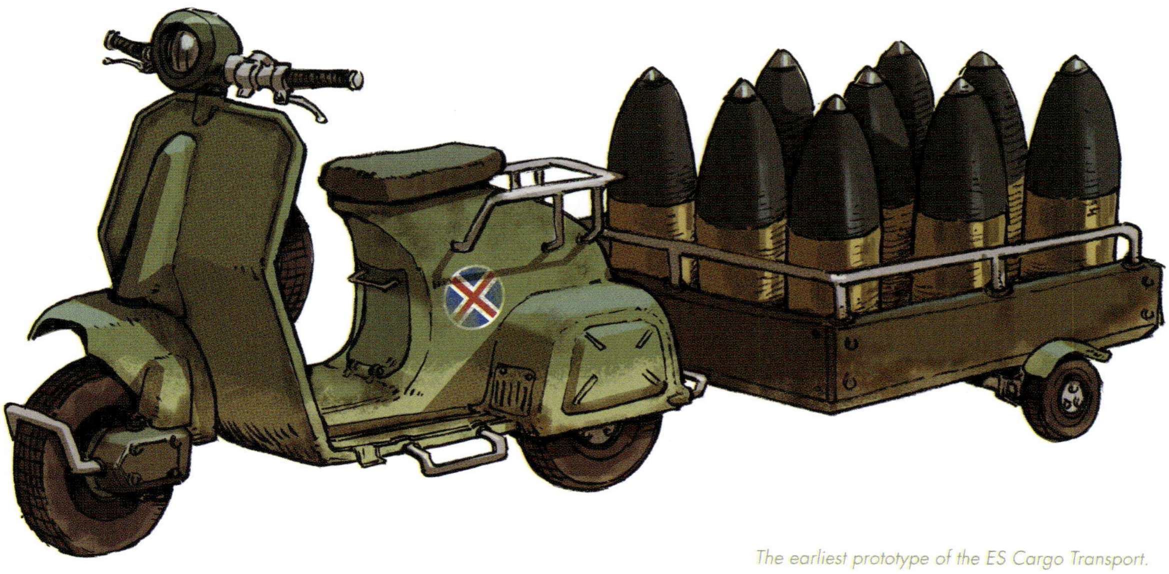

The earliest prototype of the ES Cargo Transport.

DISGUISED Q TRUCK WITH ANTI-TANKHEAD GUN

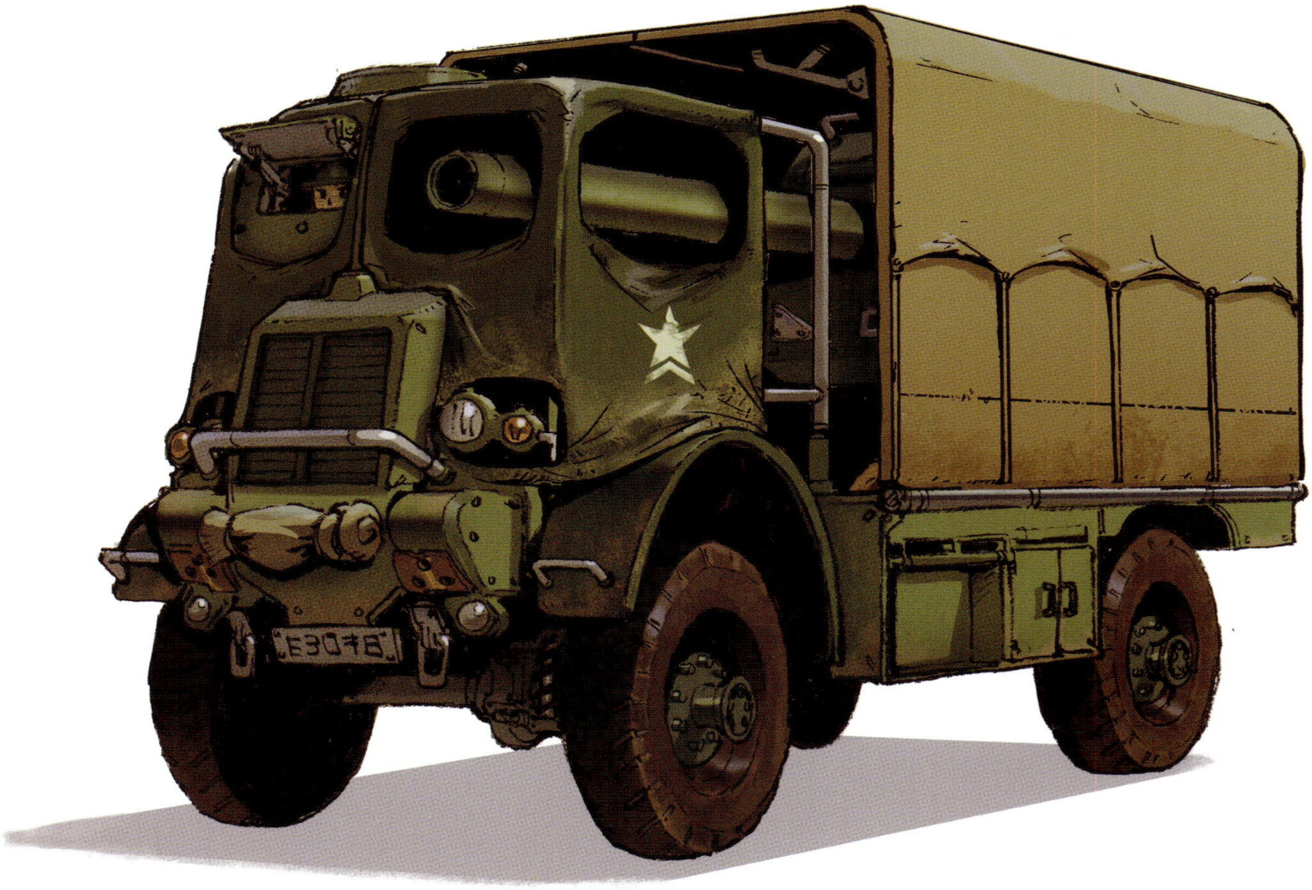

In all aspects of war, deception is key. The campaigns waged in the wide-open deserts of Alcaban were no exception. The desolate, flat landscape between the Kyzerian southern holding and Free Rozenia, in particular, saw many engagements decided more by novel tactics than by army size. In such a featureless zone, there was little opportunity for concealment or ambush. Simple observation balloons could see a force on the horizon even at distances of many miles, even further if the approaching units were producing a large enough dust cloud. Simply put, neither Alliance nor Kyzerian commanders could hide a force of any real size in the desert. Camouflage netting might have fooled casual observers, but the stark shadow of tank-sized vehicles during the day was a dead giveaway.

Nevertheless, if one cannot hide from the enemy, one can still control what he sees. Scouts and observation units would try to take stock of their targets, but were unlikely to take too close a look or loiter too long lest they be detected as well. Knowing that the enemy would be looking for recognizable tank and tankhead profiles, the Alliance army managed to fool the Kyzerian forces on multiple occasions by constructing facades using stationary structures or deflated observation balloons to create dummy tanks. Some were even mounted on simple truck chassis so that they could be moved, adding to the believability of the illusion.

The same ruse could also be employed in reverse. Given the open horizons of the area, the humble anti-tankhead gun ruled supreme. Any enemy in range could be spotted and destroyed at leisure. Getting the guns to the right position was more of a hurdle as a convoy of towed guns on the move would have been dangerously exposed. Because of this and the vast distances crossed between battle sites it became common practice to mount the guns *en portée** on the back of assigned trucks. This started out as an improvised measure but was later adopted as standard practice, with multiple armored trucks specifically designed for the purpose.

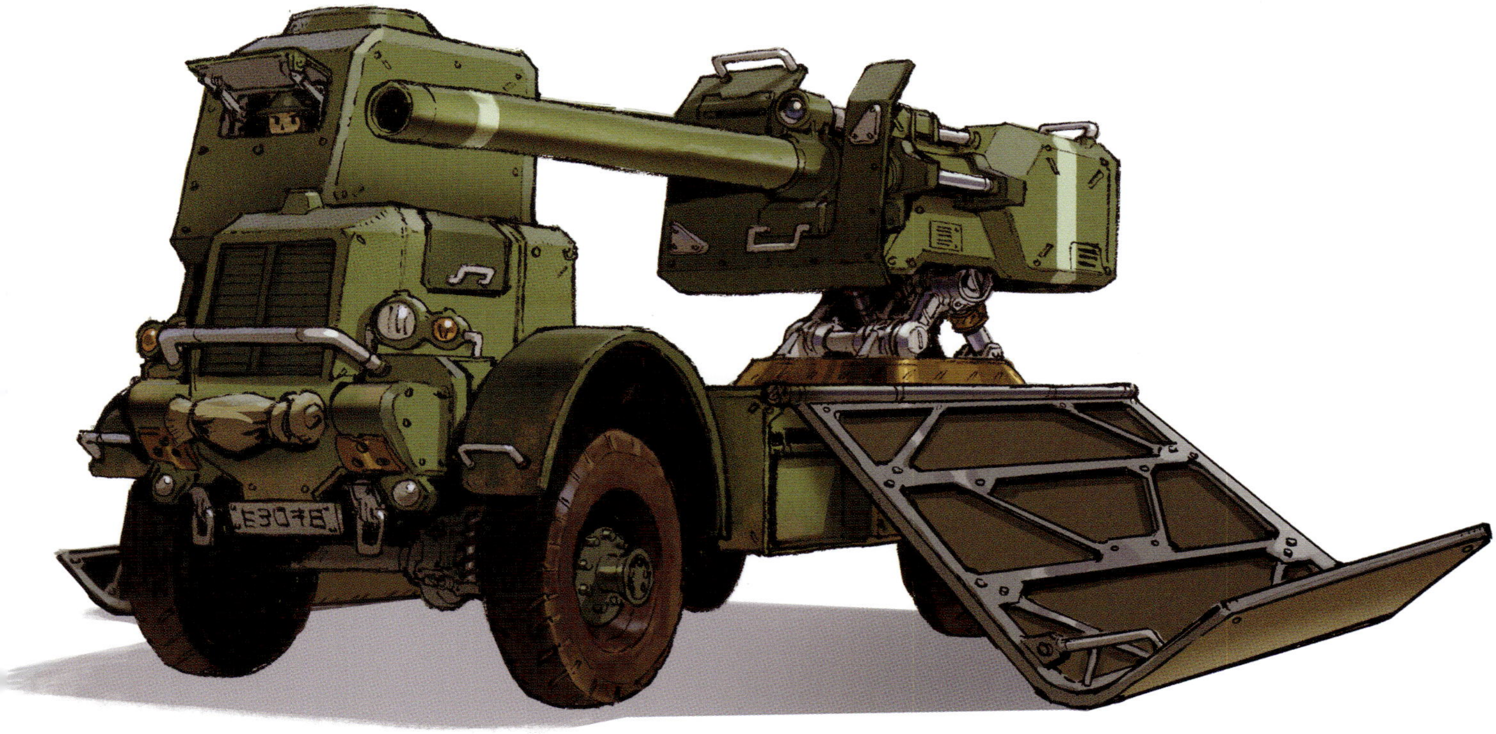

Whereas some unarmed trucks were parading around pretending to be tanks, the Q trucks instead donned soft tops which, to the eyes of enemy spotters, turned the ready-armed gun truck into nothing more than an unarmed transport vehicle. A convoy of these trucks, escorted by a few jeeps or light tankheads, would appear to be a juicy target for the Kyzerian forces, or else one that could be safely ignored in favor of higher-value Brit forces. Even more craftily, those higher-value targets might be made up only of the faux tanks! Once engaged, the soft tops of these Q trucks could be separated down the middle and folded away to the sides, revealing the anti-tankhead gun in ready position.

The first vehicles used in this way had been ad hoc field improvisations. The vehicle pictured here, however, was a late-war design purpose-built specifically for this deception tactic. Of particular note is the articulated gun base, which allowed the driver/gunner to adjust the gun's aim from the relative safety of the cabin. The offset cabin also meant that the gun could be fired in a forward-facing orientation direction before its disguise needed to be discarded. After the first shot, loader crews from accompanying support vehicles would be needed to reload the weapon. However, the first volley from a seemingly unarmed convoy of these trucks often left the enemy so surprised and shell-shocked that the crew could reload in relative safety before return fire became particularly intense.

SPECIFICATIONS

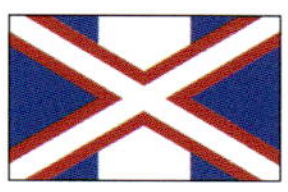

COUNTRY: Grand Britony
UNITS BUILT: 350
NO. OF CREW: 3
MAIN ARMAMENT:
1 x 120 mm anti-tankhead autoloader
ARMOR: 8–18 mm
MAX SPEED: 35 km/h

**En portée* describes the practice of carrying an artillery piece on a truck which can be fired from the vehicle or quickly dismounted and fired from the ground.

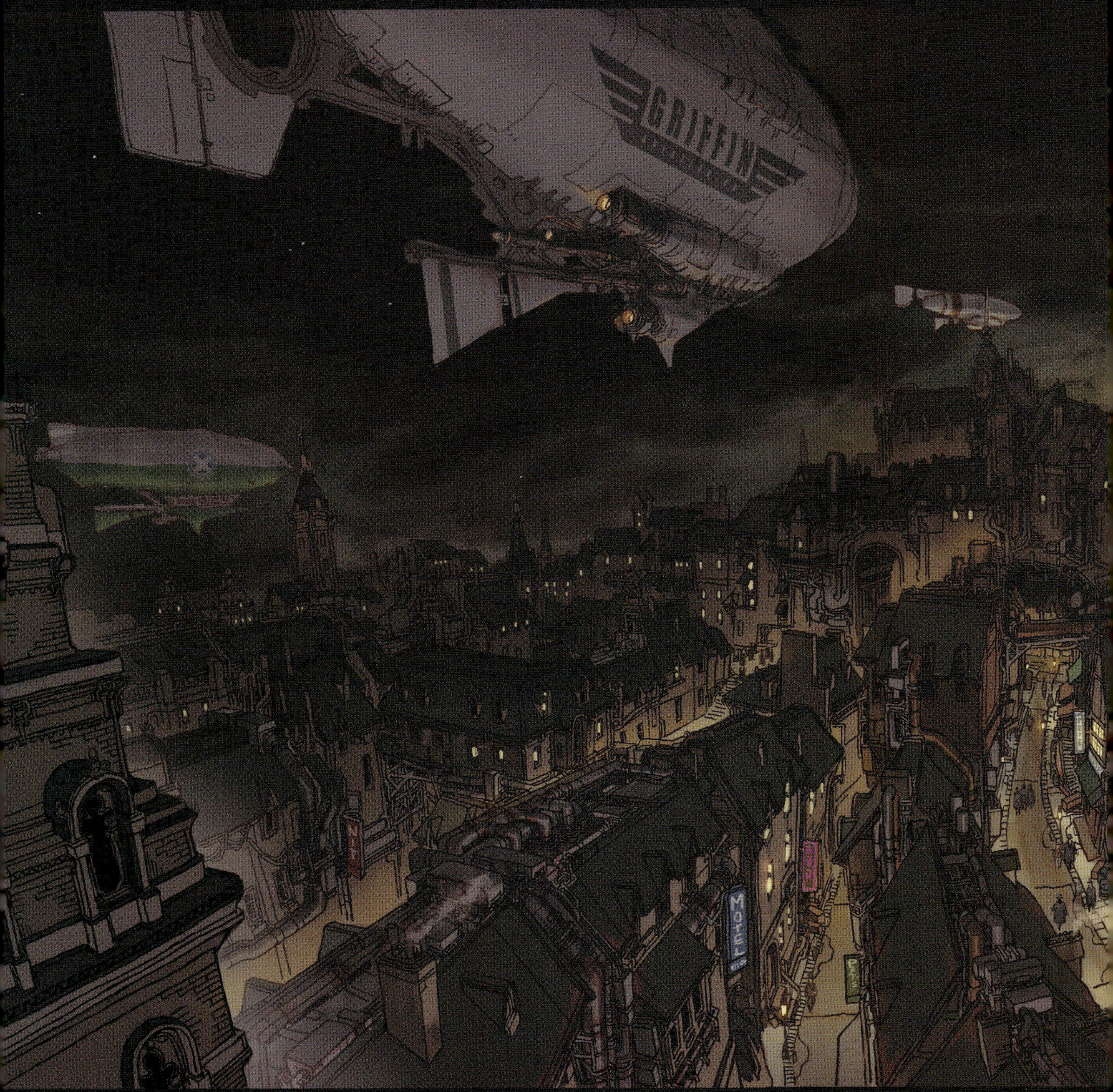

LUDENCIA

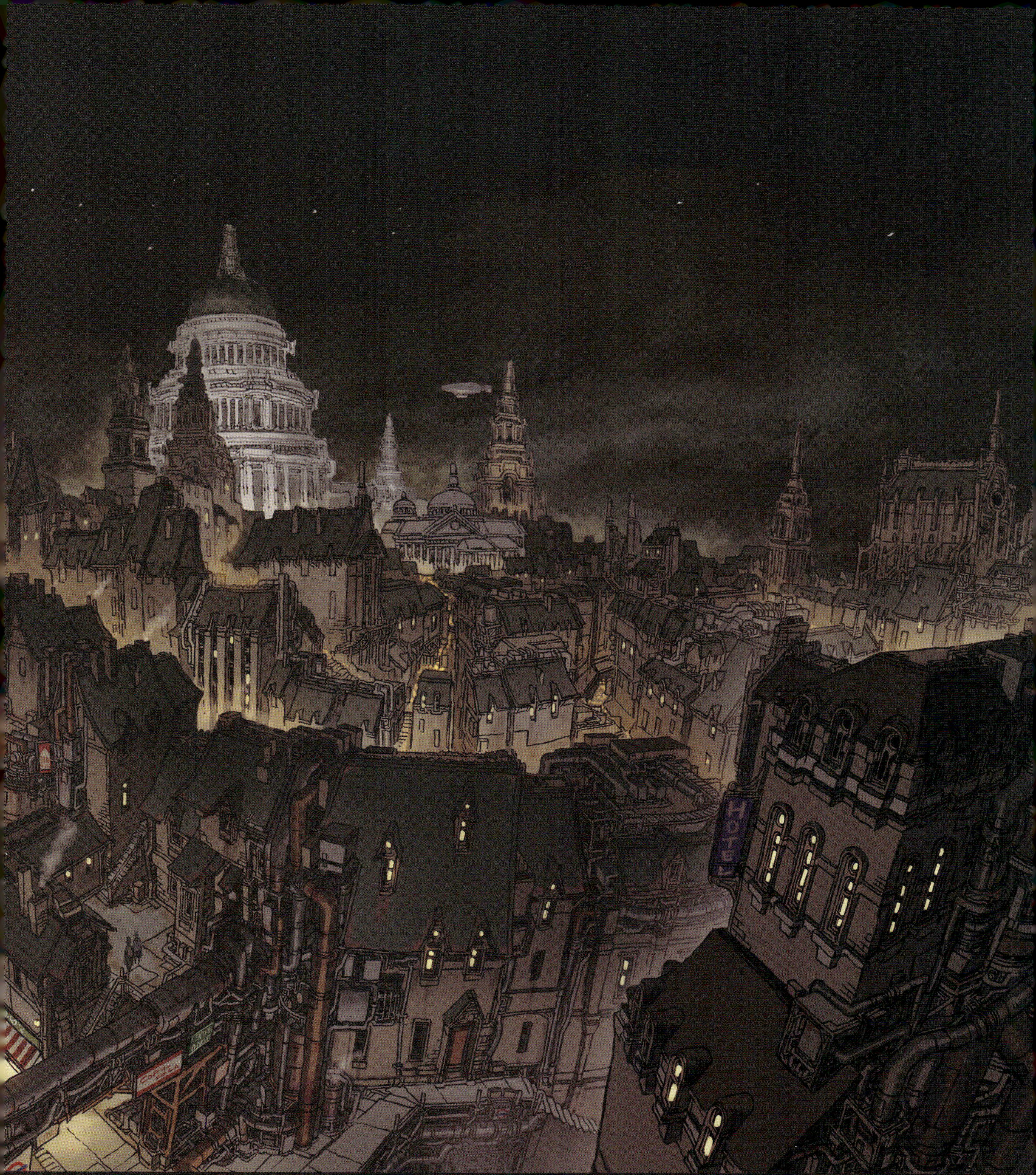

Mighty Ludencia, capital of Grand Britony and headquarters of the Britonica Alliance. The War Century, along with all its tragic consequences, brought many modernizations to the sprawling streets of this metropolis. Wartime factories and armories sprouted across the city, creating its now-famous network of exposed pipes and conveyors. Yet, she remains as shining a vista as ever. Despite so many plans and programs to shell and bomb her proud monuments, she escaped largely unscathed—especially compared to other less fortunate cities impacted by the Neverwar.

MARK I

When the Kyzerians deployed their first tankheads, the effectiveness of such a radical design came as a shock to the Alliance Army Command and soldiers alike. Brit intelligence had known for some time that a "wonder weapon" was being developed, ostensibly based on materials obtained in the Zeehond raid (WC009). Not much attention was paid to this project—mainly because most of what was known was rumor and it had yet to materialize as a credible threat. The fact that both Zeehond and the special project were known to be spearheaded by one Razivir, a priest of the Kriegsdiener religion who had long raved about the coming of "gods forged in the shape of man," reinforced the Alliance's low estimation of the potential threat. Whether warranted or based on false confidence, it proved folly to underestimate Razivir.

In the summer of WC010, walking tanks did indeed smash through Alliance defenses and made the first headway on the Front in many months. Learning what devastating advances their enemies had made, the entire Brit warmachine scrambled to obtain as much information as possible about these machines, to reverse-engineer them, and to develop their own. However, without the insight gained from the Isle à Von artifacts, it was hard to understand the mechanisms animating the few Stormwalkers they obtained. In effect, the Brits were trying to make a copy from a copy of an incomplete relic.

Understanding the complexities of bipedal design might have been beyond their capacity at the time and, hard-pressed by the advance of the Kyzerian Stormwalkers, the Brits made a radical

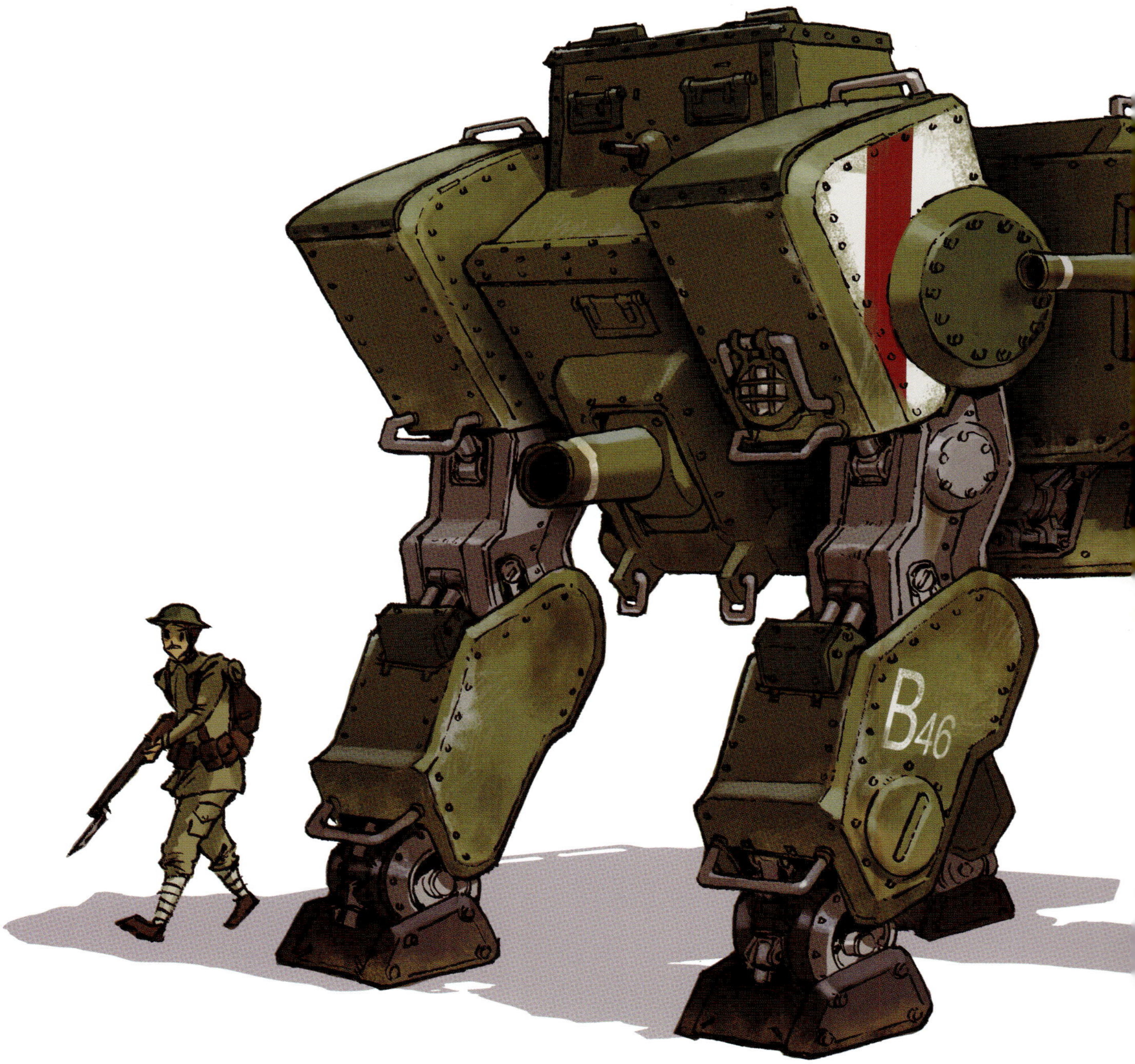

decision. While development of bipedal models continued, most of the research and development effort was diverted to focus on producing quadruped tankheads first. Nevertheless, the Tankhead Mark I was never intended as a stop-gap vehicle. Its engineers and designers tried to make it fill all the combat roles the Stormwalkers had been observed to fill, and to try to push these roles further where possible with a view to permanent utilization.

The four-legged configuration had a number of downsides. It turned more slowly than the already-lumbering Stormwalkers and its walk cycle, besides being particularly long, could not be interrupted before a full cycle had completed. In effect, this meant that once the Mk I had initiated its cycle, it would need to move all four of its legs and return to its resting position before the driver had control over the vehicle again. As this could take up to 20 seconds depending on what point of the cycle the machine was at, the Mk I could be very slow to respond to changes on the battlefield.

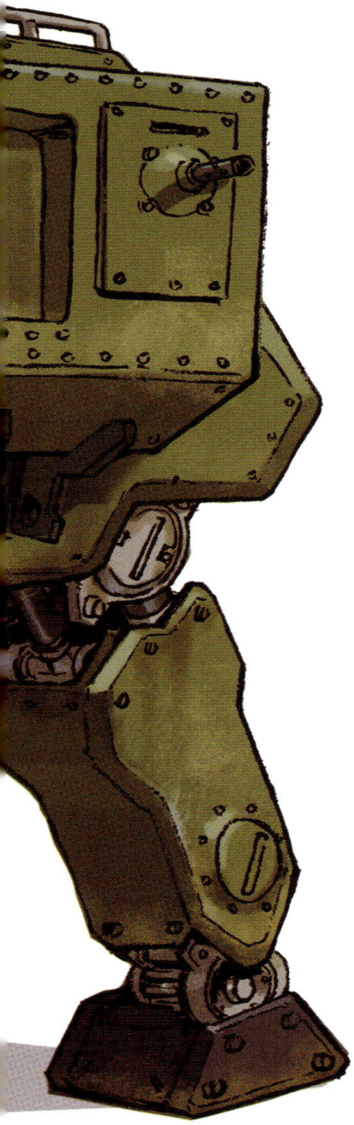

Another detrimental aspect was that the legs weren't particularly reliable. However, while some claim these early legs were more prone to failure than the Kyzerian designs, this is a flawed perception. A more accurate conclusion is that the number of legs—twice that of the Stormwalkers—and the fact that immobilization of just one of them would disable the tankhead accounted for their higher rate of failure.

Despite its heavier weight and lower height, the Brits encountered the same problems with recoil as the Kyzerians, especially since gyrojet weapons had not yet been developed, so the full recoil of launched ordnance was still being transferred to the tankhead. Nevertheless, the Brits decided to mount as large a caliber of howitzer in the lower hull of the tankhead as could be fired without toppling the vehicle. The arc of fire of this howitzer proved less than optimal and the ungainly controls of the Mk I meant that accurate indirect fire over long distances was nigh impossible. On shorter ranges, however, its large-caliber explosive shells were capable of destroying a Stormwalker with a single direct hit. Even a near miss was often enough to knock an enemy tankhead out of combat, as the shell created craters and broken ground deep enough to trip the Stormwalkers. In fact, more Stormwalkers were disabled this way than by ballistic damage, which might suggest the crews intentionally aimed at the ground under the Stormwalkers' feet. It is more likely, however, that the swimming gait of the tank simply made accurate fire difficult and shots often landed short of their intended target.

To its credit, the four-leg configuration did allow the Brits to mount more weapons on their tankheads than the Kyzerians. Twin medium cannons in the sponsons, combined with multiple machineguns, meant that a stationary tankhead Mk I could effectively become a heavily armed fortification on the battlefield. While it was prone to mobility failures, the Mk I was also relatively surefooted and could cross uneven terrain slowly but effectively, allowing it to occasionally flank enemy positions by taking routes the Kyzerians had thought untraversable.

All in all, the Mk I was a flawed machine that managed to perform quite well, especially considering the limited research and data the Alliance were working from compared to the Kyzerians. It should come as no surprise, however, that the Brits continually endeavored to replace the Mk I with new designs. The locomotion system, in particular, was something the Alliance focused on as they desperately tried to match the Kyzerian mastery of bipedal mechanics, or at least produce an alternative capable of matching its agility.

SPECIFICATIONS

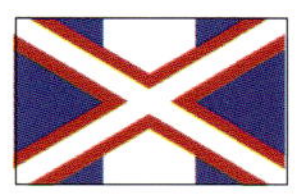

COUNTRY: Grand Britony
UNITS BUILT: 250
NO. OF CREW: 8
MAIN ARMAMENT: 2 x 6-pounder guns
SECONDARY ARMAMENT:
1 x 120 mm bunker-buster mortar cannon
3 x 7.62 mm air-cooled machine guns
ARMOR: 6–15 mm
MAX SPEED: 8 km/h

WHIPPET MK.I

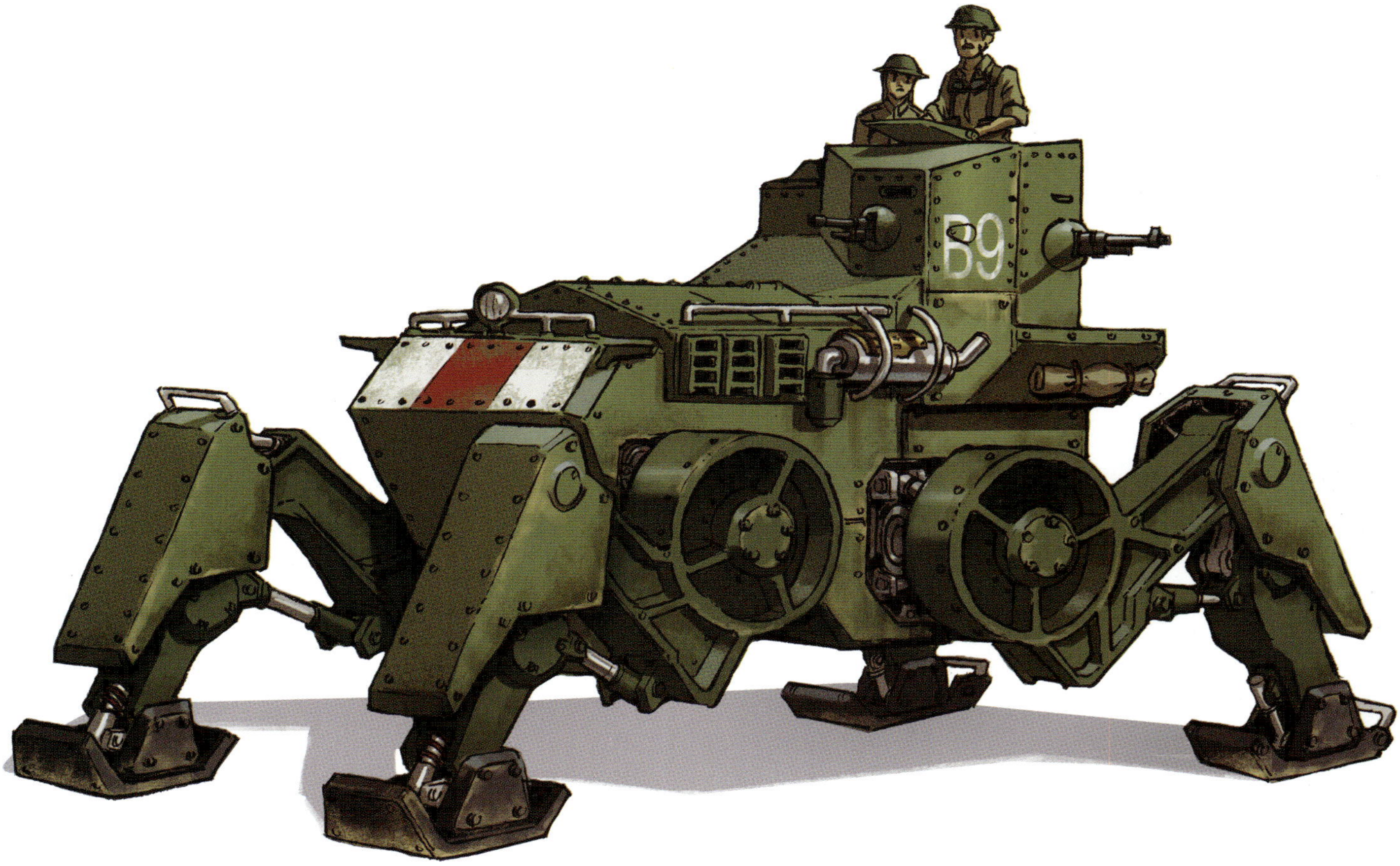

SPECIFICATIONS

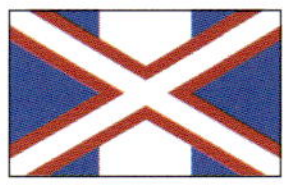

COUNTRY: Grand Britony
UNITS BUILT: 180
NO. OF CREW: 3
MAIN ARMAMENT: 4 x 7.7 mm machine guns
SECONDARY ARMAMENT: none
ARMOR: 14 mm
MAX SPEED: 15 km/h

The designers of the Whippet managed to design a very fast tank for the time. They however, did not improve upon suspension or stabilization mechanisms, making for a very uncomfortable ride for the crew.

The Whippet Mark I was one of Grand Britony's first serious attempts to create a different class of tankhead. While the Alliance's Mark I tankhead was in many ways the equivalent of the Kyzerians' Stormwalker, it should be noted that Alliance doctrine had long put a stronger emphasis on tactical maneuvering, and the mobility required for it, over power and armor. While a devastating new weapon, the Mark I tank was a cumbersome machine that the enemy would inevitably see coming well in advance. To better fit with their established style of combat, the Whippet was conceived as a fast harrier tankhead. The unit might not be able to break the enemy lines on its own, but wherever the Mark I managed to pierce the enemy trenches, the Whippet would race through the hole and raid the supply and artillery trenches behind enemy lines. The argument was made that this was a very safe place to be as, logically, any Kyzerian anti-tankhead weaponry would be found on the front line pointing outwards. The Whippet would only face small arms, which the tank was sufficiently armored against. In reality, these early designs still broke down and got bogged in terrain frequently, stranding their crews well away from support, vulnerable to enemy grenades and improvised weapons.

WHIPPET MK.II

While the Whippet Mark I was successful in many ways, its design flaws contributed to many losses. In particular the leg units' tendency to "run away" on particular downward angles led to the loss of a large number of machines during trench assaults. As the Alliance's understanding of the Kyzerian tankhead mobility mechanisms grew, the "galloper arrangement" of the Whippet Mark I came to be seen as outdated. A plan was implemented to allow engineers to upgrade tankheads in frontline workshops. This new leg configuration system worked with most of the Whippet Mark I's internal mechanisms and used the same mounting arrangement as the galloper legs. The new Mark II legs were more conventional quadruped walkers, sacrificing some speed but in exchange for much greater reliability and load capacity.

To take full advantage of these improvements, roughly half of the Whippet Mk.IIs were converted to Type Bs, carrying a medium gyrojet cannon. Gyrojet technology had come into early production in the Alliance since the Whippet Mark I's inception. Its incorporation made the Whippet Mk.II an effective combatant against most regular tanks and it could even penetrate the Stormwalker's weaker rear and joint armor if hit at a favorable angle.

SPECIFICATIONS

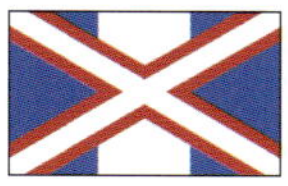

COUNTRY: Grand Britony
UNITS BUILT: 224
NO. OF CREW: 3
MAIN ARMAMENT: 1 x 6-pounder gun
SECONDARY ARMAMENT: none
ARMOR: 14 mm
MAX SPEED: 12 km/h

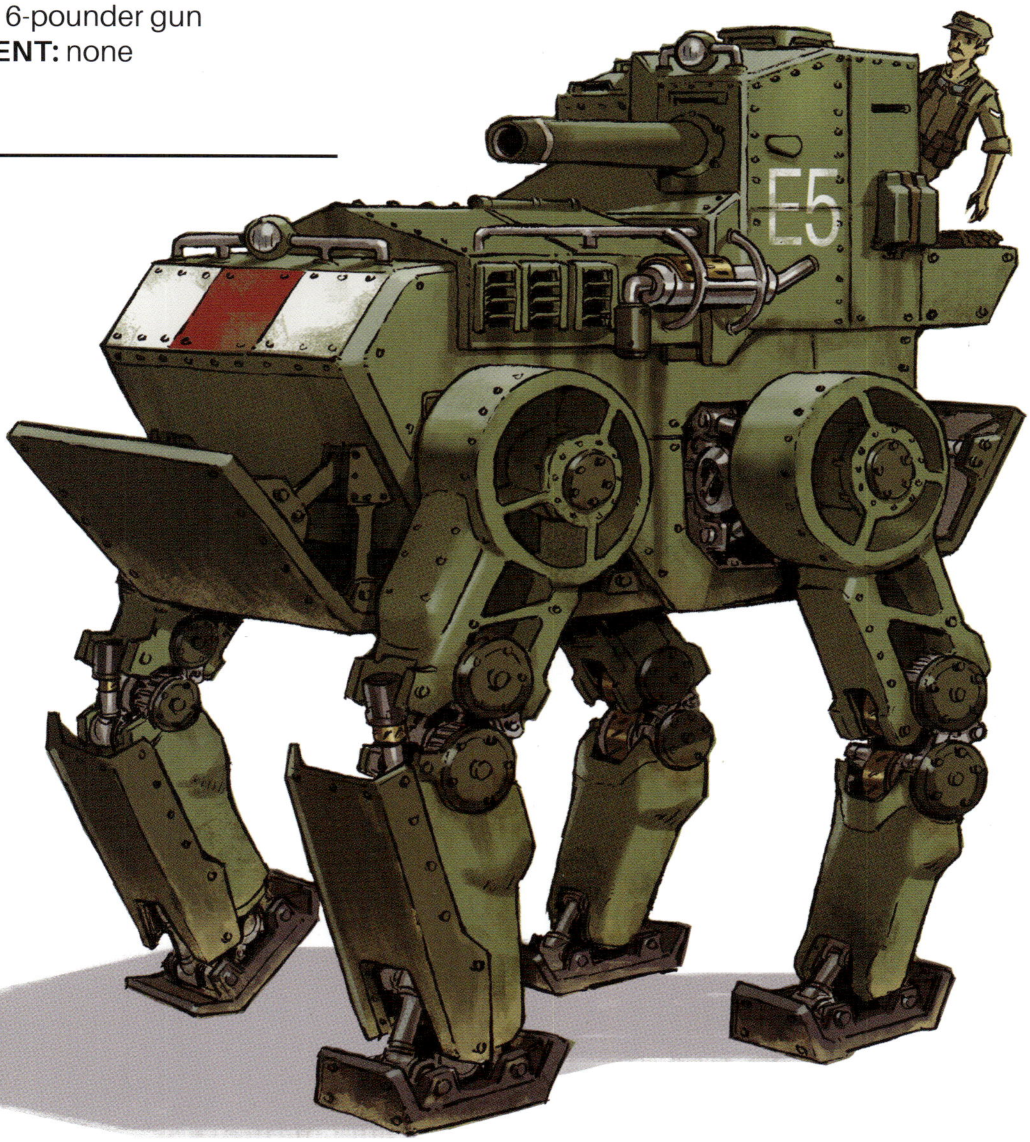

VICKERS CARDEN LOYD PERSONAL WALKER

SPECIFICATIONS

COUNTRY: Grand Britony
UNITS BUILT: 115,035
NO. OF CREW: 1
MAIN ARMAMENT:
1 x 7.62 mm light machine gun
ARMOR: 7 mm
MAX SPEED: 15 km/h

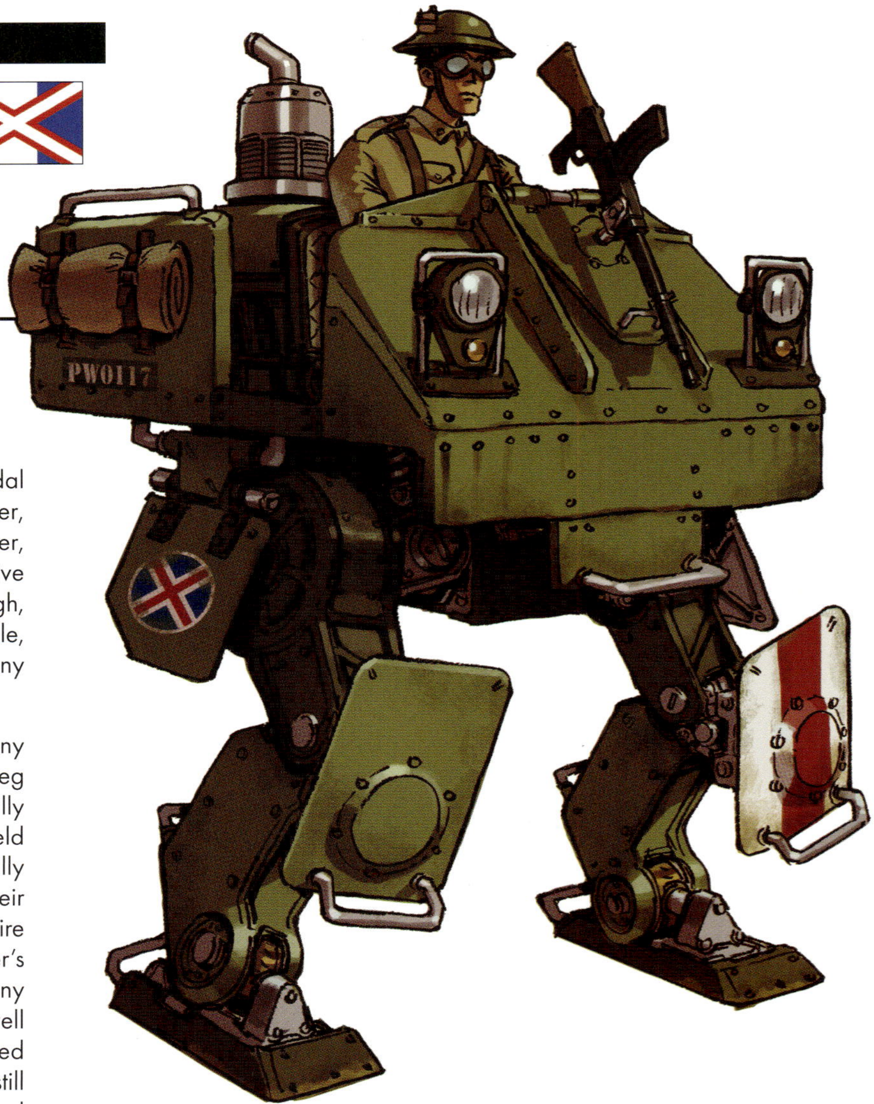

One of the most prolific and overlooked bipedal combat units of the Neverwar, the VCL Personal Walker, also known as the Universal Walker or Bren Walker, was originally designed as a light and inexpensive recon unit. During its long tenure in the war, though, it was adapted into almost every role imaginable, from amphibious assault vehicle or tank hunter to any number of auxiliary or engineering applications.

The well-rounded VCL never failed to fulfill any of the many roles it was made to fill. Its simplified leg mechanism and engine configuration were usually reliable and small enough to be serviced in the field when necessary. In this respect, the Walkers actually had an advantage over tankheads which, due to their size, required gantries and cranes, which would require more permanent workshop facilities. The Walker's remarkable speed meant it could get into or out of any position quickly, making it effective for ambushes as well as for reinforcing maneuvers. Its armor, never intended to stop any form of anti-materiel weaponry, could still stand up to significant small arms fire, and the raised position of the crew made them harder to hit, especially by infantry firing upwards from trench defense works.

The simple design of the Personal Walker meant it could be produced in any location with modest industrial capacity but it also meant it was relatively easy to refit the unit as needed for a variety of roles. The upper sections of the first series of Walkers were armed with a front pintle machine gun or, on the C variant, an anti-tank rifle. This already allowed the VCL to harass infantry or other light vehicles, but later developments saw almost any type of weapon mounted on its chassis. Most notable were the mortar carriers, which made good use of the unit's open-topped design to allow for fast reloading by the crew. The two-man variant maintained the pintle machinegun and Alliance crews of even the mortar variants are noted for never backing down from an infantry engagement at close range. This remarkable aggression and willingness to get stuck in at close quarters was fully utilized in the Wasp variant, which swapped the machine gun for a terrifying flamethrower that, contrary to popular belief, could be fired forward while running at full speed without issue.

Maybe the most famous variant comes from the Archipelago forces, who lacked an effective way to transport their 2-pounder anti-tankhead gyrojet cannons through the rough terrain of jungle islands. They devised a heavily modified version of the VCL with the engine positioned to the very front of the hull to allow for the cannon on the back to be placed in a fully articulated mount. This variant proved indispensable against the local Yamaichi forces, who wreaked havoc on heavier tankheads with lunge mines wielded by S4 Crossleys purchased from the Alliance years before.

It should come as no surprise that the VCL Personal Walker served well beyond the end of the war, continuing to see use both in auxiliary units as well as the more localized conflicts that broke out in the aftermath of the Neverwar.

VICKERS 75

A short-lived experiment into light tankhead mobile howitzers, only three prototypes were ever produced. The first, shown here, requiring a crew of four, was only capable of firing backwards from a kneeling position, though the driver needed to dismount before firing.

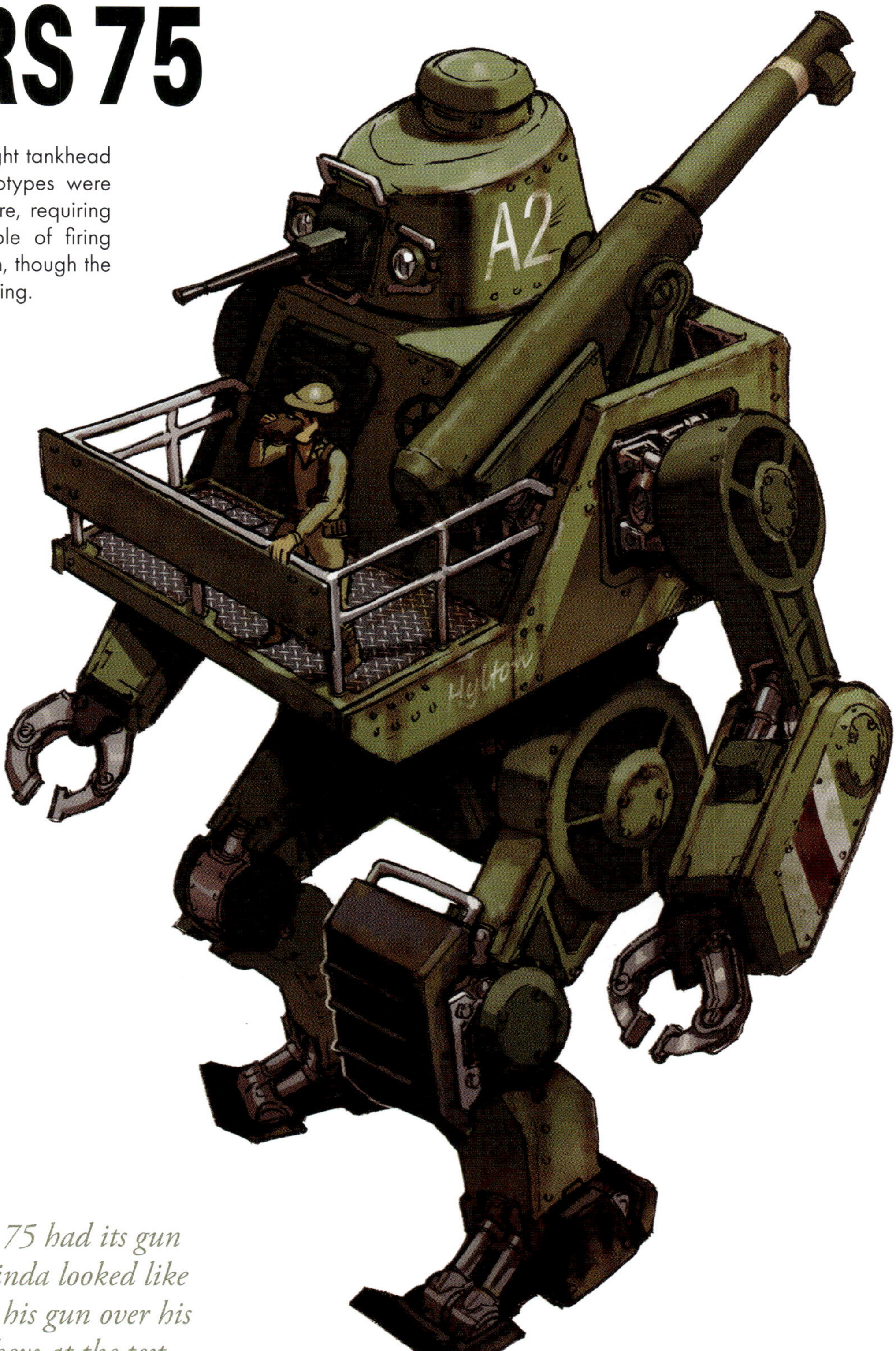

"'Cause of the way the 75 had its gun over its shoulder, it kinda looked like a guy marching with his gun over his shoulder, see? So the boys at the test range called it all sorts of names 'cause of that. Rifleman, Marcher, Beefeater, you name it. Figure we had more nicknames than actual walkers."
- Proving Grounds Overseer Teddy Glanville

SPECIFICATIONS

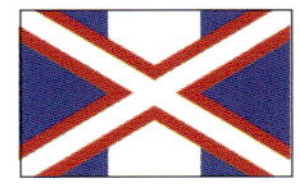

COUNTRY: Grand Britony
UNITS BUILT: 3
NO. OF CREW: 5
MAIN ARMAMENT: 1 x 6 in howitzer
SECONDARY ARMAMENT: 1 x 30-cal machine gun
ARMOR: 7–23mm
MAX SPEED: 6 km/h

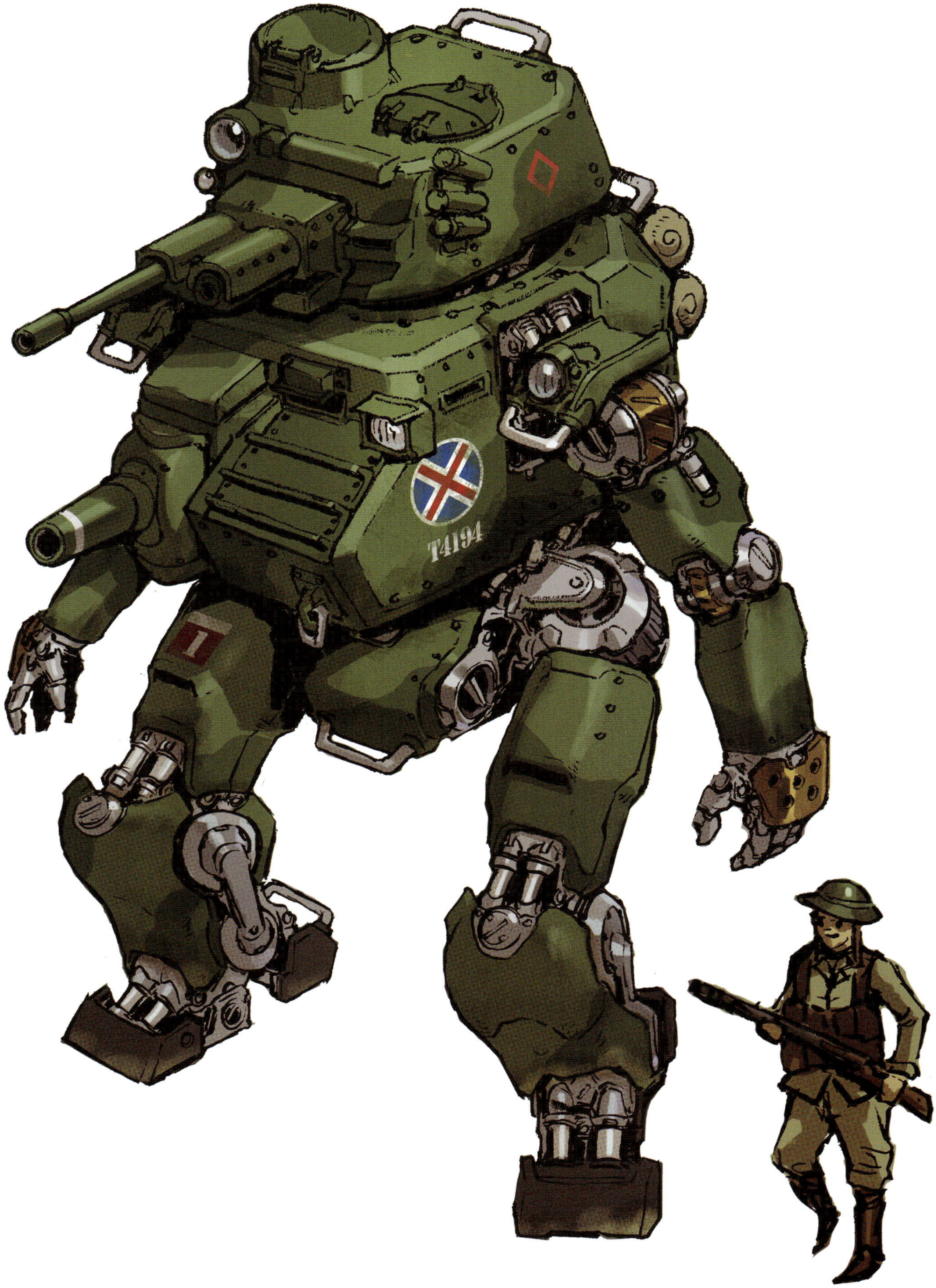
T4194
I

VICKERS MK.VI

It is very easy to dismiss the Alliance light tankheads as paralleling the early Kyzerian Panzerkopfs developed by the Empire during the same period. The truth, however, is that the two sides arrived at similar machines from very different starting points.

The development of the Alliance light tankheads started much earlier than the Panzerkopfs, long before the ceasefire agreements of WC012. But whereas the Kyzerians had a solid grasp on tankhead mobility from the outset from their study of the Isle à Von artifacts, the Brits were working from secondhand knowledge gathered from captured Stormwalkers. Although the early Brit tankheads were quite cumbersome compared to the already-lumbering Kyzerian counterparts, the developmental focus on improving mobility was apparent with units like the Whippet Mk.I.

Alliance Command wanted to fill a gap in their ranks between the fast wheeled tanks and the heavy tankheads, leading the Britonicans to develop the first light tankheads. These tankheads focused on the mobility aspect that Alliance forces had traditionally favored in war. The light tankhead line, exemplified by units like the Vickers Mk.VI, was the first to gain fully bipedal motion. By the time the ceasefire came into effect and production of heavy tankheads was prohibited, the Alliance had a very solid grasp on light tankhead design. To Alliance Command, the Vickers Mk.VI was the pinnacle of this unit type. It was, in their estimation, the perfect machine for fast anti-infantry response and, most importantly, a scouting machine capable of traversing steep and rough terrain with minimal risk of falling over.

Something that set the Mk.VI apart from earlier light tankheads was a short thumper cannon in a hull-mounted semi-swivel. This weapon was not a gyrojet but a short howitzer designed mainly for attacking fortifications, though it did see its fair share of anti-infantry and even anti-tank use. The official designation of the gun was "short thumper cannon." This was shortened to "stumper cannon" by crews, which eventually led to the entire short tankhead becoming affectionately nicknamed "Stumpy."

For a short period after the ceasefire, the Mk.VI was produced in greater numbers than any other Alliance tankhead. It performed outstandingly during those early days of renewed conflict. Soon, however, it would be outpaced by new, more agile medium and heavy tankheads. For the Alliance, though, such vehicles all used technological insights gleaned from the light tankhead program.

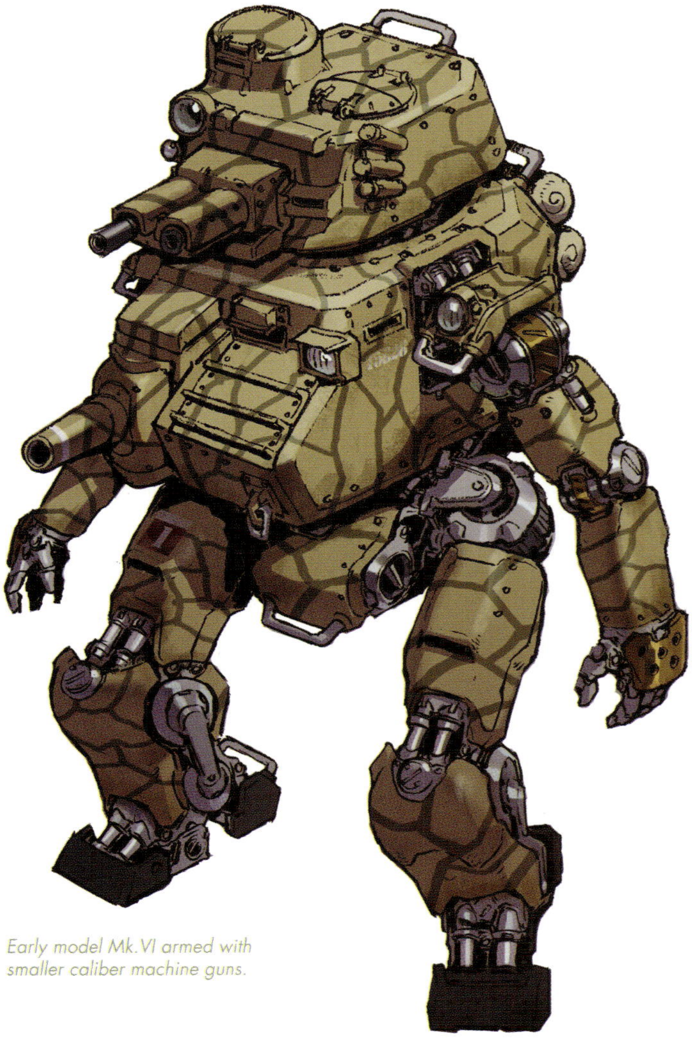

Early model Mk.VI armed with smaller caliber machine guns.

SPECIFICATIONS

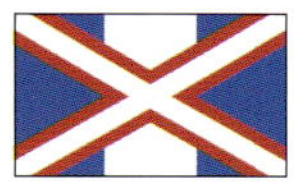

COUNTRY: Grand Britony
UNITS BUILT: 3,282
NO. OF CREW: 3
MAIN ARMAMENT: 1 x 6 in short thumper cannon
SECONDARY ARMAMENT:
- 1 x 15 mm heavy machine gun
- 1 x 7.92 mm machine gun
- 3 x smoke grenade launchers

ARMOR: 4–14 mm
MAX SPEED: 56 km/h

"The 6 inch stumper gun is firing, the 15 mm machine gun is firing, but somehow it's facing towards the rear. The 7.92 mm has been overheating and smoking for over an hour. I try to say 'Steady advance' on the A channel, but the driver – who is on the B channel – is backpedaling. I take a peek out the top of the turret, spot about a dozen enemy tankheads straight ahead, and just then someone hands me a ham sandwich."
-Lieutenant Keaton Gallarde, Britonican Tankhead Commander

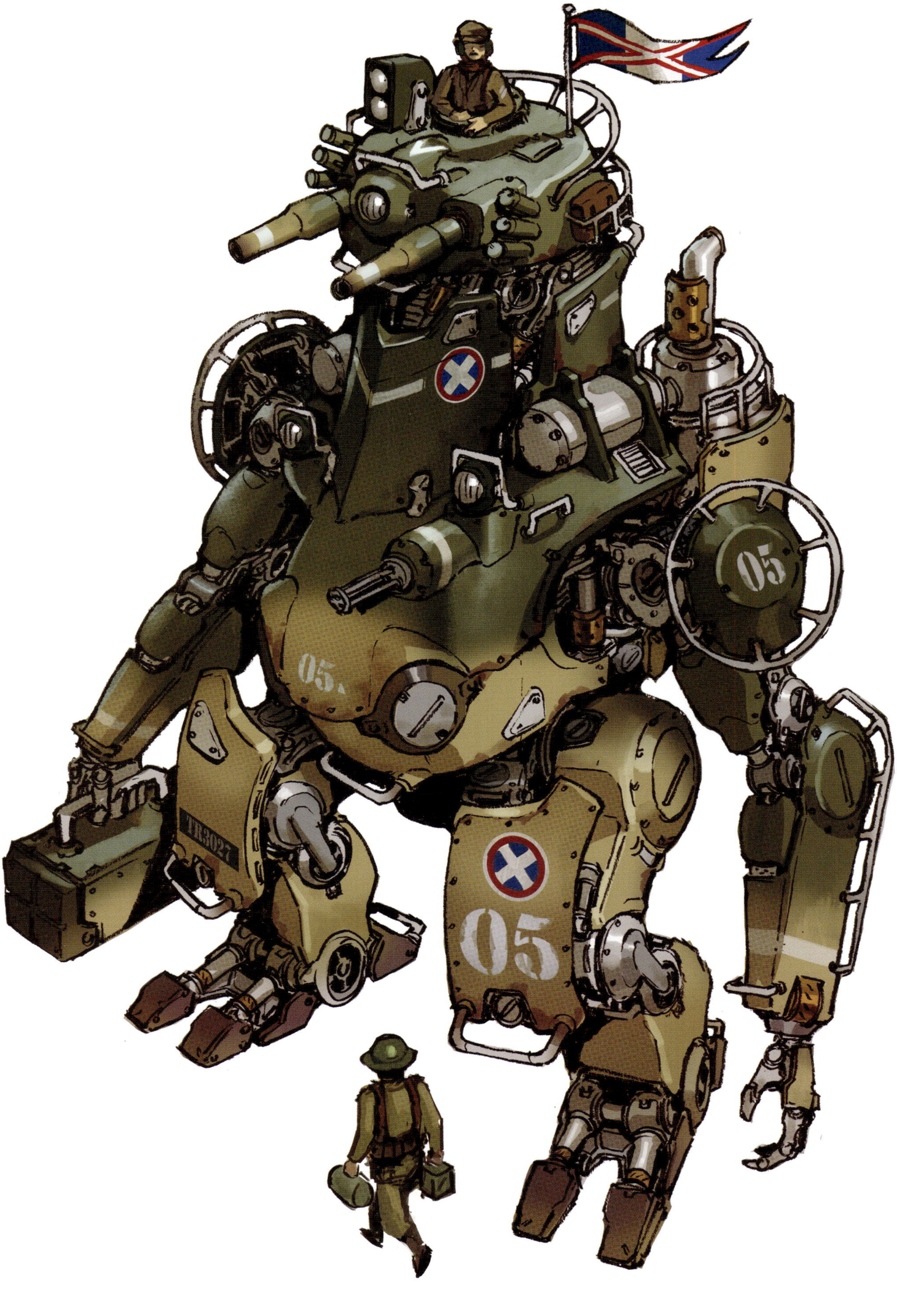
05
05.
TR3027
05

XM1 UGLY DUCKLING

The Vickers Mk.VI "Stumpy" finally provided the Alliance with a reliable bipedal mechanism that could match Kyzerian tankheads in speed and mobility. In light of this success, the Alliance was desperate to start production on medium and heavy tankheads capable of the same agility as the Vickers Mk.VI. Sadly, these ambitions overshot their technological capabilities at the time.

While Stumpy's legs did indeed provide good bipedal motion, it soon turned out the design had size limitations. Scaling it up too much resulted in catastrophic trips and failures during testing. Although solutions to this problem were a primary goal at the time, it was decided that production on something heavier than the Mk. VI should start as soon as possible. The XM1 design team were given very specific speed, armor and weaponry minimums, and were told to work around the limitations of the legs as best they could, as long as the XM1 could walk and be ready for production within months.

The upper height limit of the initial leg mechanisms was already known, so the XM1 design team expanded in the only direction they could, making the stance as wide as it was feasible. This allowed for a bigger torso capable of containing the power plant and weaponry requirements, if only just, with the engine sticking out in a large lightly-armored backpack.

This wide, ungainly leg configuration led to what could only be described as a waddle when running. All tankheads exhibit a back and forth swaying motion when moving at speed, and all need to find some way to compensate if they are to fire accurately on the move. In the case of the XM1, its side-to-side sway, as well as its back-and-forth tilt, was too extreme. The range of motion required from a stabilizer able to compensate could not fit into a turret head of any practical size.

In order to fulfill its intended role as the Alliance forces' main battle tankhead, a more conventional turret head was required. By this point, the XM1 prototypes had acquired the "ugly duckling" nickname among engineers for its waddling gait. One researcher took inspiration from the moniker and proposed reverse-engineering not from human biomechanics but from those of animals, ducks and pigeons in particular. A self-balancing head-bobbing mechanism was developed. This pneumatic stabilizer would, when operating as intended, temporarily "freeze" the head in place, staying still as the body moved below it. Then the head would snap back forward. This "bobbing" turret head created a brief window during which effective, accurate fire was possible.

Overall, the largest downside to this set of mechanisms was its height. The bobbing turret head's pendulum-like counterweights required long arms to swing on, and the mechanism ended up lifting the head of the XM1 up further. This made the design top-heavy, a shortcoming eventually compensated for by making the body lower and wider, and extending the arms. While restoring some equilibrium, this potbellied, gangly-armed assembly engaged the waddle further, even, some would say, to comedic levels.

Less than one hundred XM1 Ugly Ducklings were eventually produced, and only half of them are confirmed to have actually seen combat. While considered by many to be a pinnacle example of outside-of-the-box thinking, its unconventional proportions were inferior to the humanoid tankheads that soon became viable for the Alliance. However, the height of the Duckling, a major flaw in a frontline combat role, was looked upon more positively once the unit was relegated to a secondary line position, as the high turret could function as a mobile watch and guntower. Thereafter, the Duckling became a very popular machine for guarding ammo dumps and other important locations and facilities.

At the height of the war this led to a situation—unfortunate for Alliance operators but fascinating for historians—where Kyzerian tankhead crews came to understand that a Duckling head poking above trees or ridgelines likely marked the position of a valuable target. When communications about this were intercepted by Alliance codebreakers, Army Command came up with a devious counter. They positioned XM1 units as bait to draw Kyzerian raiding forces into ambushes. This strategy became known as the "Sitting Duck Gambit" and, while very effective before the Kyzerians adjusted, it led to the loss of the vast majority of functioning Ugly Ducklings. Of the few units remaining today, none ever saw combat, but were used for training or else never fully assembled, production having shifted to newer tankheads during the process.

SPECIFICATIONS

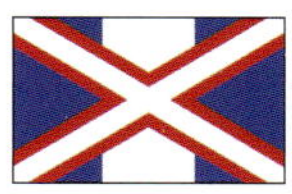

COUNTRY: Grand Britony
UNITS BUILT: 93
NO. OF CREW: 4
MAIN ARMAMENT: 2 x 37 mm gun
SECONDARY ARMAMENT:
1 x 7.62 triple-barreled heavy machine gun
6 x smoke grenade launchers
ARMOR: 15-30 mm
MAX SPEED: 27 km/h

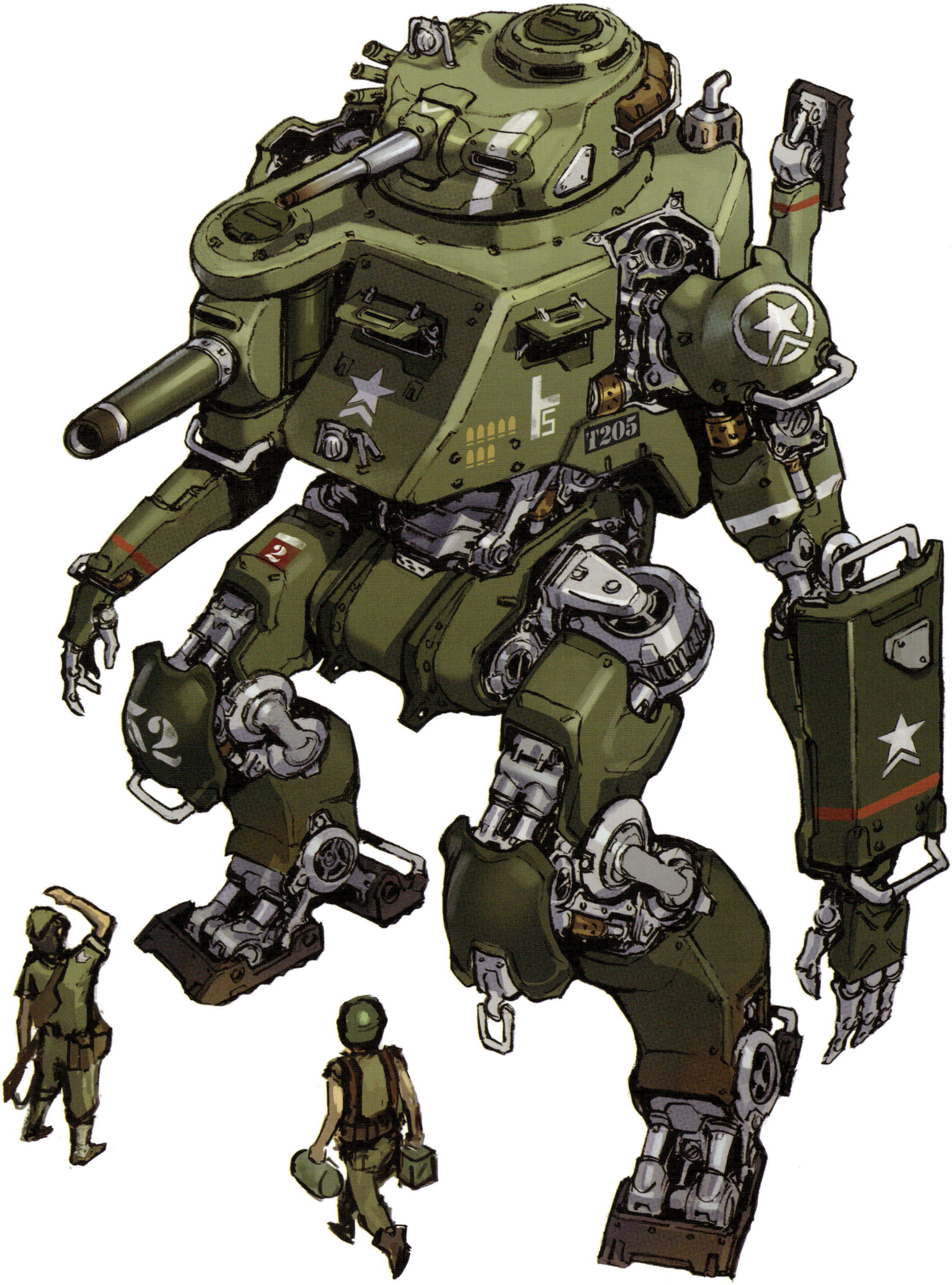
T205
2

M3 GRANT

During the ceasefire period of WC013 to WC015, the Liberterrans were generally optimistic that diplomatic talks between the Alliance and Kyzeria would lead to long-lasting peace. This has been attributed, both contemporaneously and today, to the nation's distance from the conflict and to a generally positive mindset stemming from the economic boom they were experiencing at the time.

When war did start up again in WC015, the Liberterrans quickly realized they lacked knowhow and experience in the production of tankheads. Looking back, it can be easy to dismiss this disadvantage as inconsequential. After all, the continental Grand Britony army came out of the ceasefire with a tankhead force majorly composed of the Vickers Mk.VI, which was arguably obsolete when the fighting began in earnest. However, Grand Britony seriously doubted that the Liberterrans could produce a tankhead equal to the Mk.VI, let alone surpass it.

The Britonicans would be proven half right. In particular, mounting the tankhead's main weapon on its turret and compensating for the recoil with neck actuators and fast-deploying rear stabilizers were engineering challenges that required more trials and testing than there was time for, now that fighting had resumed. A solution was sorely needed, as the Liberterrans weaponry had significantly developed. The high power gyrojet cannon was quickly settled on as the weapon of choice for this project, but had to be mounted on the hull to circumvent the recoil issue. The workaround had the added benefit of providing additional space to incorporate a cutting-edge stabilizer system. Unfortunately for historians, frontline reports as to the system's effectiveness vary wildly. Some operators claimed the system made the weapon the most accurate gun they ever crewed, even while moving, while others habitually shut the system down, complaining that it actually decreased the weapon platform's effectiveness.

To speed up development, the lower mobility assembly was largely copied from one of the few tankheads developed by the Liberterrans during the ceasefire, while its hull was completely redesigned. This actually complicated weapon placement, as the leg actuators used an unconventional mechanical linkage that extended from the pelvic joint up into the center of the body. This meant the weapon had to be laterally displaced, leading to the M3 Grant's recognizable asymmetrical design.

SPECIFICATIONS

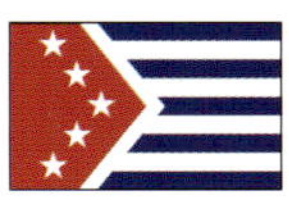

COUNTRY: Liberterra
UNITS BUILT: 1,058
NO. OF CREW: 6
MAIN ARMAMENT: 1 x 105 mm gun
SECONDARY ARMAMENT:
- 1 x 37 mm gun
- 3 x smoke grenade launchers

ARMOR: 14–51 mm
MAX SPEED: 50 km/h

Like the mobility assembly, the turret head weapon was retained from the previous ceasefire designs. While undeniably outdated, the relatively rapid-firing light anti-tankhead gun did provide the tankhead with a more effective secondary weapon than many of its contemporaries and allowed it to deliver some firepower in the arc of fire its main weapon could not cover.

Another area where the M3 profited from technological progress was the engine, which was both reliable and powerful for a tankhead its size. In fact, the M3 remained one of the fastest medium tankheads produced during the Neverwar despite its short development time and largely improvised design.

The M3 was surprisingly effective when it first debuted on the battlefield. Its mobility made it a useful strike tank during the early fighting of this period, when battlelines were quickly being redrawn as the balance of power shifted. During its initial engagements, it could outrun and outgun almost all Kyzerian tankheads, though it did not perform as well against the new designs that started to appear immediately after.

By WC043, the M3 had been rotated out of frontline combat in the Southern Front, replaced by more modern units like the M4 Sherman. The M3 had a much longer lifespan in the campaigns against the Yamaichi. Because the Dominion lagged behind in tankhead production during this period, the otherwise-outdated M3 was still an effective combatant. Its lightweight design also made it easier to transport during the island-hopping campaigns that front became known for.

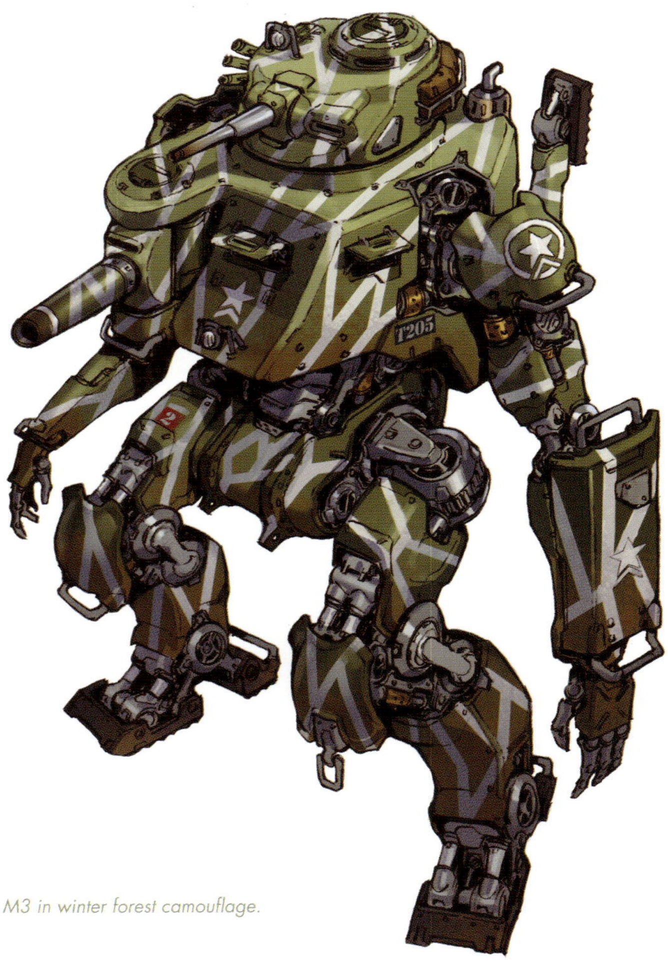

The M3 in winter forest camouflage.

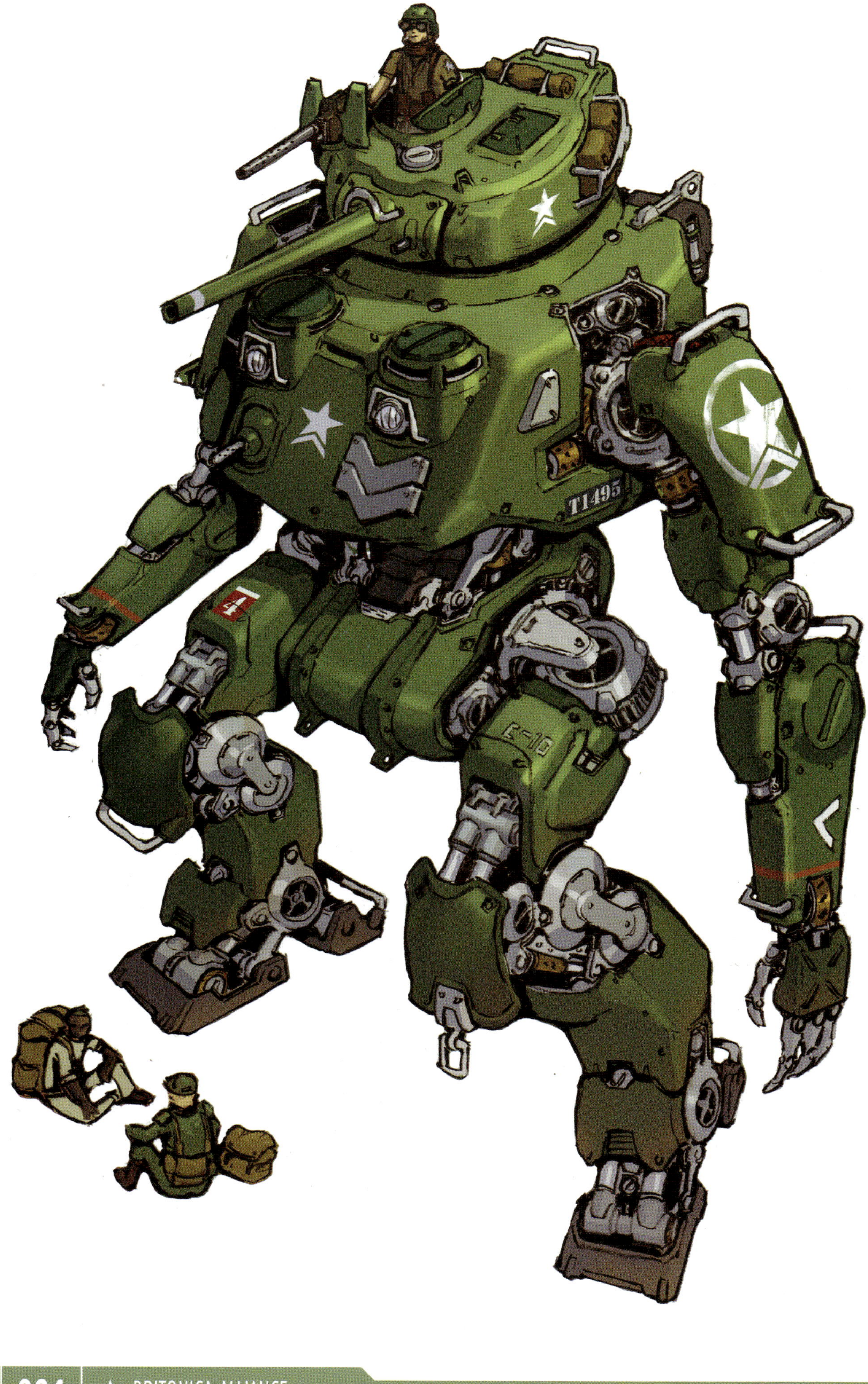
T1495
4

M4 SHERMAN

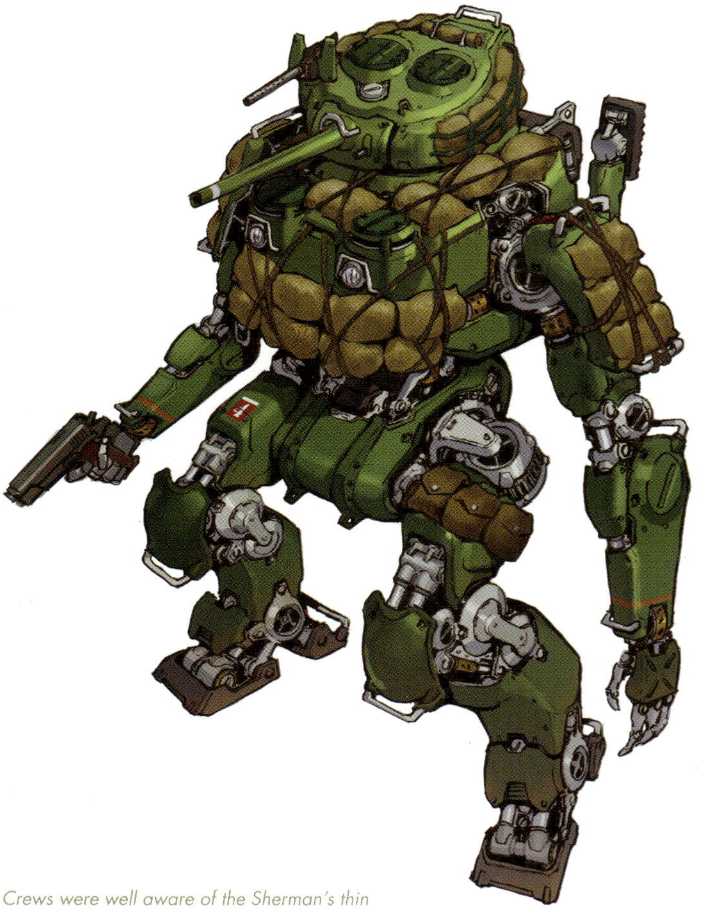

Crews were well aware of the Sherman's thin armor and occasionally bolstered its defense with sandbags or wooden logs. Note the 76 mm handheld pistol cannon, which was used in dense urban combat.

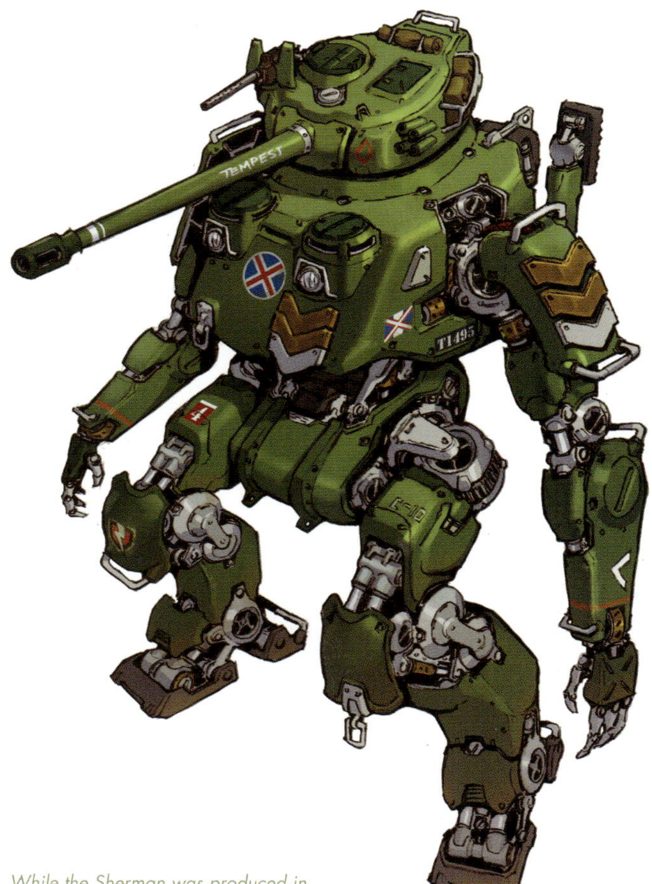

While the Sherman was produced in Liberterra, the entire Alliance made use of it, often with their own bespoke modifications. Grand Britony's Sherman Firefly, fitted with a powerful 3-inch gun, was arguably the most effective of all M4 variants.

The M4 Sherman was, in many ways, a greatly redesigned M3 Grant. It used many of the same components in its manipulator arm and leg assemblies. Even the hull was similar in design, though it featured a single main gun in a regular turret rather than the asymmetric body-mounted cannon of the M3. Its new main gun was chosen not so much as an anti-tankhead weapon but for its effectiveness against infantry and other light targets. The task of anti-tankhead combat would be relegated to specialized tankhead-hunter types and anti-tankhead gun units. That, at least, had been the strategy in mind when the Sherman was developed. As Liberterra was still not expecting any combat on their own home territory, there were strict design limitations put in place for the M4 to make sure it could be transported by train and be fitted into other existing infrastructure, and thus could be rapidly deployed to any front in the three-sided war. The result was that the Sherman was a bit narrower and taller than the designers would have preferred, but that was considered a reasonable tradeoff for speed and ease of reallocation.

While Alliance strategic doctrine required enemy tankheads to be engaged by specialized tankhead destroyers, in practice the M4s that encountered opposing armor had little choice but to immediately defend themselves. While the Sherman could hold its own against the main Kyzerian tankheads in terms of weapons, its comparatively weaker armor meant that it would have to rely on teamwork and surprise attacks to come out the victor. This was where two of the big advantages of the M4 became apparent: first, it was designed from the ground up to be produced in large numbers; second, it was quite easy to maintain in good condition, especially compared to its Kyzerian counterparts. As Kyzerian tankhead numbers were reduced and replenished more slowly, the Sherman was more likely to be accompanied by another operational unit (or several) that could help it take down its opponent.

CONTINUED>>

M4 SHERMAN (CONTINUED)

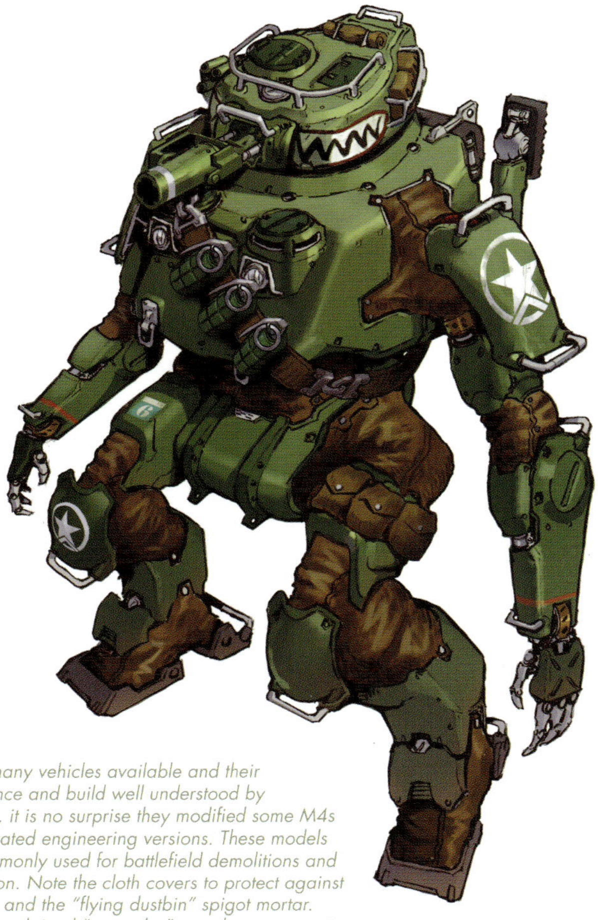

With so many vehicles available and their maintenance and build well understood by engineers, it is no surprise they modified some M4s into dedicated engineering versions. These models were commonly used for battlefield demolitions and construction. Note the cloth covers to protect against fragments and the "flying dustbin" spigot mortar. The tankhead-sized "grenades" seen here were not adopted for wide use, though, as operating them reliably with the tankhead's manipulators was found to be tricky—sometimes fatally so.

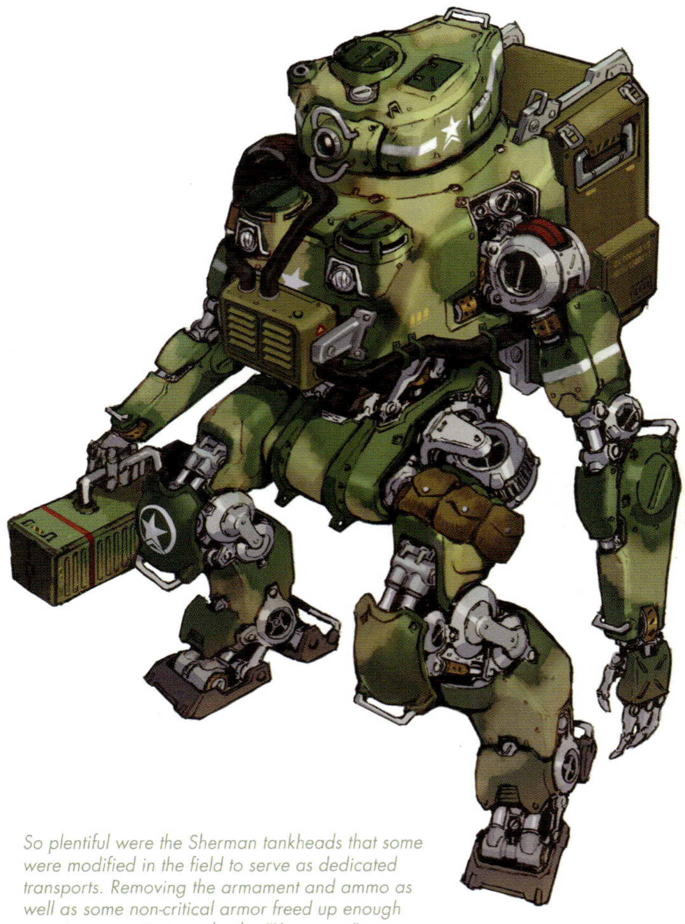

So plentiful were the Sherman tankheads that some were modified in the field to serve as dedicated transports. Removing the armament and ammo as well as some non-critical armor freed up enough carrying capacity to make the "Kangaroo" a versatile battlefield transport. The model shown here has been further modified with a front-mounted radiator to cool its beleaguered engine.

The balance would soon shift again, however, when a new generation of Kyzerian tankheads like the Tiger made their presence known on the battlefield. As mentioned previously, the M4's main gun was never intended specifically for anti-tankhead use and it was immediately apparent that the M4 stood no chance of penetrating the anterior armor of the Tiger tankheads. Faced with such an adversary, once again teamwork and superior tactics were the only way for the M4 to survive.

The ease of maintenance of the Sherman should not be underestimated. After all, the tankhead that works will always win against the supposedly superior one that doesn't.

While the Kyzerians and the Yamaichi strove to completely reinvent the tankhead over and over, the Liberterrans never stopped producing or improving the M4. Nearly seventy thousand Shermans were produced throughout the Neverwar. Of that staggering number, there were differences and augmentations made over the years, leading to many variants such as the anti-aircraft M4AA(20Q) Skink, the M4 Kangaroo transport and the M4A3 AVAE 230 engineering type.

Whenever there was a need for a new type of combat vehicle in the Alliance forces, it was only natural for Liberterran engineers to consider how a modified M4 might fill the role. Likewise, every conceivable type of weapon system had been tried out on a Sherman at one time or another. Sometimes these experiments never moved beyond the prototype, but others led to some innovative variants. For instance, the M4AA Skink replaced the mainline model's turret to create an anti-aircraft variant while others, like the M9X MacArthur, greatly changed hull design to fit a 155 mm self-loading recoilless gun system. These adaptive

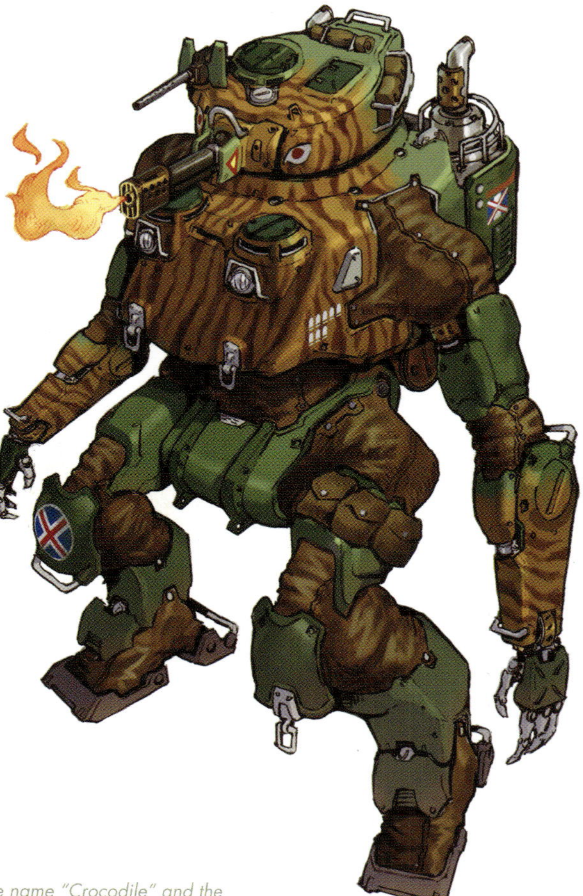

The name "Crocodile" and the flamethrower of this M4 variant reference the Britonnian flame tank, which Liberterran troops greatly respected. Note the reconfigured back-mounted power plant, which housed a large fuel bowser.

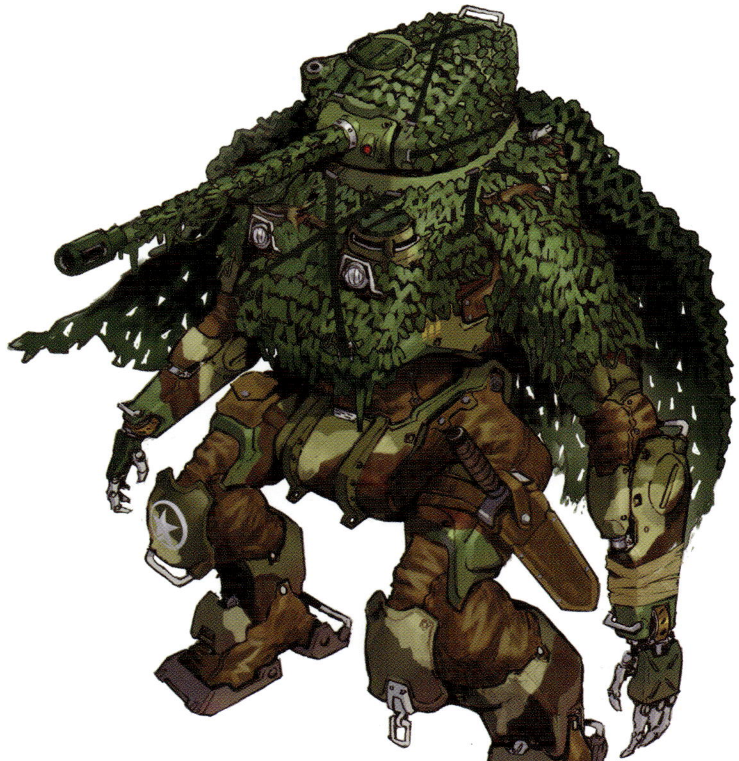

The Firefly was a devastating tank hunter but it still had the light armor of the regular Sherman. This meant that Firefly crews had to rely on stealth and speed to strike first, retreating quickly if the target tankhead was not destroyed in the initial strike. Note the engineering joint covers used here to allow the tankhead to submerge into boggy water without jamming its mechanisms, allowing the unit to utilize more unconventional hiding spots.

changes had the added benefit that the crews were already familiar with their components and thus would have an easier time maintaining them.

With huge production numbers, it also became possible to equip other forces in the Alliance military with the M4. The beleaguered Rozenians valued it because its armor was superior to their own T-34. To their dismay, the somewhat tall Sherman with its narrow ground contact would sink in and get stuck and sometimes even fall over in the Rozenian mud. Brit formations also made great use of the M4. Due to early, heavy losses following the end of the ceasefire, Liberterran Shermans were used to reinforce weakened units. They would gradually become more and more common in Britonican forces. In fact, by the end, they were the most numerous tankheads in both nations' militaries. Brits also developed a number of their own variants—most notably the Sherman "Firefly," which mounted a highly effective quick-firing 17-pounder cannon, turning the Sherman into a deadly tank hunter quite capable of destroying even the once feared Tiger tankhead.

SPECIFICATIONS

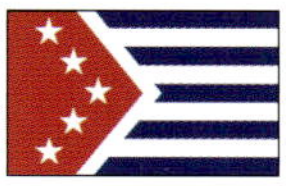

COUNTRY: Liberterra
UNITS BUILT: 69,234
NO. OF CREW: 5
MAIN ARMAMENT: 1 x 75 mm gun
SECONDARY ARMAMENT:
2 x 30-cal machine guns
1 x 50-cal heavy machine gun
ARMOR: 12–177 mm
MAX SPEED: 48 km/h

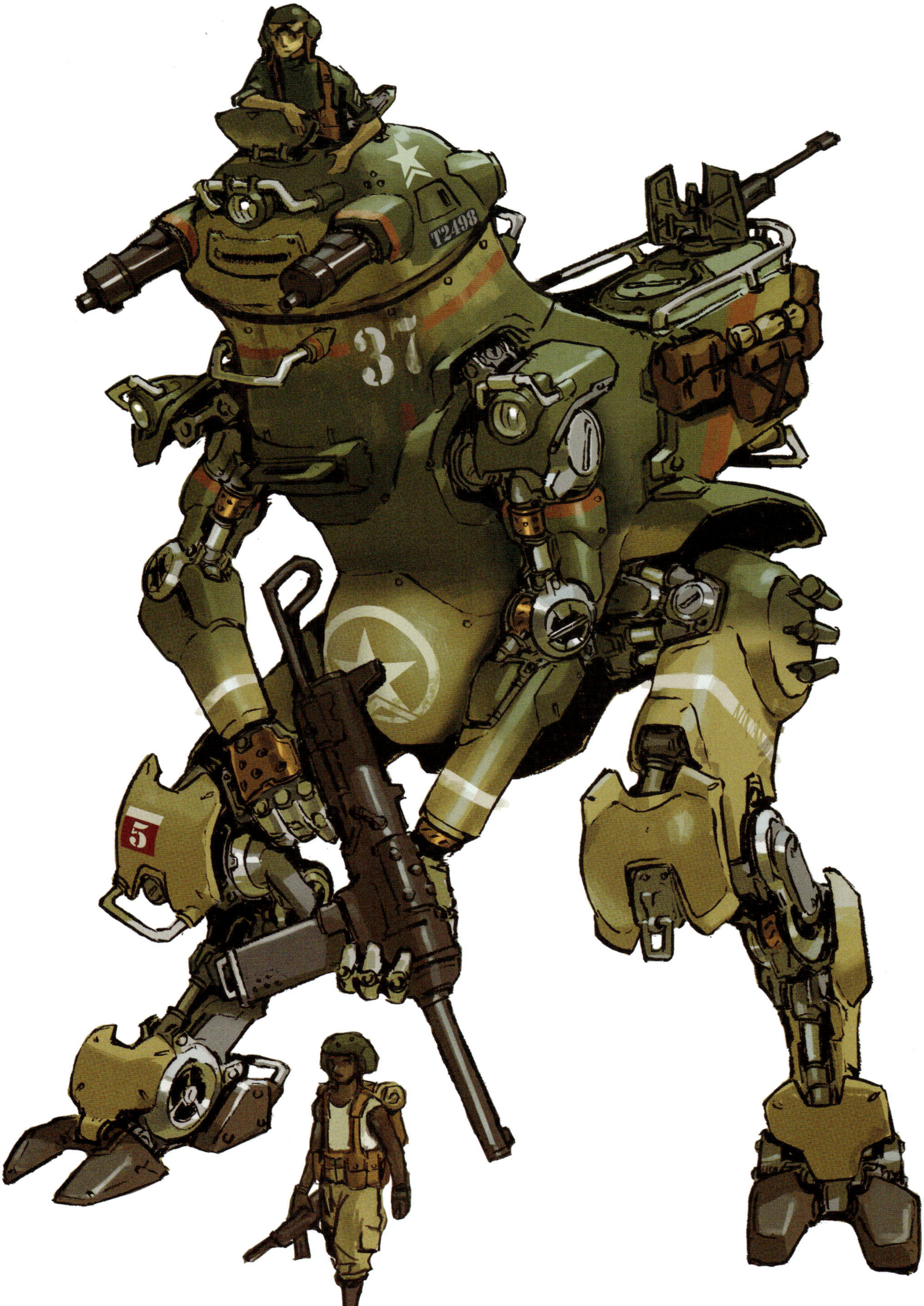
T2498
37
5

S4 CROSSLEY

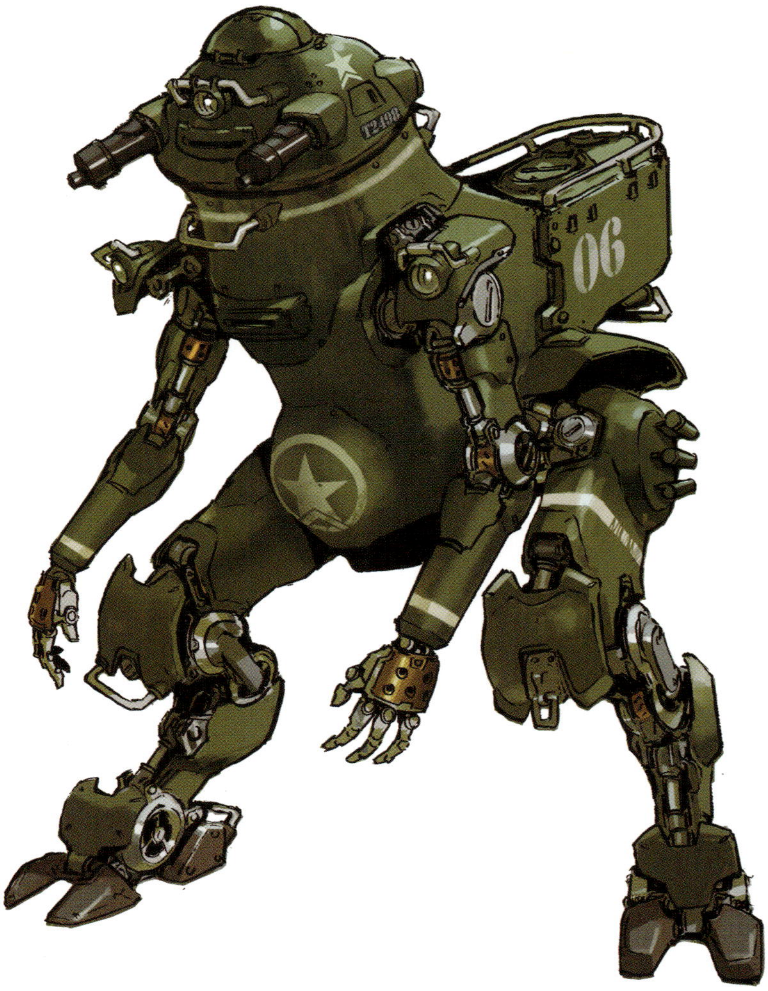

A stock Crossley without any additional armament.

Mirroring the development of the Kyzerian Späher scout tankhead to a significant degree, the Crossley came about during the ceasefire period of WC012. With nations being forced to build lighter tanks under the stipulations of the ceasefire agreements, the idea of a light scouting tankhead suddenly became enticing to development branches. Thus, the S4 Crossley was born.

It is notable that, during this much calmer ceasefire period, a few dozen Crossleys were sold to the Yamaichi Dominion early in WC014, before the troubles in the Darumeshia Archipelago escalated that year. To get around restrictions on exporting combat vehicles during wartime, these vehicles were sold without any armor or weaponry. The Yamaichi armored and armed their Crossleys differently from the Alliance models in several notable ways, further adding to the general confusion that surrounds this enigmatic tankhead.

The Crossley featured a digitigrade leg configuration, much like the Späher Panzerkopf, but unlike its Kyzerian counterpart, the Crossley lacked rebound spring suspension that would have enabled sprinting. Instead, the Crossley's designers opted for a more circular pattern of leg movement reminiscent of walking birds. This led to the unit being nicknamed "Ostrich" during action in the Alcaban desert theater, where it was used by the Auxiliary Desert Patrol Unit.

While never intended as a front line tankhead, the Crossley carried ample small-caliber firepower in its stock configuration, making it suited for raids against unarmored targets.

SPECIFICATIONS

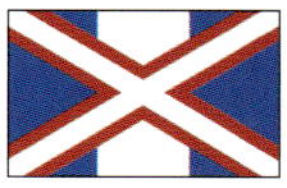

COUNTRY: Grand Britony
UNITS BUILT: 3,850
NO. OF CREW: 4
MAIN ARMAMENT: 2 x 7.7 mm machine gun
SECONDARY ARMAMENT:
- 1 x handheld 75 mm rapid-fire cannon
- 1 x 50-cal heavy machine gun
- 6 x smoke grenades

ARMOR: 20 mm
MAX SPEED: 95 km/h

CONTINUED>>

S4 CROSSLEY (CONTINUED)

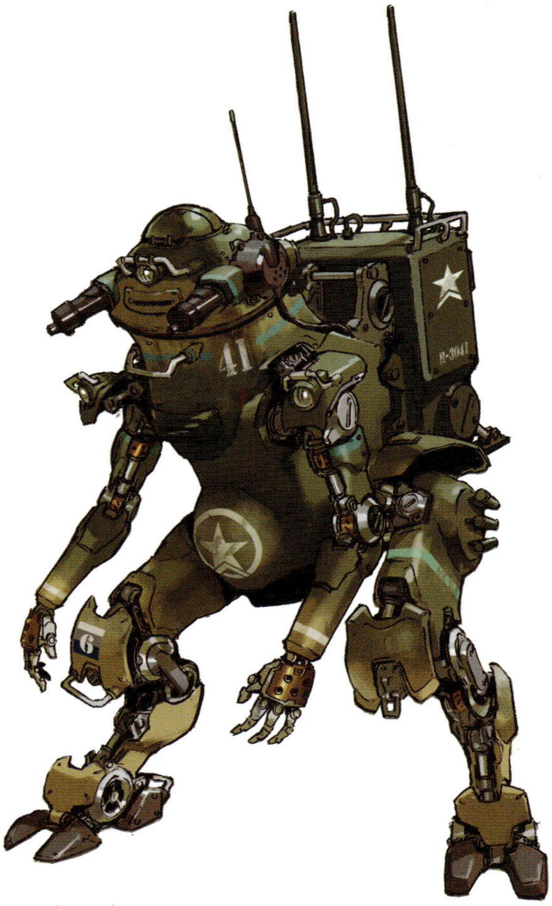

Due to its agility and modulability, the Crossley was indispensable as a radio vehicle for scout units moving through rough terrain.

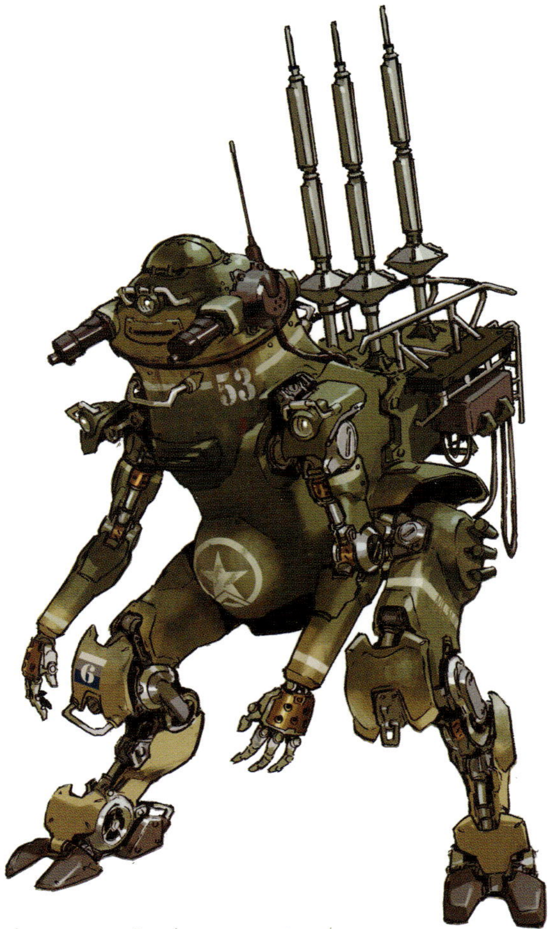

Late in the war, some Crossleys were equipped to function as radio relays. Their combat use had declined over time but, by placing one of these units at a high point, communications throughout the area could be greatly improved.

However, its real strength lay in its modular backpack system and overall uncomplaining, adaptable design. The renowned Auxiliary Desert Patrol Unit, better known as the "Desert Rats," successfully demonstrated how this machine could be adapted for specialized use by its crew. The Desert Rats became famous for strapping as many guns, fuel, and water containers as possible to their Crossleys when crossing the vast deserts of Alcaban. They also showed that the Crossley's notably dextrous arms could be used to wield all sorts of weapons, including many for which it had not been designed. Most tankheads had trouble balancing while walking if they were holding weapons in their hands, but the Crossley's peculiar gait, combined with the counterbalance of its backpack, alleviated this problem. The rapid-fire handheld weapon known as the M3 "Grease Cannon" was a late war addition semi-officially developed for the Crossley by the Liberterran Army Division. This weapon system would continue to see use in a number of conflicts after the war.

The backpack mount was first designed to allow for an optional secondary gunner position. Even before the Crossley was fully deployed, however, engineers realized that the machine had an adequate load capacity to mount a variety of alternate systems in this position. Most official modules designed for the Crossley were of types one would expect for a light scouting vehicle: additional observation or communication equipment, command positions, and transport. A light mortar was also experimented with, with mixed results. Field modifications extended these options for carrying additional cargo and for performing simple engineering jobs. More famous now are the offensive modifications that equipped the otherwise lightly-armed tankhead with flamethrowers or rocket launchers. In the latter half of the war, anti-air-weapons modules were developed to help combat the rising threat of jetheads.

The tendency to improvise additions or substitutions was mirrored on all fronts where the Crossley was deployed, both by the Alliance as well as the Yamaichi (and even the few captured by the Kyzerians). This fact has led to great confusion as to which weapons were developed for Crossley use through official channels and which ones were improvised in the field. The same can be said about its modular backpacks, which were modified in the field just as often, if not more.

M4F CRAB MINEFLAYER

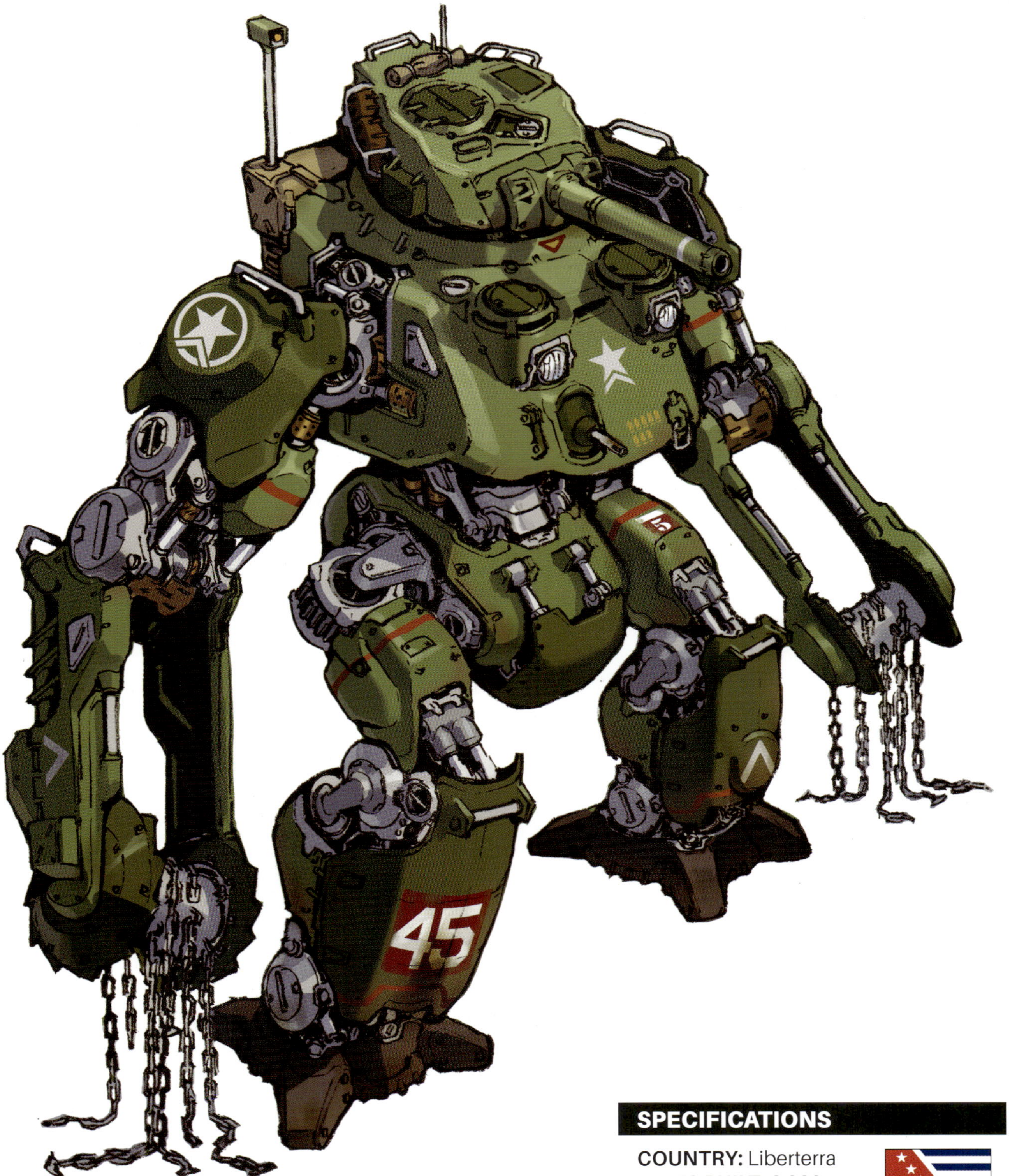

A specialized obstacle-clearing variant of the Sherman, the M4F Crab was notable for its reinforced and padded feet, as well as for the shoulder-mounted telescoping beacon to help mark its position in the inevitable dust clouds. But most notable, of course, are the chain flails, which rotate at high speed and "plow" the ground in front of the tankhead to safely detonate mines and clear a path for other vehicles and tankheads.

SPECIFICATIONS

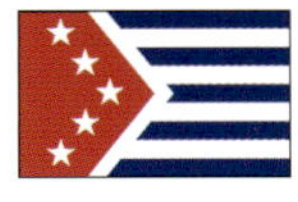

COUNTRY: Liberterra
UNITS BUILT: 2,000
NO. OF CREW: 5
MAIN ARMAMENT: 1 x 75 mm gun
SECONDARY ARMAMENT:
2 x 30-cal machine guns
ARMOR: 13–178 mm
MAX SPEED: 20 km/h

ROZENIAN TANK FACTORY

The discovery of the Isle à Von artifacts, followed by the production and rapid innovation of tankheads, led to many new technologies and techniques that soon were also adopted in industrial processes. The factories that resulted were simply in another class compared to those built before the Neverwar.

The factories in Rozenia produced mechanical cogs of war, of both machines and men.

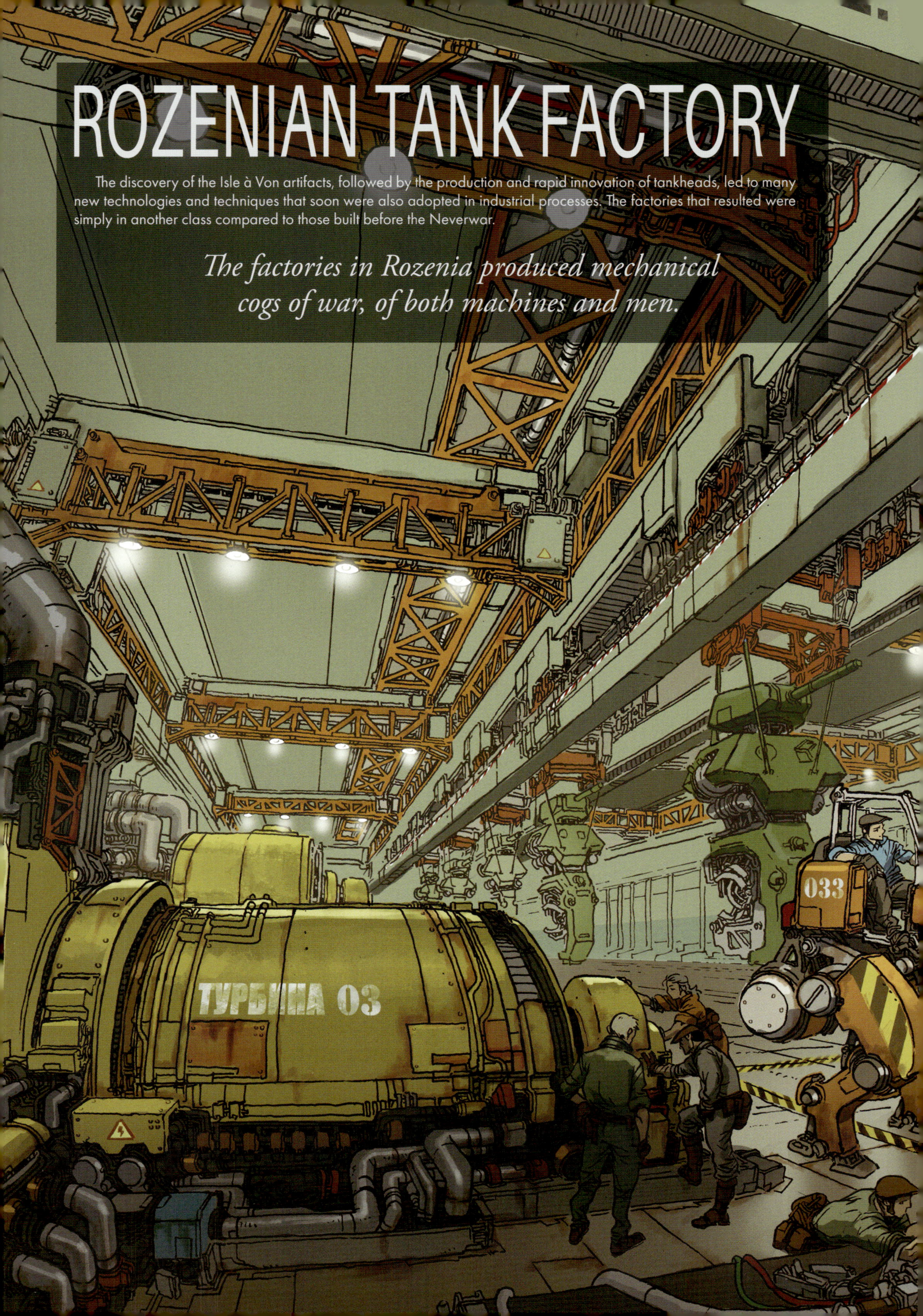

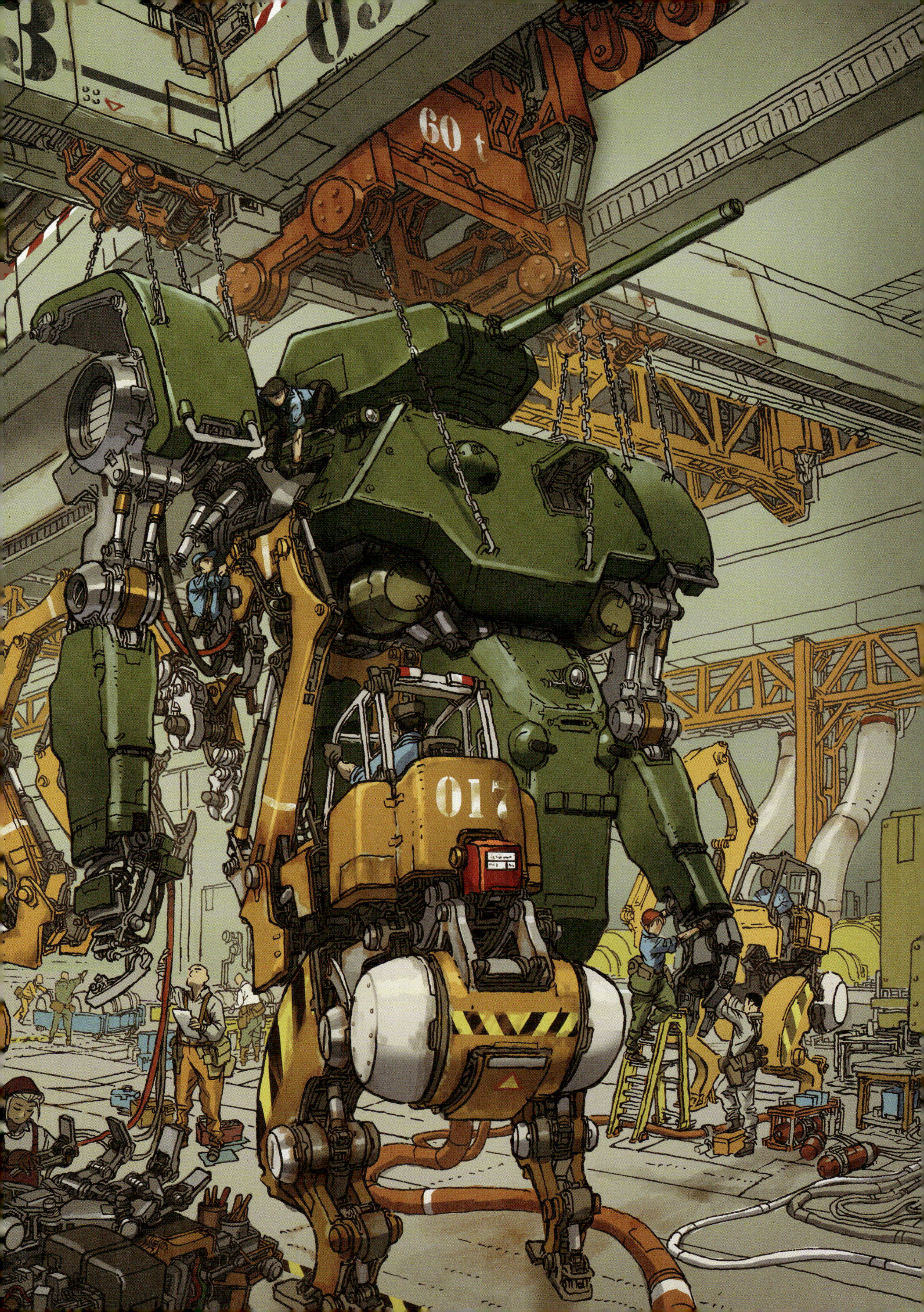
60 t
017

TK-1 TANK KONG 1

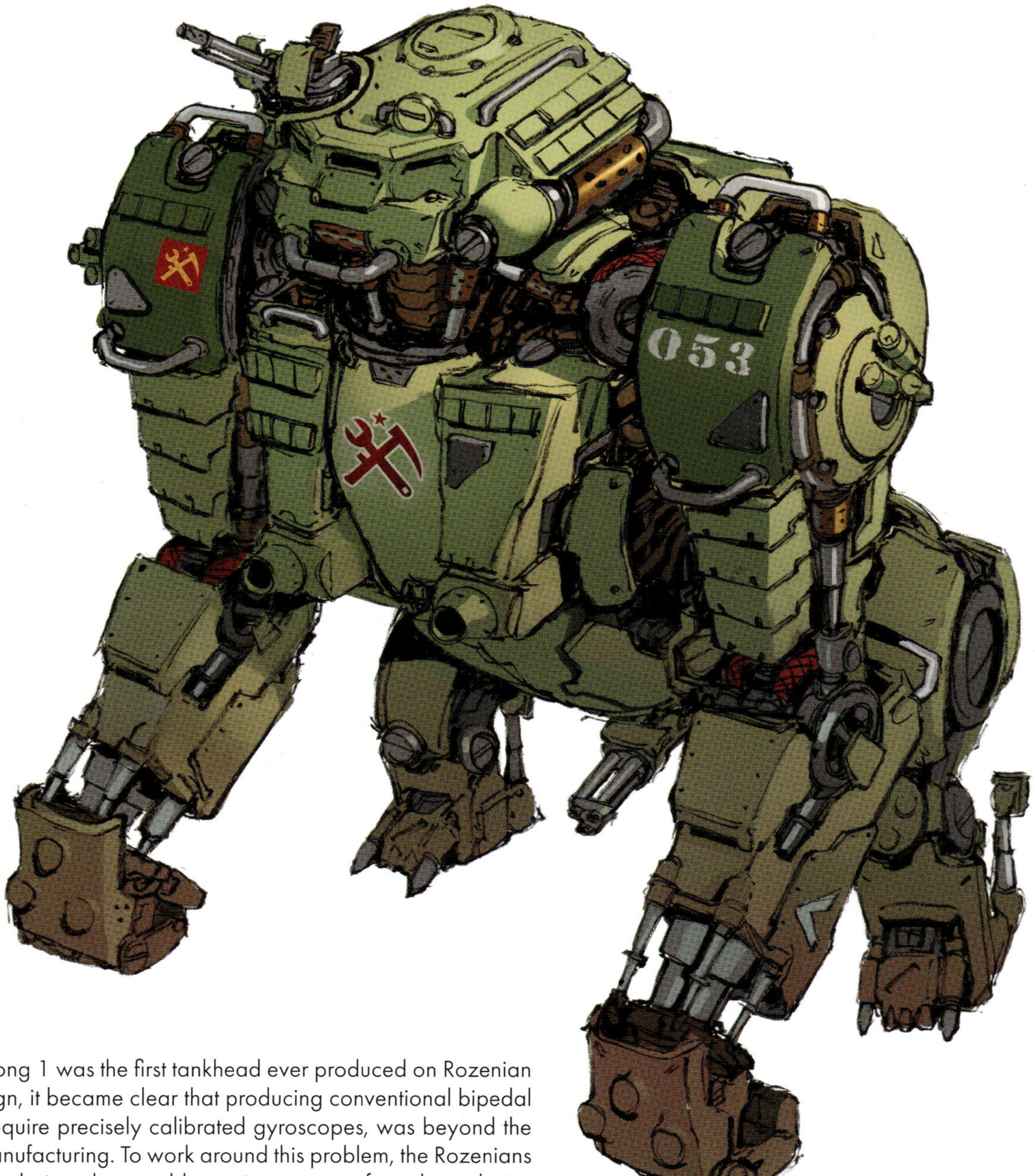

The TK-1 Tank Kong 1 was the first tankhead ever produced on Rozenian soil. Early in its design, it became clear that producing conventional bipedal tankheads, which require precisely calibrated gyroscopes, was beyond the capacity of local manufacturing. To work around this problem, the Rozenians produced a hybrid design that could use its arms as front legs during movement to augment its stability.

Early prototypes proved that this configuration was not only steady but also capable of transporting great weight. On the other hand, it suffered from a very low top speed. Capitalizing on these qualities, the Kong became a trusted workhorse of engineering divisions on the Rozenian front. Most models were equipped with twin smoothbore mortars, used for demolishing targets or firing smoke rounds, and an underslung machine gun, which has received a plethora of pejorative nicknames over the years.

The Kong's unconventional configuration did result in a number of troublesome quirks. For example, the hydraulic expansion cylinders located on either side of the turret head would get dangerously hot during operations. Nevertheless, the raw power of the frame made the Kong a beloved design among engineers and it was still in use well over a decade after the war.

SPECIFICATIONS

COUNTRY: West Rozenia
UNITS BUILT: 5,600
NO. OF CREW: 5
MAIN ARMAMENT: 2 x 100 mm mortar
SECONDARY ARMAMENT:
- 1 x 20 mm multi-barreled rotary cannon
- 1 x 7.62 mm heavy machine gun
- 6 x smoke grenade launchers

ARMOR: 16–100 mm
MAX SPEED: 32 km/h

TK-1 TANK KONG 1 (VIKTOR MAIMIRSK CUSTOM)

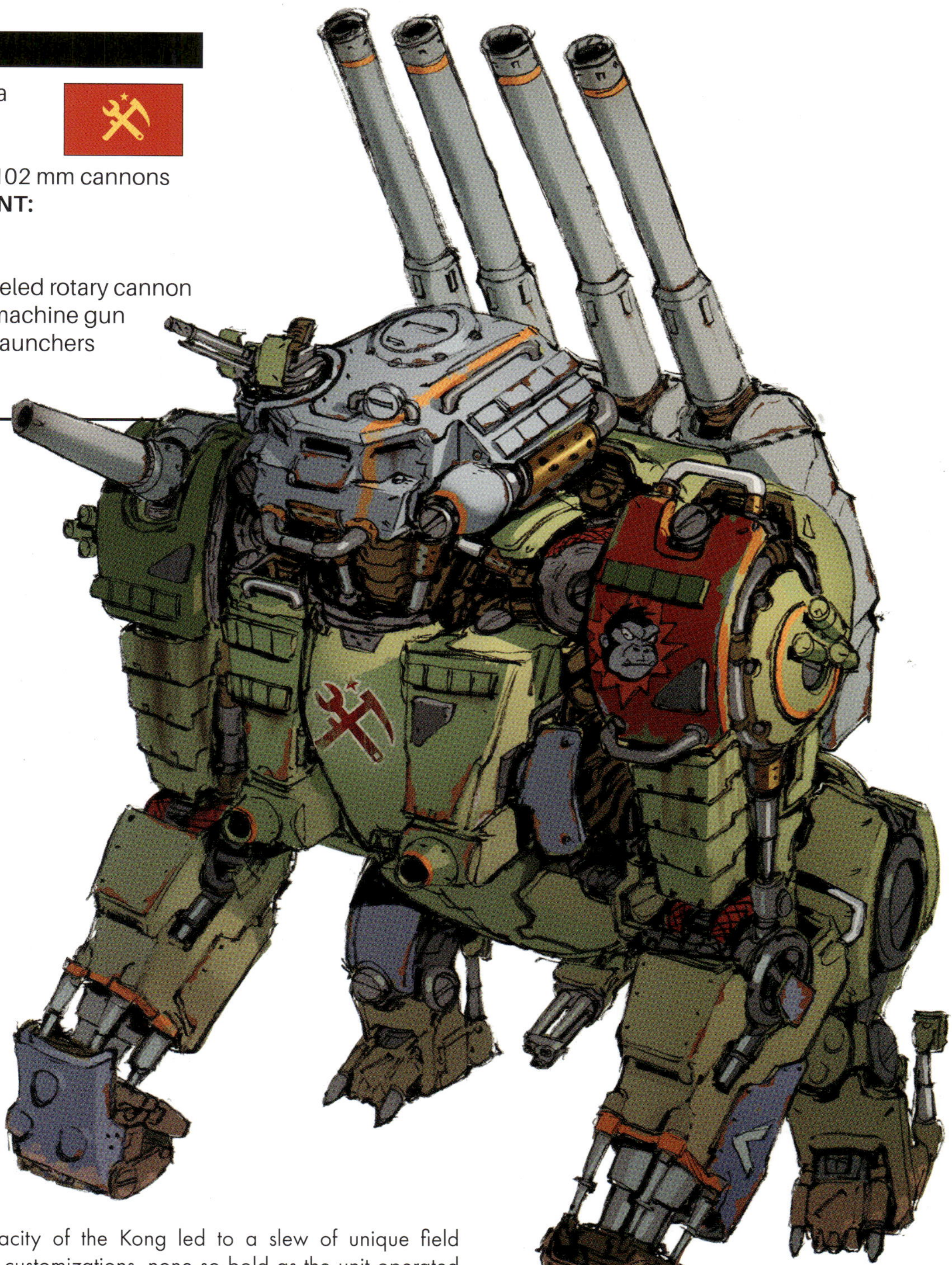

SPECIFICATIONS

COUNTRY: West Rozenia
UNITS BUILT: 1
NO. OF CREW: 5
MAIN ARMAMENT: 4 x 102 mm cannons
SECONDARY ARMAMENT:
- 2 x 100 mm mortars
- 1 x 75 mm cannon
- 1 x 20 mm multi-barreled rotary cannon
- 1 x 7.62 mm heavy machine gun
- 6 x smoke grenade launchers

ARMOR: 16-120 mm
MAX SPEED: 27 km/h

The tremendous load capacity of the Kong led to a slew of unique field adaptations and personalized customizations, none so bold as the unit operated by the famous West Rozenian ace, Commander Viktor Maimirsk. The primer-gray head with orange stripe had already become a famous icon when Maimirsk led a raid on a Kyzerian cruiser docked on the north Rozenian coast for refits. After capturing and scuttling the vessel, Maimirsk found that one of its turrets had been removed to the docks for maintenance. Maimirsk had it fitted onto his personal Kong unit. It is still unclear how these weapons were reloaded in the field.

T-34 GURKO

SPECIFICATIONS

COUNTRY: West Rozenia
UNITS BUILT: 84,070
NO. OF CREW: 4
MAIN ARMAMENT: 1 x 85 mm gun
SECONDARY ARMAMENT:
- 3 x 7.62 mm machine guns
- 1 x handheld 50 mm rapid-fire sub-cannon
- 6 x smoke grenade launchers

ARMOR: 20–90 mm
MAX SPEED: 53 km/h

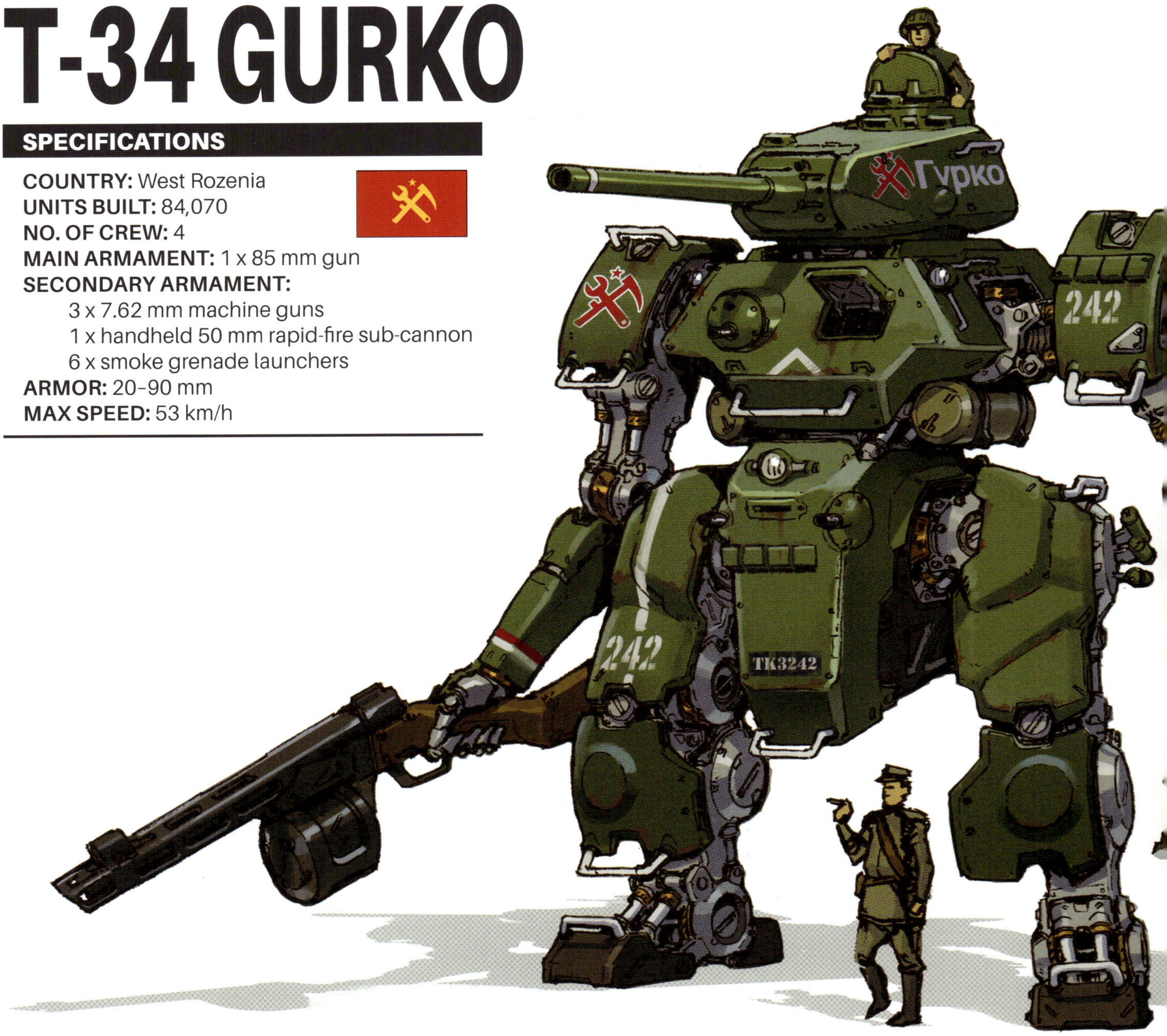

In WC001, Rozenia faced unrest following the assassination of their exiled crown prince and his Britonican wife, heralding the inauguration of the War Century. In WC004, they faced civil war after the death of their tsar, which ignited the Neverwar in earnest. In the fall of WC005, Rozenia would become the focal point of the war between the Alliance and the Empire in the latter's Nolburg Plan. When the ceasefire agreements of WC012 were signed between the Alliance and the Empire, it marked the first time that Rozenia would not be an active battlefield in more than a decade.

Respite from war did not lead the Rozenians to idle, however. A fervor to retake the entirety of their homeland burned strong in the Western Rozenians and, with the looming threat of a new conflict ever present, they set about rebuilding their war industry. The ceasefire document stipulated strict weight limits on production of all military vehicles in signatory nations. As a result, the Rozenians suddenly found themselves on a much more even playing field with both their allies and enemies, who were now limited in what tankheads they could produce. With tankhead technology provided by the Alliance and captured from the Empire, the West Rozenian provisional cabinet immediately funded a project to design and produce a vehicle that pushed the upper limits of the weight provision. Perhaps the Kyzerians would have considered the project provocative if it had been conducted under the auspices of the Alliance. In the event, though, their reaction seems to have been one of dismissal or even contempt.

Typical of the resourceful Rozenians, the tankhead they produced, the T-34 Gurko, was a fascinating mix of cutting-edge technology, ingenious design, and outright jerry-rigging. The angled armor on the hull was unique for its time. On the other hand, the turret and hull armor were very crude, with imprecise castings, pockmarked surfaces, thick welds, and misaligned rivets. These visual imperfections were simply aesthetic, however, and had no significant impact on functionality. Everything that needed to fit, fit; and, ultimately, the only places where corners had been cut were where there had been room for it.

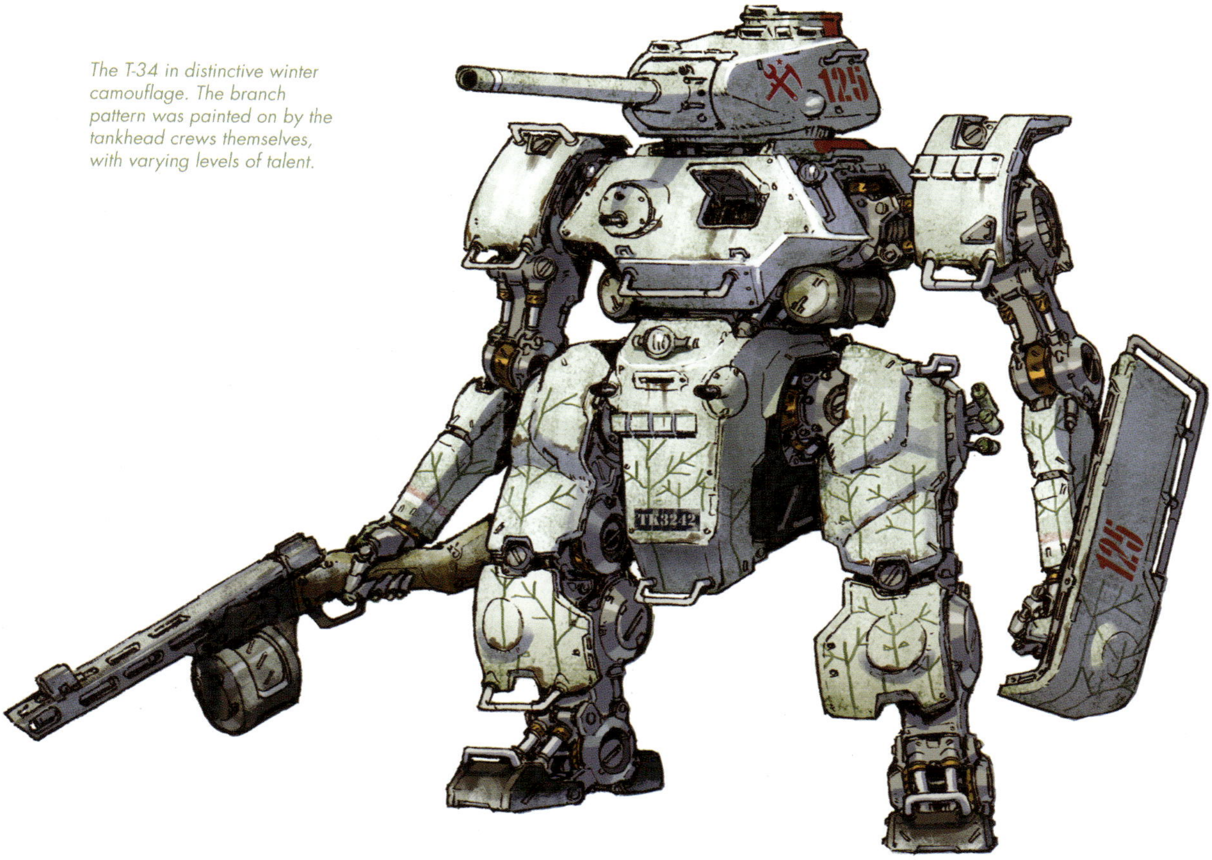

The T-34 in distinctive winter camouflage. The branch pattern was painted on by the tankhead crews themselves, with varying levels of talent.

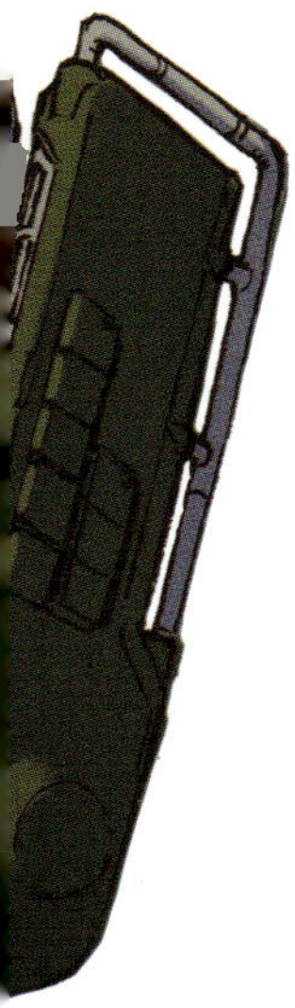

*"If it fits, it's good enough.
If it doesn't fit, make it."
- West Rozenian Chief Tankhead
Engineer Anto Dyatshi*

Due to the decentralized nature of Rozenia's manufacturing industry, parts for the new T-34s were fabricated by many small factories and workshops throughout the country before being assembled at designated plants. In Kyzeria or the Brit countries, where industrial logistics had already been refined over a number of years, all parts would have been labeled and numbered prior to shipment, variations clearly marked, and accordingly sorted at receiving plants. The Rozenians took a more pragmatic approach, stacking parts by type in the order they were received. Since newer parts were stacked on top of older ones, many late-model T-34s actually incorporated older components.

The T-34s saw combat for the first time in WC014, when New Kyzeria invaded northern Rozenia. During these first fights a number of flaws became apparent immediately. Most crippling was the fact that engines in the early models had a tendency to overheat and their internals quickly wore out due to the lack of proper air filters. To extend the tankhead's longevity, T-34 crews were not allowed to train in their designated tankheads for more than a couple of hours at a time.

Kyzeria's excessive pride in what it saw as its superior engineering acumen had clouded that nation's judgment when it had first heard about the T-34 project. When the Kyzerian Panzerkopf III encountered the T-34 in the field, however, they found the "rough and tumble" Rozenian tankhead to be a hard nut to crack, able to weather multiple direct hits. The T-34s, on the other hand, managed to cripple or destroy Panzerkopf IIIs with surprising regularity.

In light of its success, the T-34 would be produced in staggering quantities for the duration of the war. The unit's design only saw a few significant upgrades, most effective of which proved to be the 85 type, which received a much more powerful high-velocity gyrojet cannon. This allowed the T-34-85 to effectively menace Kyzerian Tigers, despite the latter having superior armor.

Notably, the T-34 was also one of the few tankheads to be seen regularly wielding handheld weapons. This concept had been proposed even at the beginning of tankhead development, but testing had always revealed the same problem—specifically, a debilitating tradeoff in balance. On the move, a tankhead uses its arms to balance itself, not unlike a human. Handheld weapon systems led to unevenly distributed weight, which in turn resulted in more crashes and an overall reduction in handling. Compounding the issue, the arm movement was normally controlled by the driver or commander, who did not have the time or angle of view to carefully aim handheld weapons in a combat situation.

Despite all of this, the Rozenians designed the T-34 with an accompanying handheld rapid-fire cannon, using a simple, heavy shield in the opposite hand to account for the weapon's weight. Why, when all other armies had decided to scrap handheld weapon systems, did the Rozenians persist? The reason was actually bureaucratic. There was a loophole in the wording of the WC012 ceasefire agreement: tankhead handheld weapons were counted as independent weapons, like field guns, rather than parts of a tankhead's armament. Therefore, they did not "count" toward the machine's official loaded weight, allowing the Rozenians to skirt the ceasefire weight limits.

KV-2

The KV-2 is arguably more famous than its unimpressive older brother, the KV-1 medium tank, despite performing equally poorly, if not worse. In fact, the KV-2 owes its notoriety to its iconic design—specifically its big, square turret head. The KV-2 utilizes many aspects of the T-34's design, most prominently its arms and pelvic gearing. It was heavily armored in most areas and was especially strong against frontal assault, impervious to all but point-blank fire. The hefty recoil from the howitzer in the oversized turret demanded a stable chassis to fire from, accounting for the KV-2's thick and stumpy legs when compared to the T-34.

SPECIFICATIONS

COUNTRY: West Rozenia

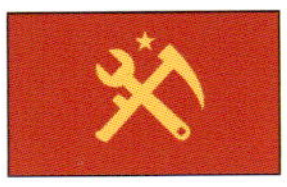

UNITS BUILT: 420
NO. OF CREW: 6
MAIN ARMAMENT:
1 x 152 mm howitzer
SECONDARY ARMAMENT:
3 x 30-cal light machine gun
3 x smoke grenade launchers
ARMOR: 61–109 mm
MAX SPEED: 27 km/h

Produced in the always-overstretched Rozenian tank factories, the KV-2 was designed with an engine and gearing system almost a decade out of date. There were persistent reports that drivers of the KV-2 regularly used a mallet to engage the leg gearing. The underpowered engine was simply not able to meet the demands placed upon it by the heavy armor and weaponry. It was slow and ponderous, deadly weaknesses in a combat unit and ones that enemy forces took advantage of. As the war developed, the Rozenians made multiple attempts to shore up the KV-2's rearward defenses with additional armor, but this inevitably slowed down the machine even further. While combat records for the KV-2 show passable numbers, in the larger picture its lack of speed meant it was rarely where it needed to be on time, and on a strategic level it failed to have much impact on the war.

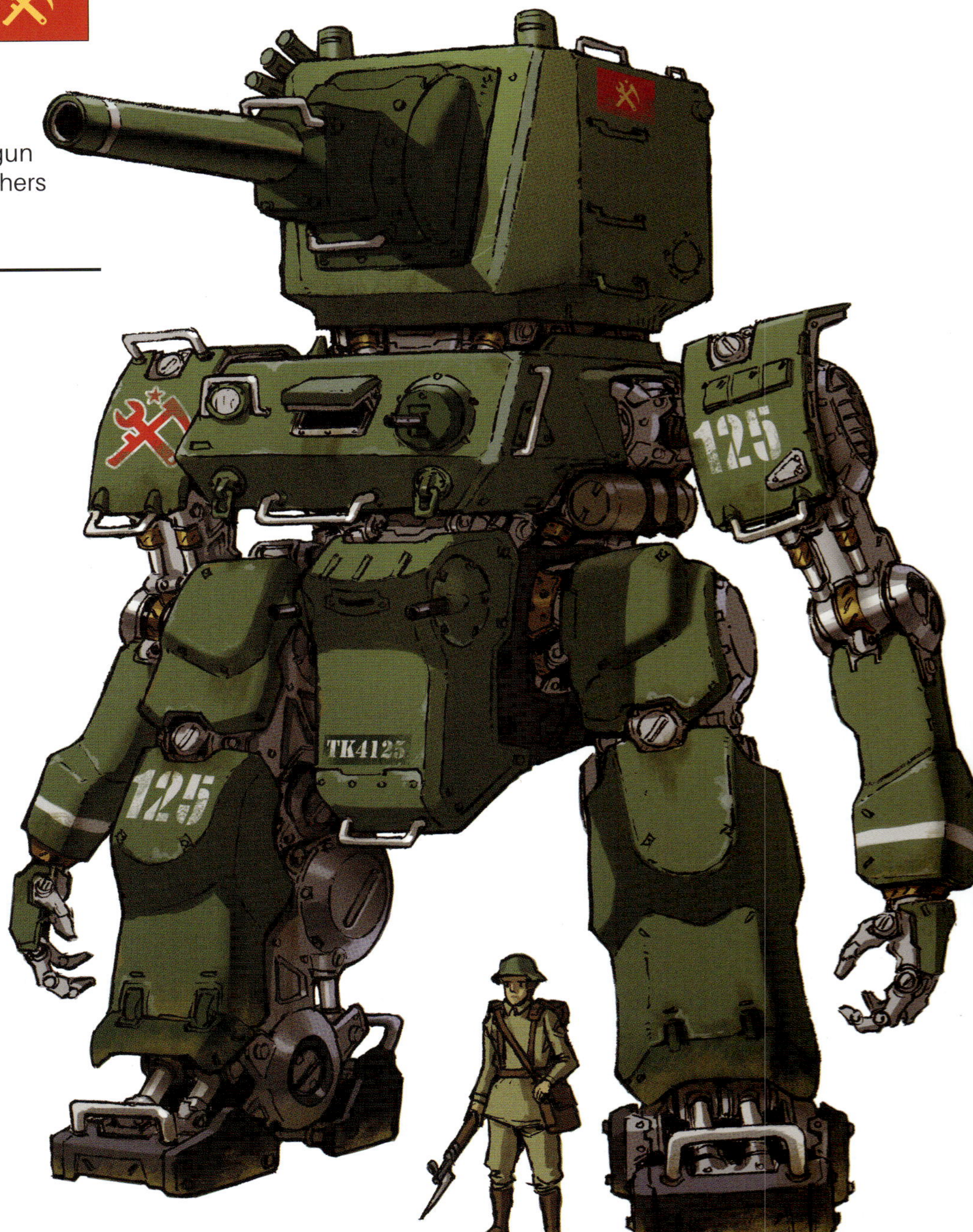

KV-2P

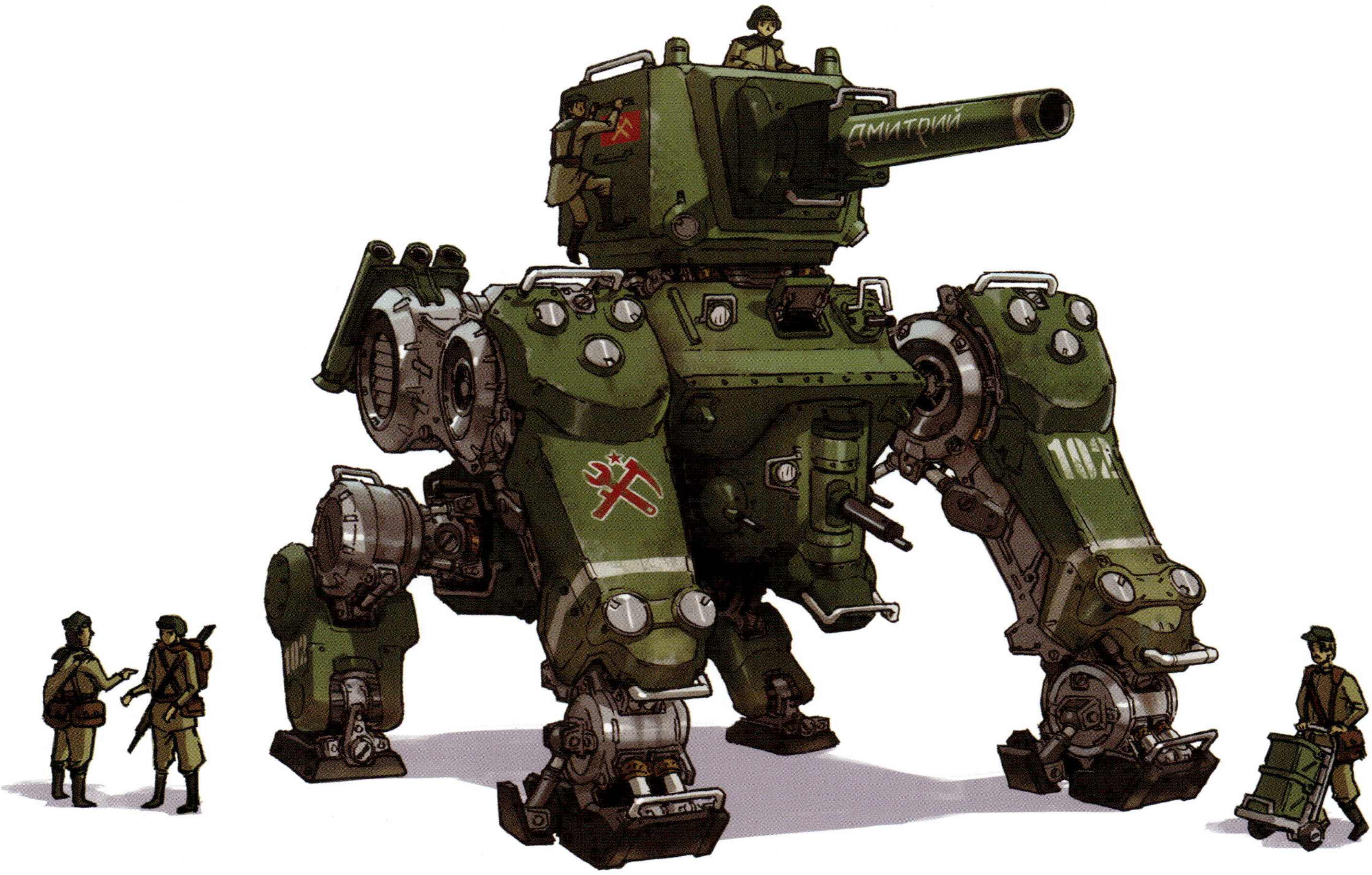

Historians speculate the KV-2P began its life as an extensive field modification with reversed TK-1 Tank Kong 1 legs and arms mounted to the KV-2 hull. This P variant became a more common sight when West Rozenia lost the TK-1 factory, leaving them with unused TK-1 limbs but no ability to fabricate the matching torsos. Meanwhile, the KV-2 factories suffered from supply chain issues leading to a similar, but opposite—and thus complementary—problem. Combining the mismatched components was a solution every bit as inelegant as it was practical, and wholly Rozenian in character.

In theory, the 2P aimed to compensate for the KV-2's underpowered engine by utilizing the arms for movement as a set of secondary legs, much like the Tank Kong 2. As a result, the KV-2P had a higher top speed compared to the KV-2, though its main advantage was its much improved stability. The tankhead could also carry additional armor or armament without significantly impacting its agility.

Records of these improvised machines are exceedingly rare. Existing photographs show a wide variety of alternative configurations. In all likelihood the factory would simply build and equip them with whatever material was available at the time and field modifications would diversify them further with time. However, their loadout seems to have been standardized late in its production lifetime, as the last batch were all equipped with the heavy front machine gun on the lower torso and a rocket artillery mount on the back.

SPECIFICATIONS

COUNTRY: West Rozenia
UNITS BUILT: 38
NO. OF CREW: 6
MAIN ARMAMENT: 1 x 160 mm howitzer
SECONDARY ARMAMENT:
- 3 x 7.62 mm light machine gun
- 1 x 7.62 mm heavy machine gun (water-cooled)
- 3 x 82 mm rockets

ARMOR: 30-95 mm
MAX SPEED: 27 km/h

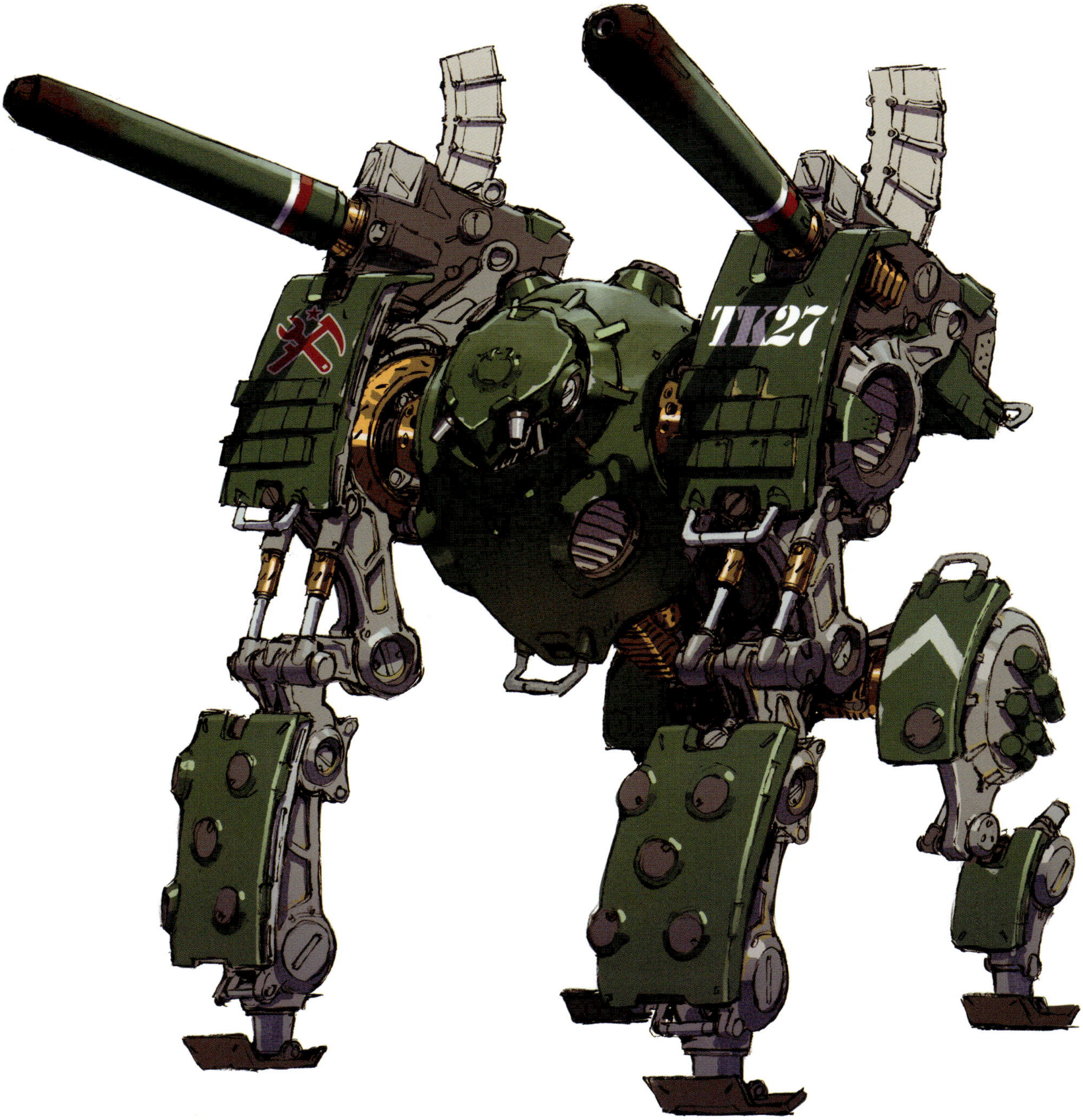

TK-2A TANK KONG 2 (ARTILLERY TYPE)

During the mid WC040s, the West Rozenian war ministry was united in a concerted effort to produce a new but still iconically Rozenian tankhead, one that would showcase the beleaguered state's potential for producing cutting-edge weaponry. Though the choice to base this new design on the original Tank Kong was apparently self-evident at the time, today some wonder why the T-34 or its derivatives were not chosen as a base. The question may be considered moot, however, as the TK-2A Tank Kong 2 ultimately shared little with its predecessor besides its posture and name. In an effort to compensate for the Tank Kong 1's shortcomings in speed with a light frame, only the forward-facing segments of the Tank Kong 2 were armored, while the manipulators of the fore limbs were done away with entirely.

The skeletal look and vestigial feet of the Tank Kong 2 has led some to believe it must have performed poorly. This was not the case. While unconventional, the highly-optimized mobility system of the Kong 2 was, at the time it was introduced, considered

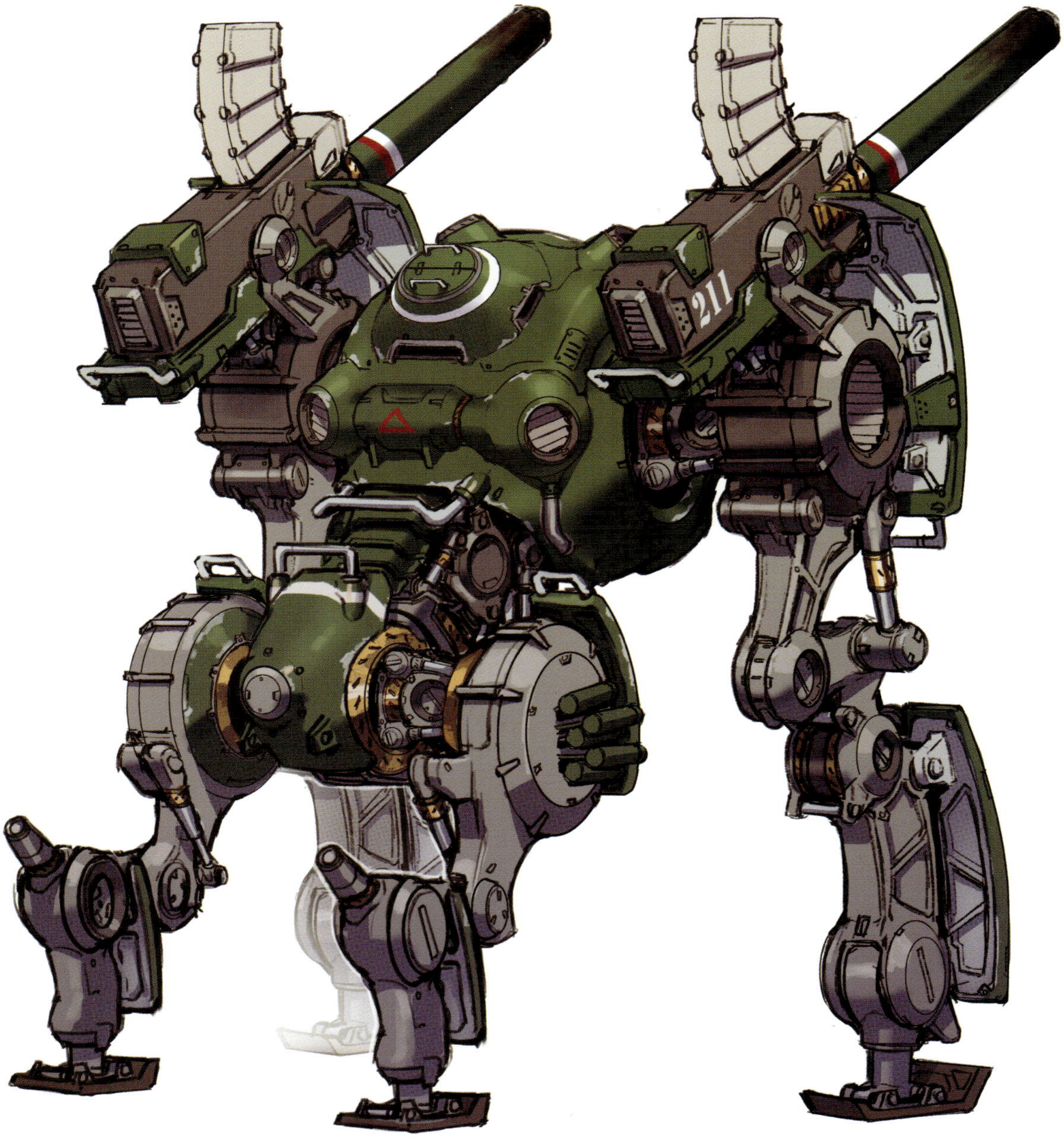

REAR VIEW

SPECIFICATIONS

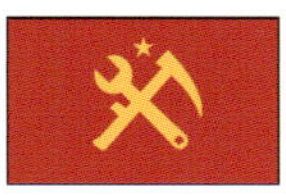

COUNTRY: West Rozenia
UNITS BUILT: 752
NO. OF CREW: 4
MAIN ARMAMENT: 2 x 30 mm rod cannons
SECONDARY ARMAMENT:
2 x 12.7 mm machine guns
10 x smoke grenade launchers
ARMOR: 16–70 mm
MAX SPEED: 35 km/h

indisputably advanced. Still, it was not without shortcomings. It was unwieldy in turns and, unsurprisingly, fatally vulnerable to ambushes or rear attacks in close quarters combat, this latter fact making it impractical for urban fighting.

A small number of variants were produced, with many more never progressing beyond the blueprint phase. Pictured here is the so-called "artillery type." The name is not particularly fitting as its weapon system was incapable of indirect fire. It utilized a complex semi-hydraulic pressure system and fired a long but small caliber rod at extreme velocity. The solid projectile used pure kinetic force focused on a small target area to improve armor penetration.

CONTINUED>>

TK-2A TANK KONG 2 (CONTINUED)

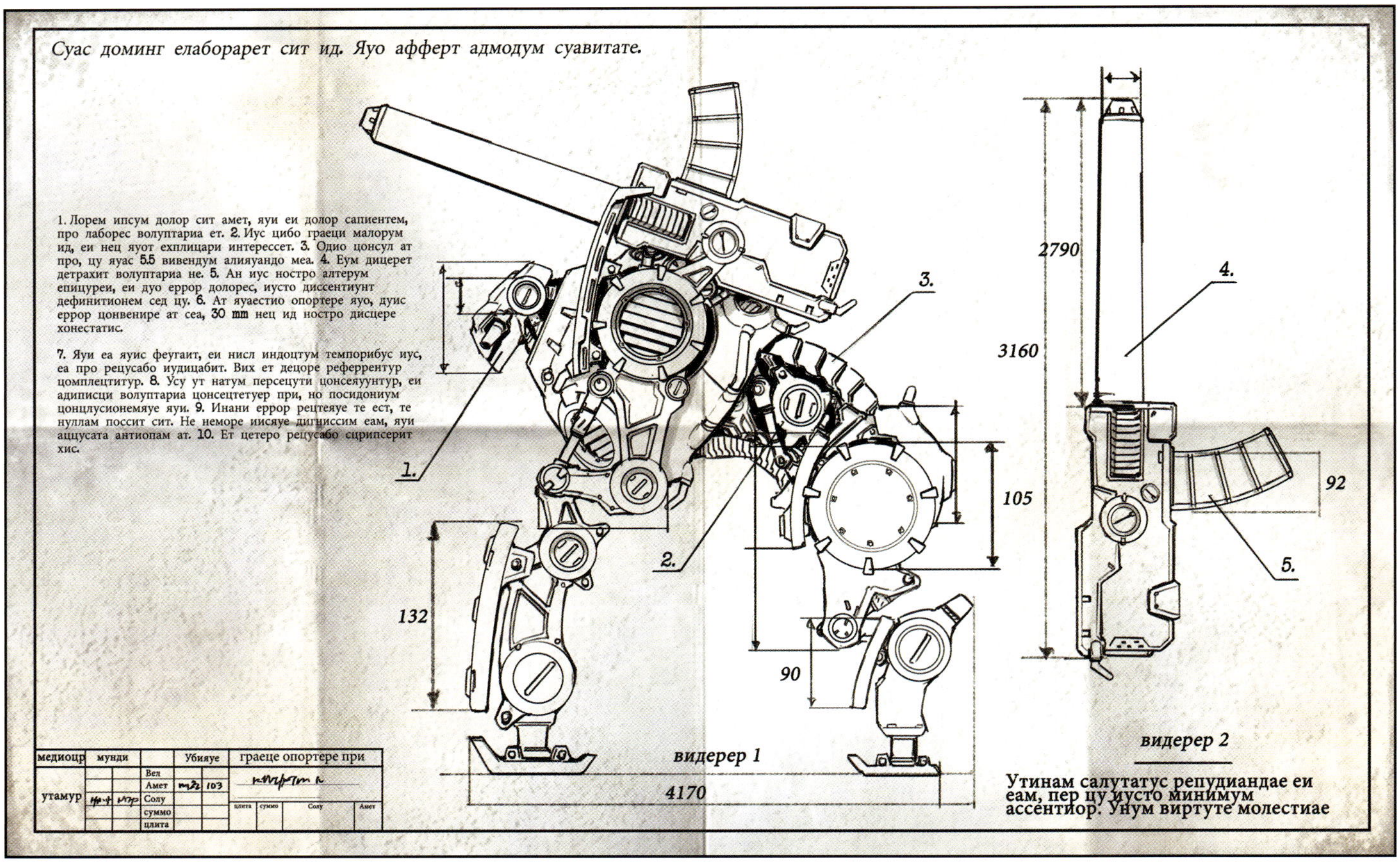

Construction blueprint of the Tank Kong 2 (Artillery Type).

More refined and more extensively-tested versions of weaponry like this would see further use after the Neverwar but, in the few occasions the Kong 2 saw combat, it delivered varying results. The projectile lacked the power of a more conventional charge gyrocannon, but still had a high rate of fire and packed a notable punch. In some reported cases the rod projectiles performed remarkably well. On the other hand, in at least two known instances, discharging entire magazines produced no operational effect on the unit's target.

Ironically, the same high rate of fire which accounted for much of the TK-2A's offensive capability could also be considered a design flaw. Because the ammo reserve was externally mounted and the manipulators had been removed to reduce weight, reloading required the assistance of other tankhead units. Since the Tank Kong 2 often ran out of ammo in prolonged battle, this could mean a major detriment to squad effectiveness.

On a more positive note, the artillery-type weaponry was very quickly found to be an effective anti-air system, as it could deliver a fast burst of multiple rounds on target. The high velocity of the projectiles reduced the degree to which gunners needed to lead the target, resulting in more accurate fire. Whereas the force of the projectiles was inadequate against many armored targets, it wreaked havoc on the much more lightly armored jetheads of the late war.

The final chapter in the Tank Kong 2's saga would come in the autumn of WC047. The West Rozenia province of Novgryov elected a new Magistrate who immediately initiated a change of alliance to East Rozenia. In effect, this marked the loss of the Bol-Obez Tank Kong factories, those that were producing the Tank Kong 2s. The production of the Tank Kong 2 continued but subsequent units were painted Kyzerian blue-gray and armed with different weaponry. These changes suited the Tank Kong 2 quite well and the Kyzerian variants had a reasonably good track record during the closing years of the war.

ALLIANCE ENGINEERING VEHICLES

The development of tankheads as the predominant battlefield unit brought with it a need for advancement in other areas of military technology, especially in the field of maintenance. The evolution of tankheads was, by and large, predicated on the development of advanced joint mechanisms. The advancement of this technology led to a new generation of engineering vehicles that were capable of faster and more efficient maintenance and field repairs of tankheads.

These vehicles varied in form and function, often combining new tankhead mechanisms with conventional crane or lift designs. Where they really stood apart was in their inclusion of light, highly articulated tool arms. Miniaturizing the humanoid appendages of tankheads into ones that were still two to five times the size of a human limb resulted in powerful yet nimble manipulators. As with full-sized tankheads, these arms took training and practice to control, as their interlinked naturalistic movement could catch inexperienced operators off-guard.

It was advances made in these tool arms that would lead to experiments in smaller tankhead-like combat vehicles such as the Alliance's Vickers Carden Loyd Personal Walker and the Yamaichi Type 22/N1 Teisatsukani Scout Crab.

VICKERS ENGINEERING WALKER

The Vickers Engineering Walker eventually led to the creation of the Vickers Carden Loyd Personal Walker.

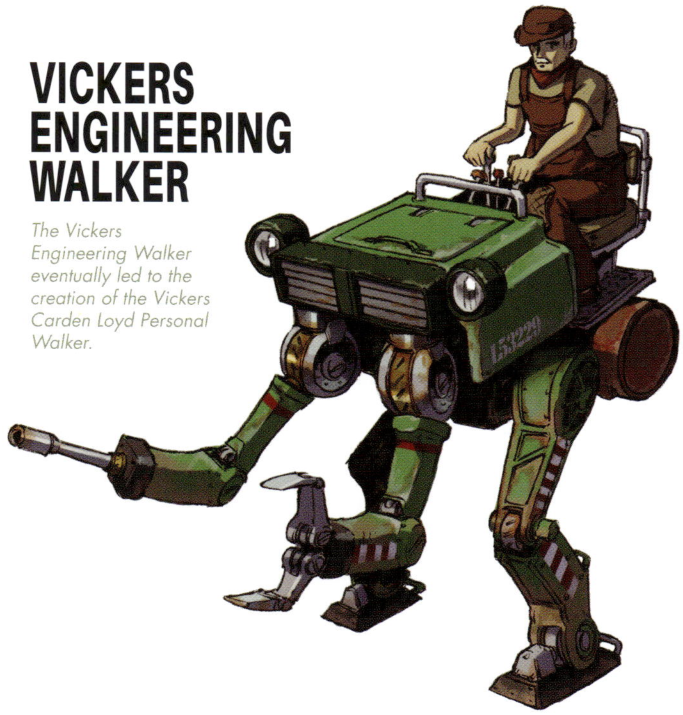

The sheer height of tankheads meant that repair crews needed to find ways to take their thankless job to a whole new level.

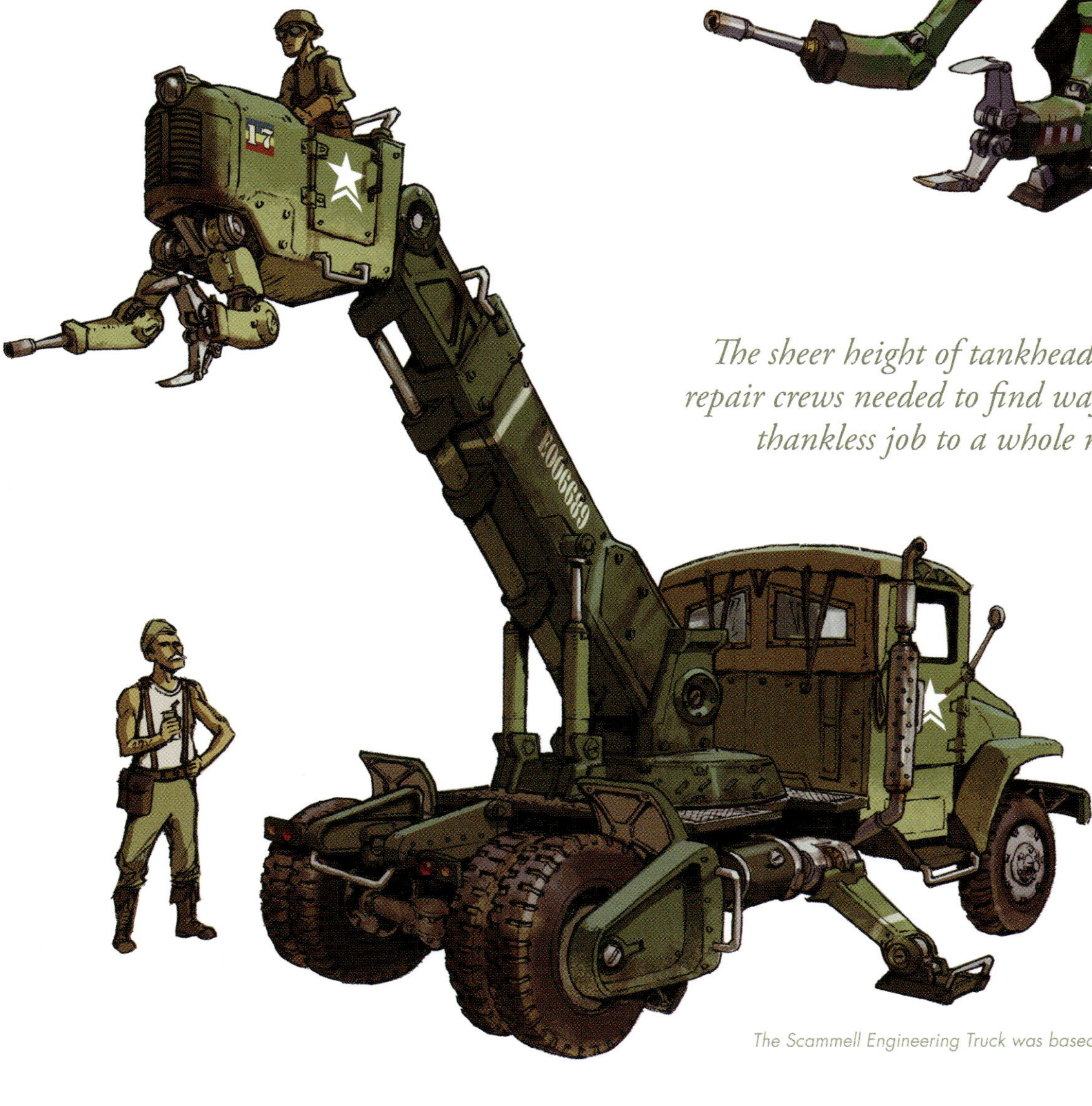

SCAMMELL ENGINEERING TRUCK

The Scammell Engineering Truck was based on the Scammell Explorer Truck.

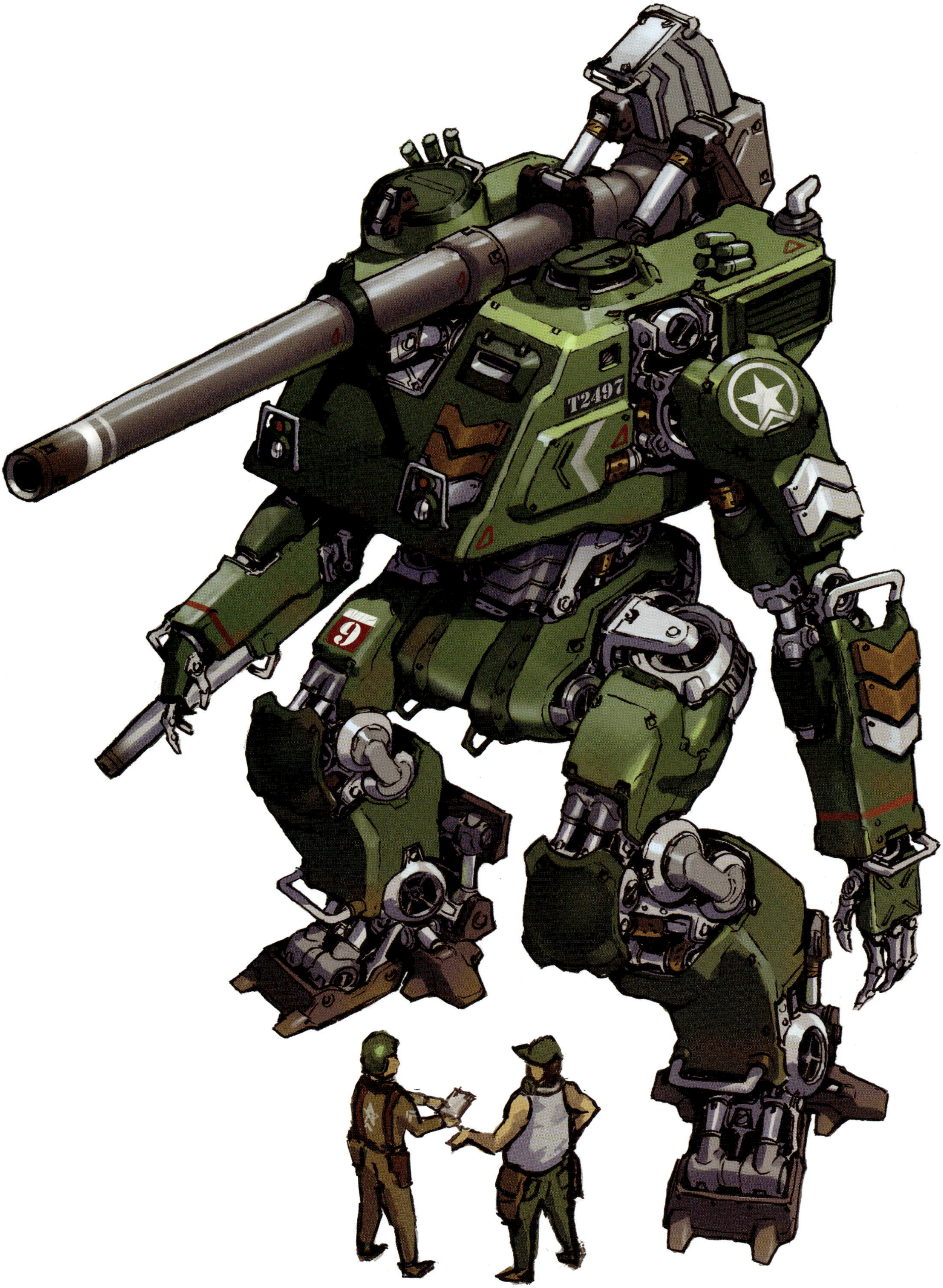
T2497
9

M9X MACARTHUR

The genesis of the M9X MacArthur was a reaction to the effectiveness of the upgraded Kyzerian destroyer tankhead Sturmgeschütz "StuG" III in WC016. It was clear to the Alliance that it needed its own destroyer tankhead in the field and a number of smaller refits were experimented with in an attempt to fill this role.

To make the M9X as powerful as possible, the Liberterran development team procured an experimental battleship autoloader gyrocannon from the navy. This huge weapon required a stretched M4 hull to accommodate it and, even then, most of the gun's operational components had to be mounted on the hull's exterior.

The fixed main gun was expected to be cumbersome when adjusting to acquire targets and its tremendous recoil would prohibit rapid-fire without deploying the M4's backpack-mounted stabilizer legs. A limited ammo capacity could also limit the M9X's effectiveness. To compensate for these weaknesses, the M9X was equipped with a 50 mm gyrojet cannon mounted on the arm. While high caliber, the shells for this weapon were kept light to prevent the weapon from unbalancing the machine. Most of the round's weight was propellant, leaving little room for explosive or general momentum when hitting the target.

During trials, the Liberterran Arms Committee was still quite positive about the arm-mounted gun and considered implementing it on further designs, but lackluster results in the field led to them abandoning the feature. This was partially due to the division of crew tasks in the M9X. The arms were only controllable by the driver, who used them to maintain equilibrium while the tankhead was on the move. In trials, when the unit was stationary, drivers capably operated the cannon. In the field, however, drivers were usually too preoccupied with displacing or evading incoming fire to use the arm cannon to an efficient degree.

The entire design of the M9X ended up getting a significant overhaul midway through development when the navy abandoned their gyrocannon in favor of yet another prototype, this one a recoilless cannon. As this weapon was roughly the same as the gyrocannon in terms of mass and dimension, the M9X also adopted the new, superior design. Instead of discharging its propulsion into the chamber, the new rifle used a rear-vent expulsion mechanism so that its recoil was effectively negligible. The M9X had previously incorporated the M4's stabilizer legs, but these were removed and replaced with smaller heel plates. This modification compensated partially for the added weight of the extended hull and allowed the M9X to achieve speeds close to those of the M4.

While a marvelous design overall, the size and complexity of the main cannon remained a problem. Due to the positioning of the magazine, in order to reload the weapon the gunner had to pop open the top hatch, look backwards at the magazine, and instruct the driver how to operate the arms. While never a simple operation, the crews did become more adept with experience. However, another similar problem arose: the exposed cannon with its open back was particularly prone to fouling, and the autoloading mechanism would often jam. This meant that the gunner would need to crawl out on top of the hull to rectify this problem with a tanker bar and brute force—often under fire, on the move, or both.

While its main gun could deliver heavy anti-armor firepower quickly, and rapid-fire was much simpler to achieve than with regular tankheads, the M9X was not a success. When it worked, it performed remarkably, but inevitable failure made the M9X very unpopular with its crews, who felt they were operating an unreliable "deathtrap." It was replaced by the M10 later in the war.

"Reloading this stupid thing is like trying to undo a knot behind your head, going on your buddy's instructions while he's getting shot at."
- M9X Driver Brandon Lu

SPECIFICATIONS

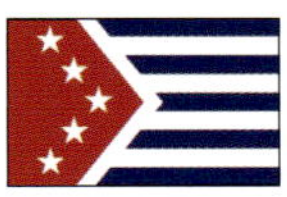

COUNTRY: Liberterra
UNITS BUILT: 622
NO. OF CREW: 8
MAIN ARMAMENT:
1 x 155 mm self-loading recoilless gun
SECONDARY ARMAMENT:
1 x 90 mm arm-mounted gun
7 x smoke grenade launchers
ARMOR: 12–20 mm
MAX SPEED: 36 km/h

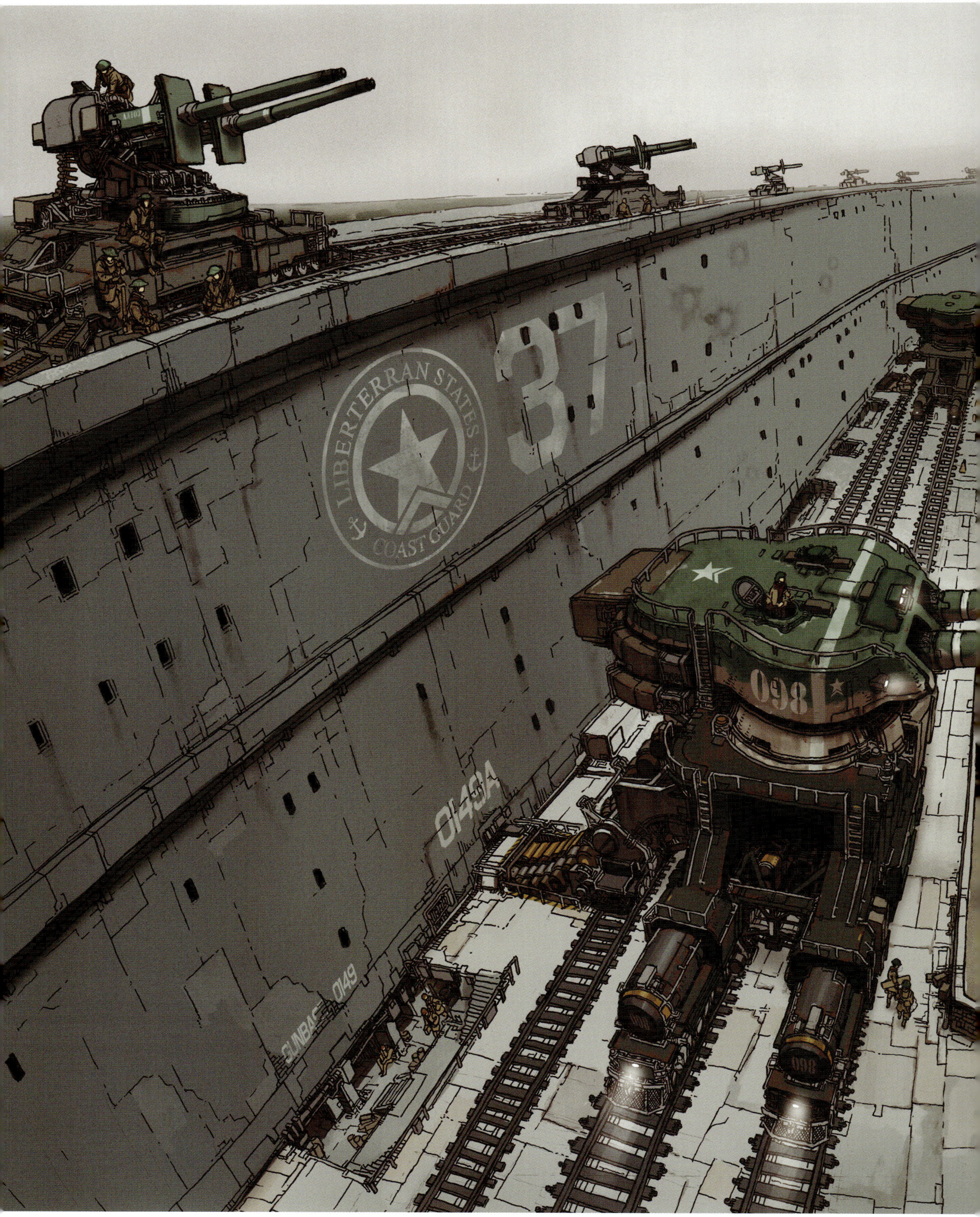
LIBERTERRAN STATES
COAST GUARD
37
098

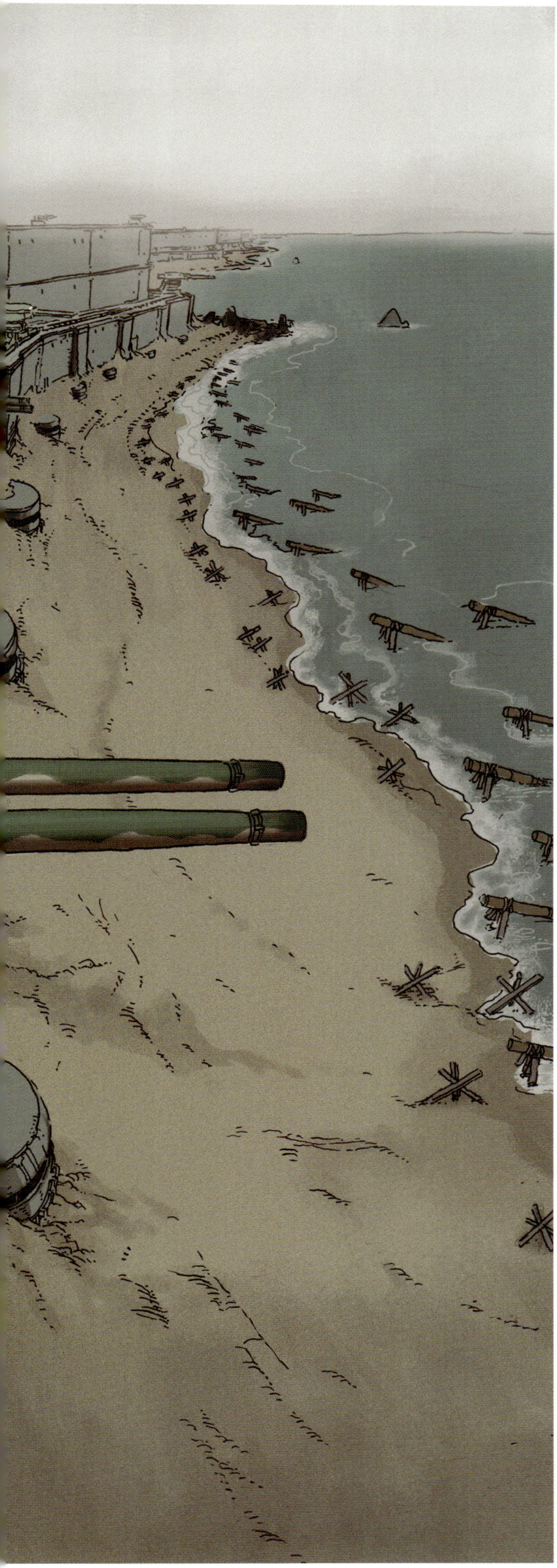

LIBERTERRAN COASTAL WALL

Even before Liberterra joined the war as a combatant nation, it was a major factor in the global balance of power. Any country or coalition bolstered by this economic and industrial juggernaut would have a significant advantage over its rivals. Yet, when the Liberterrans did officially join the Neverwar as members of the Alliance, it was not without some reticence. There arose in its populace a fear that their territories, never before exposed to the bloodshed and horror of modern warfare, might become targets of retaliatory strikes.

As a direct consequence of this unease—and much like the war itself—Liberterran coastal defenses continually evolved. Fear over the threat of invasion grew and took on more substantial shape when the enemy began to deploy walking battleships. Liberterra immediately looked to strengthen its coastal defenses, which at the time amounted to simple spotter bunkers equipped with surplus naval guns. These bunkers and their armaments were reinforced as the enemy's naval capabilities changed, opening up new potential routes of attack over the oceans.

Further expansion of the coastal defenses stalled during the ceasefire period (WC012–015). Many in Liberterra hoped peace talks would take place, freeing the world from the unrelenting grip of the global war. This optimism pushed the Liberterran economy into a boom as the country shifted its wartime production toward less martial markets. Many innovations entered daily civilian life at this time. Urban construction projects sprang up in the thousands, powered by new engineering vehicles based on tankhead technology and fueled by an urban migration of the Liberterran population.

However, a coalition of politicians and engineers from both the military and civilian sector believed that war would erupt again, and devised a plan for the ultimate coastal defense system. Mainstream political thinking and press coverage opposed the plan, but when the Kyzerians began to mobilize once more, the coastal defense system quickly gained support.

Some now argue that the government contract for construction of the wall, rather than any defensive advantages it provided, was the project's main benefit. As the economy shifted back to a wartime footing, civilian construction ground to a snail's pace, threatening to cause widespread unemployment in the civilian construction sector. Labor unions across the nation began to protest and in one case a protest turned violent. The government and company managers, alarmed by the real danger posed by striking workers piloting mechanical construction walkers, pushed the defense wall plan through the budgeting process with increased urgency.

In the spring of WC041 the Liberterra Defense Committee proposed the Coastal Defense Wall Spending Act, which would be approved by all governing bodies later that year. It would be the largest infrastructure project in that or any other nation's history. By providing enough jobs to employ those construction workers who had lost jobs in the turn to a war footing, the Coastal Defense Wall acted as a decompression valve for the fomenting unrest. But while the project was a boon for the fortunes of Liberterran workers, averting crisis, it is also true that the ambitious plan might not have been possible without such a large available workforce.

While the wall's design remained fundamentally the same as that described in the initial proposal, arming it presented a logistical challenge. The Liberterrans

CONTINUED>>

LIBERTERRAN COASTAL WALL (CONTINUED)

were already producing shells and artillery pieces at unprecedented speed. However, most of this production was going to a general effort to modernize the navy and little could be spared for coastal defense. Strategic analysts calculated the amount of firepower that would be needed to adequately defend each section of wall. Their conclusions caused anxiety throughout the defense department. Even if they halted navy modernization immediately, they would only be able to arm a third of the total length of the wall.

To this day, it is unclear who came up with the solution to this conundrum, which now seems so obvious in hindsight. Still, it cannot be overstated how forward-thinking the entire project became once the idea was implemented. Inspired by the moving cranes used on docks, the idea was proposed to mobilize the guns of the coastal wall by putting them on rails. Skilled labor and modernized machines for laying track had become widely available following the stagnation of civilian infrastructure projects, which made the solution not only elegant but also feasible. In the end, the idea was received so positively that not one set of tracks but two were installed, the lower for the large naval guns and the upper for the anti-air barrage cannons. Large crates of shells could be transported along with the air defense guns on the dual tracks atop the wall. Each gun carriage was equipped with a small diesel engine, allowing them to link up like trains and move to intercept points along the wall at an impressive pace.

For the naval guns, however, the logistics were more challenging. The lower portion of these gun platforms was entirely taken up by a modified pair of massive LSATC S160 Class locomotives, necessary for transporting the large naval guns, and there was little track capacity left for the transportation of ammunition. Storing the large shells with the guns was not an option, as a single well-aimed or even lucky shot could result in detonation and loss of the entire platform.

Taking inspiration from recent innovations in metropolitan subway systems, a third rail line was enclosed within the wall itself. This allowed for the transportation of personnel and equipment on much faster carriages on the tracks inside the wall, and, during combat, special ammo transportation carriages would be deployed to stock the ammo of the large guns, keeping the ammo safe from incoming fire. These carriages would be connected to hoppers that ran on their own narrow-gauge tracks immediately behind the large guns, and would connect to the internal ammo carriages with flexible feed chutes through ports in the wall. When a gun reached its destination, the hopper could hoist shells up to the turret on a continual beltlift using power sourced from the locomotives.

The battle effectiveness of this weapons system proved itself in many defensive actions, of which the repulsion of the Kyzerian "Operation Oberherr" is superlative. On the morning of the invasion, the northern extremity of the Coastal Defense Wall and the West Tramonic Sea was enveloped in heavy fog. In these conditions, elements of a wider Kyzerian invasion fleet, including the carrier Elbe, became separated from the main force, drifting towards the wall's northernmost span. However, the fog also masked the carrier detachment long enough for it to approach within striking distance, undetected. Eager to capitalize on what was interpreted by the Kyzerian commander, Captain von Vonozak, as a tactical advantage, he scrambled all his combat jetheads. The small force was spotted at the last moment, however, by a coast guard PT boat, and mobile gun platforms were moved north on the rail system. Meanwhile, the Kyzerian pilots were disoriented by the fog, so that when they suddenly found themselves above the imposing bulwark they were unprepared for the fierce barrage that welcomed them. The entire flight was downed.

Still, the Liberterran victory in the northern skirmish almost spelled defeat in the south. One of the Kyzerian jetheads that was gunned down crashed into the rails, cutting the mobile air defense guns off from the south after the brief northern action was decided. Luckily, a group of unproven Skink anti-air guns was able to be moved into the gap from a nearby factory and testing site in Hutchinford. They performed well, and the Kyzerian invasion was ultimately turned back.

The Kyzerians never learned how close they had come to punching through the Liberterran defenses that day and they would never again attempt a straightforward assault against the coastal defense wall. Most sections of this mighty fortification still stand today, well after the war for which they were constructed. An imposing mark, perhaps forever scarring the once beautiful coast and reminding subsequent generations of the tremendous effort, resources, and willpower invested into the conflict, even so far from the frontline.

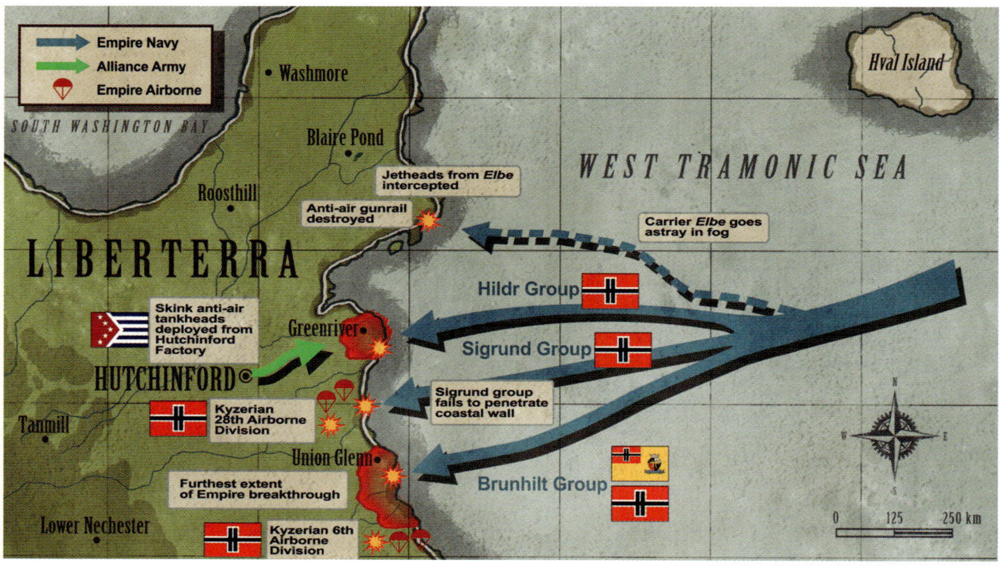

A map showing the Kyzerian plan of attack and the furthest extent of their breakthrough.

M4AA(20Q) SKINK

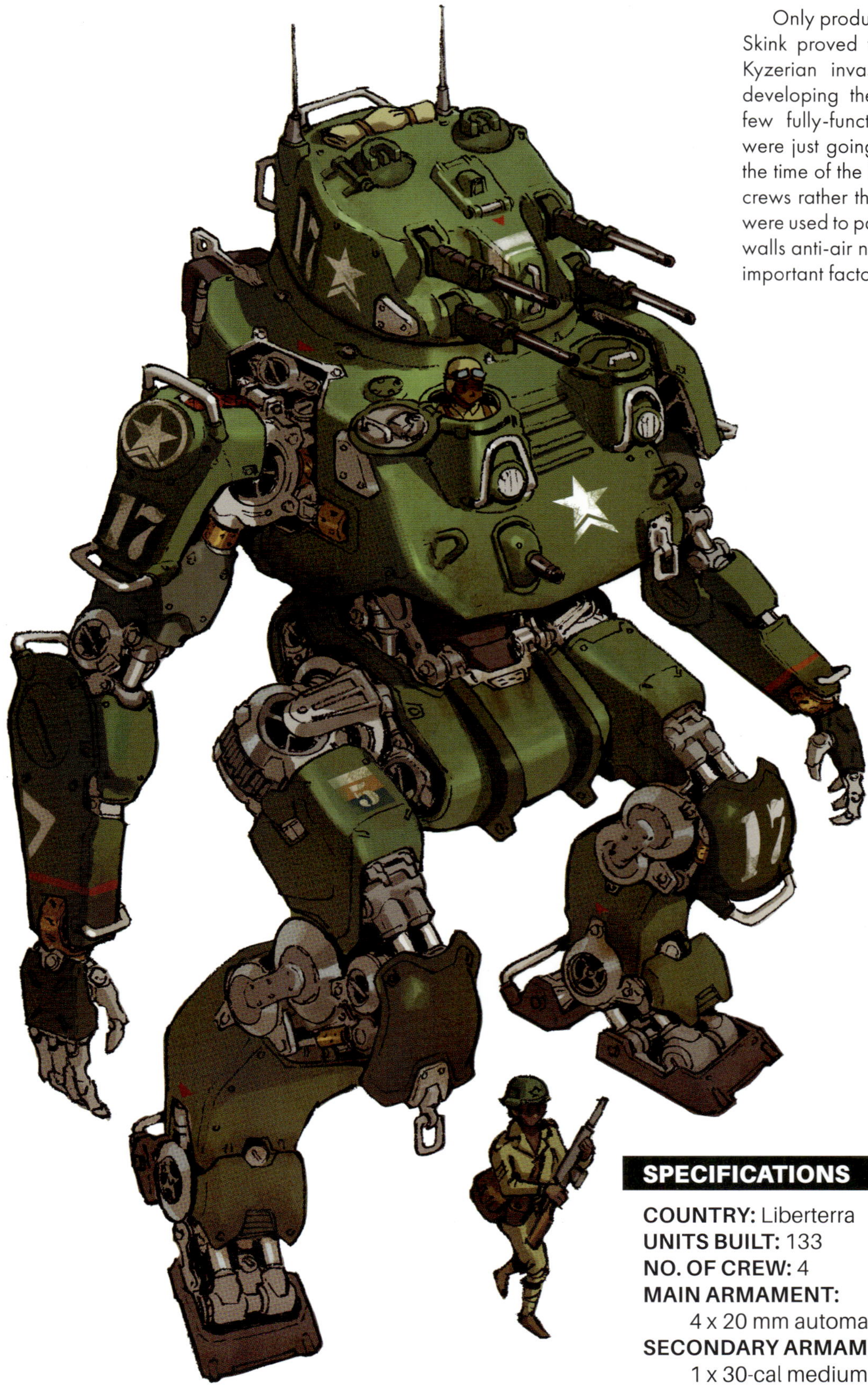

Only produced in limited numbers, the M4AA Skink proved very effective during the ill-fated Kyzerian invasion of Liberterra. The factories developing the new Skink design only had a few fully-functional models completed, which were just going through proving ground trials at the time of the invasion. Driven by testing ground crews rather than trained soldiers, these vehicles were used to patch holes in the Liberterran coastal walls anti-air network, thus becoming a small but important factor in this hard-fought battle.

SPECIFICATIONS

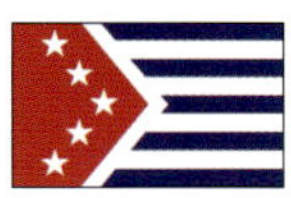

COUNTRY: Liberterra
UNITS BUILT: 133
NO. OF CREW: 4
MAIN ARMAMENT:
4 x 20 mm automatic anti-aircraft cannons
SECONDARY ARMAMENT:
1 x 30-cal medium machine gun
ARMOR: 50 mm
MAX SPEED: 56 km/h

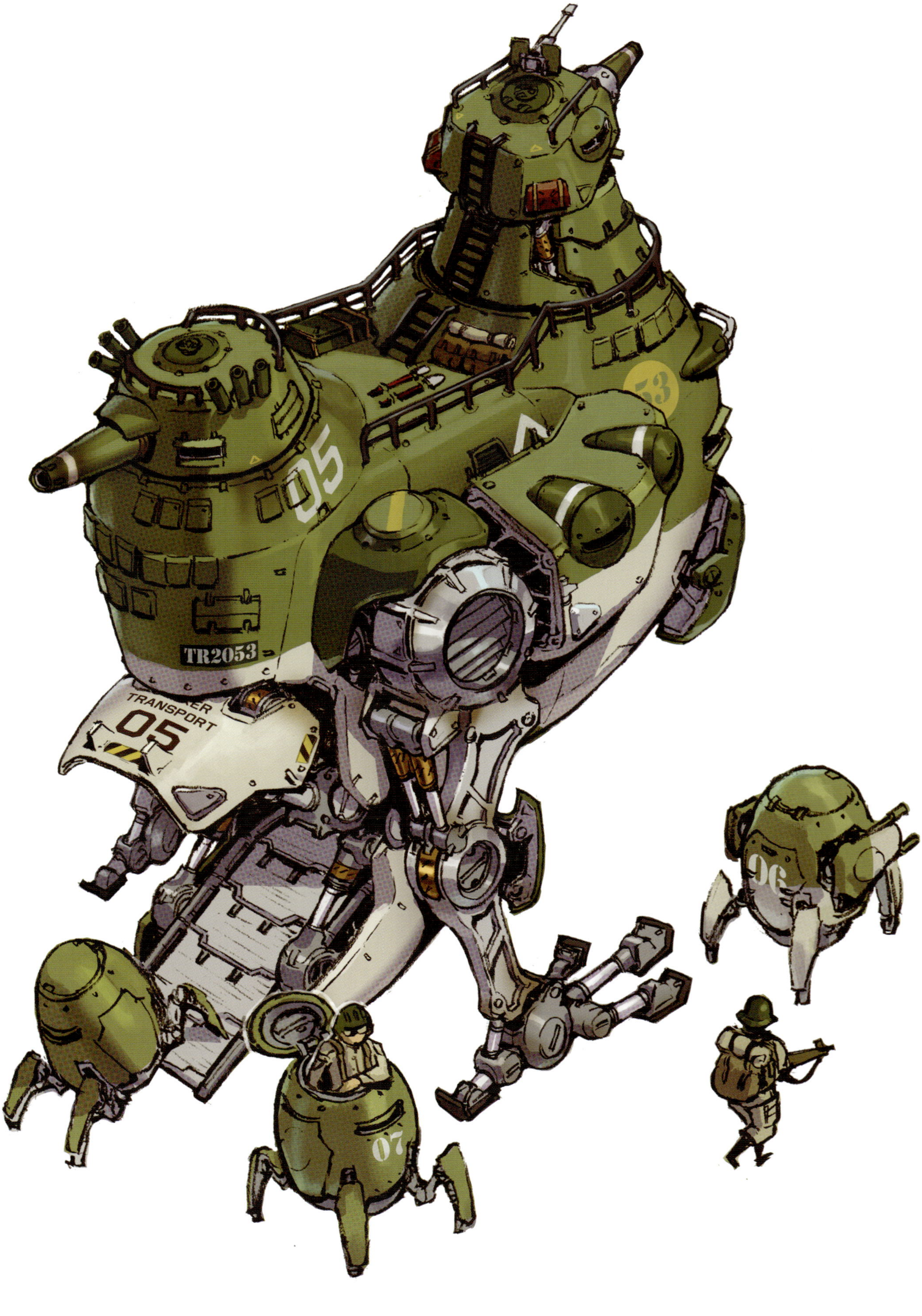
05
53
TR2053
TRANSPORT
05
06
07

FT3 HEN TRANSPORT

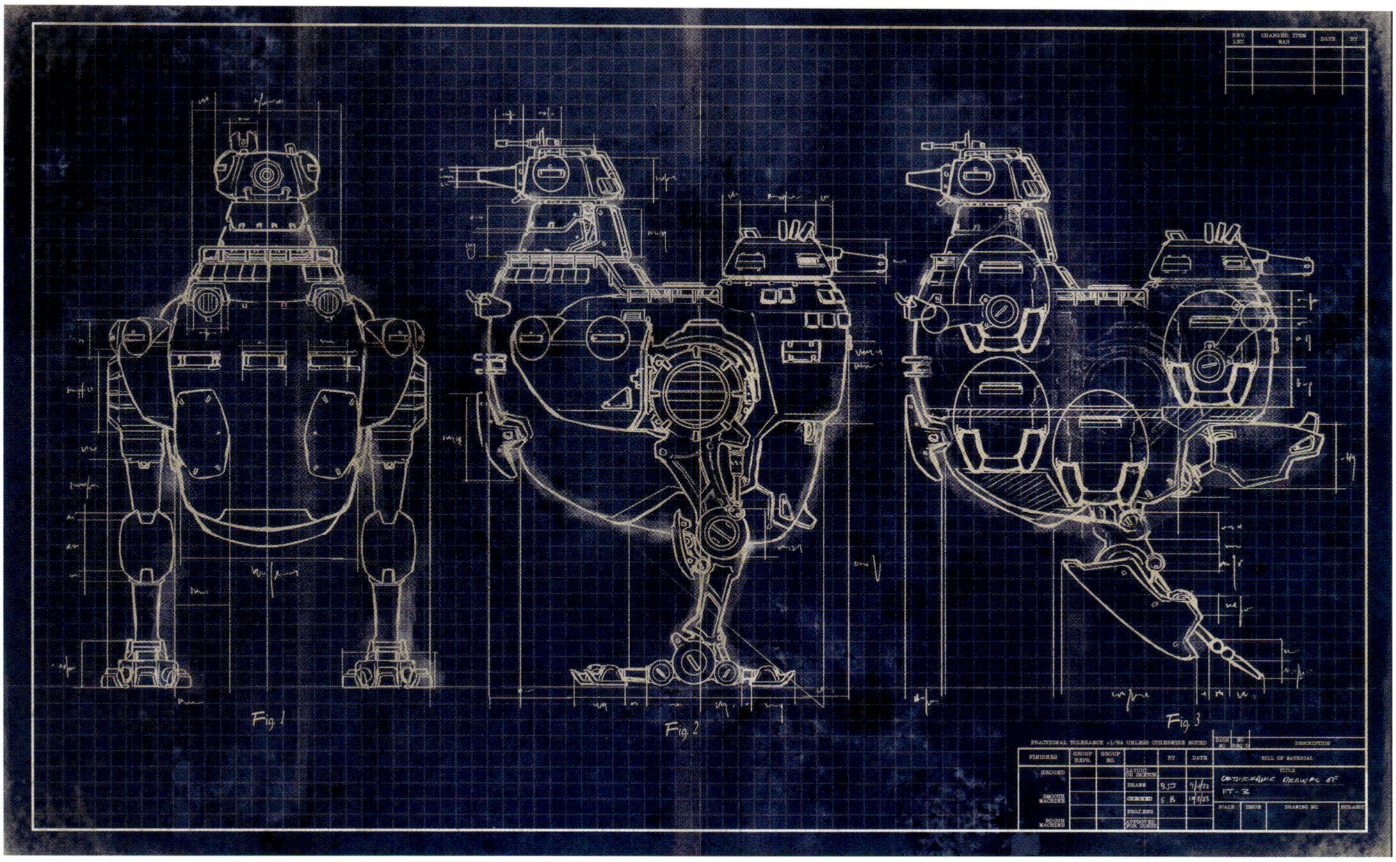

Blueprint of the approved FT3 design.

SPECIFICATIONS

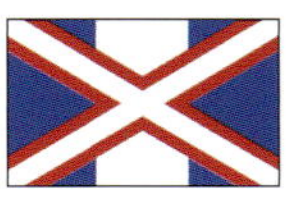

COUNTRY: Grand Britony
UNITS BUILT: 279
NO. OF CREW: 10
EGG TANKETTE CARRIED: 4
INFANTRY CARRIED: 12
MAIN ARMAMENT:
1 x 75 mm mortar cannon
SECONDARY ARMAMENT:
1 x 50 mm gun
1 x 30-cal machine gun
6 x smoke grenade launchers
ARMOR: 5–30 mm
MAX SPEED: 20 km/h

As has been the case in every prolonged modern conflict, the technology available at the final hour was remarkably different from that of the opening campaigns. Despite this evident fact, military theory and praxis lagged behind as commanders and troops alike preferred to stick to proven methods over new or even experimental ideas.

While the birth of the tankhead and its supplemental technologies were, unequivocally, the military breakthroughs of the century, it is easy to underestimate how much this technology matured in the early years of tankhead design. The power-to-weight ratio of engines and transmissions, power plants, fuel efficiency, and, above all, the design understanding of articulated limbs evolved at a tremendous rate as every nation and workshop raced to keep a step ahead of the opposition or industrial competitors.

The idea of large, even gigantic walking warmachines was no longer a marvel by the last year of the Neverwar. The mechanisms capable of transporting such titans had been fully refined and even scaled down to create smaller tightly-packed machines, demonstrating a mechanical understanding that early Neverwar engineers would never have thought possible. The most famous example of this evolution would be the FA5 "Rooster" artillery platform, which packed an entire battery worth of cannons into a single well-armored vehicle.

CONTINUED>>

FT3 HEN TRANSPORT (CONTINUED)

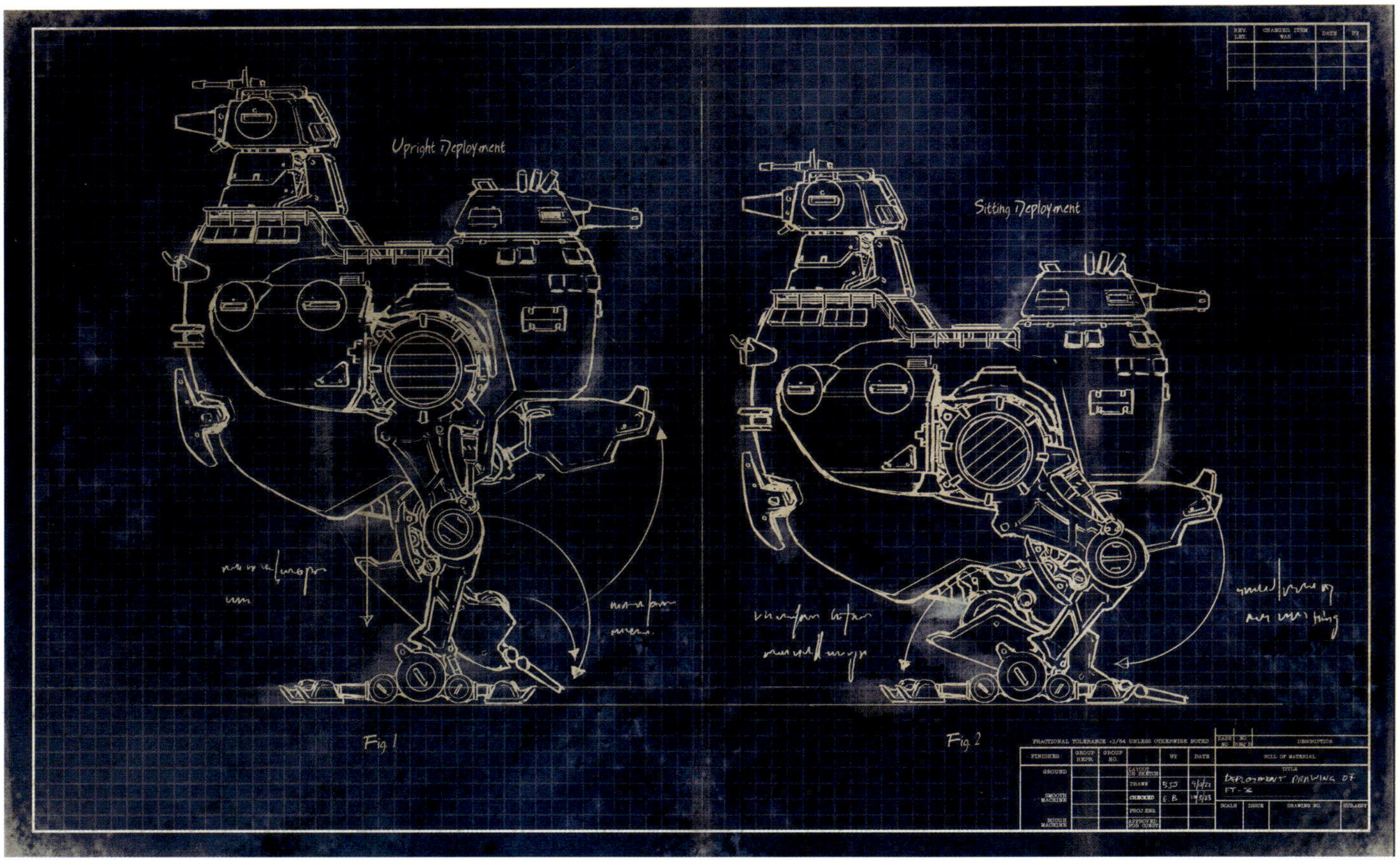

Blueprint showing the ability of the FT3 to deploy its "Egg" tankettes in either a standing or "roosting" position.

Before this catastrophic level of firepower was packed into the Rooster's reverse-jointed frame there were other models designed. One in particular would have a much bigger impact on warfare than its cannons ever could. The so-called "Egg" tankettes, engineering support vehicles built to support larger units, were the first to incorporate scaled-down limb articulation in single-pilot, fully-enclosed mobile weapon pods. The Eggs were about as fast as infantry on foot but entirely immune to small arms fire and, with four mounted heavy machine guns, four times more deadly. The design was even small enough that it could operate in interior spaces after passing or forcing its way through doorways or walls, changing the way fortification assaults could be carried out.

Of course, such concentrated power left little room for ammo or fuel, meaning the Eggs were entirely unsuited for prolonged engagements without support. Their weight and speed also meant they were unable to keep up with a mobile vehicle assault force, so every pod would require a dedicated truck to transport it, plus another vehicle to transport needed supplies. The Hen transport tankhead, up to this point somewhat of a solution looking for a problem, offered an opportunity for a novel combination of capacities perfectly suited to the conundrums posed by the Egg tankettes.

The Hen's basic design was developed before the Egg. It was a very tough and reasonably fast tankhead capable of transporting significant cargo loads. In fact, even at its max volumetric infantry load, it would still be utilizing only half of its carrying capacity in terms of weight. Using the Hen as a transport for heavier materials made little sense, though, as regular trucks could transport those materials faster at a fraction of the production cost. The notion of a heavily armored logistics vehicle had some merit (and the Hen would see use in this role a few times during the war) but was also judged to be an application too specific to merit funding.

Marrying these two weapon platforms, however—the miniaturized and fuel-hungry Egg tankettes and the almost overly powerful transport Hen—resulted in a system that would have revolutionized the war had it not come to an end very soon after the designs were finalized. The first model of Hen transport that did see limited use was an unorthodox machine in terms of appearance. It was equipped with a short range gyrohowitzer and mortar cannon intended to provide fire support to its passengers. Inside its hold, the Hen could carry four Egg assault tankettes, along with supplies and maintenance crews. A cramped hotbed sleeping compartment was also included, although most crews preferred to camp out on the top deck of the Hen's exterior hull.

The first preliminary production models of this combat unit were cautiously deployed in field trials and proved very promising. Urban combat, a nightmare for attacking forces since the birth of warfare, was solved with the combination of the Hen Transport and Egg Tankettes. The tall Hen required cover in order to reach a build-up area, though its armor provided some protection en route.

VICKERS "EGG" TANKETTE

A closer look at the Vickers "Egg" tankette with its quad 50-cal heavy machine guns.

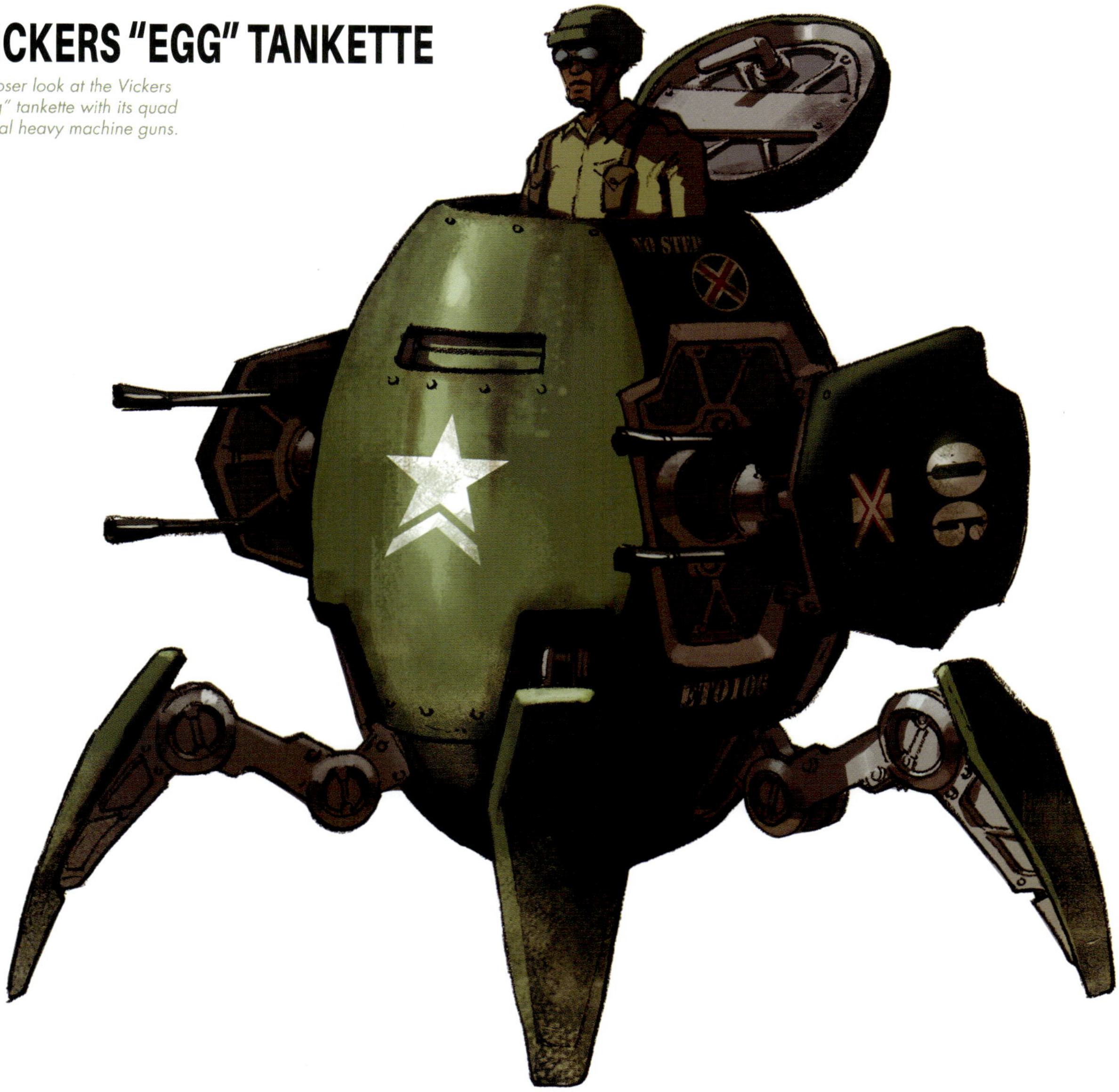

Once among the buildings, regular tankheads that would lead urban assaults became vulnerable as infantry could strike at their legs from any angle and retreat before the tankhead could counter. The Hen, on the other hand, would unload its deadly cargo of Eggs into the streets. The armored walking pods could sweep buildings ahead of the advancing tankheads with impunity, blowing away any opposition. If they encountered fortified enemy positions, the Hen's bunker buster weapons would be called in. It was reported that the Eggs were also very resistant to the shrapnel the Hen's explosive weaponry produced. This meant that supporting fire could be called in frightfully close to the tankettes' position with limited risk of friendly fire casualties, something that would be unthinkable with regular infantry.

All in all, the Hen and Eggs displayed a tactical synergy rarely seen prior to their introduction or since. The fact that they were only introduced shortly before the Grandmaster Gamble (WC049) brought a sudden end to the Neverwar meant that the effectiveness of the combination would never truly get a chance to prove itself in relation to the overall conflict. The spirit of their design lives on, however, in the smaller, more agile power armor we see deployed in modern times—machines more suited to the blindingly fast and confused urban skirmishes that have marked terrestrial combat since the end of the Neverwar.

When Kyzerian Command asked their engineers which came first, the chicken or the egg, they weren't trying to be funny.

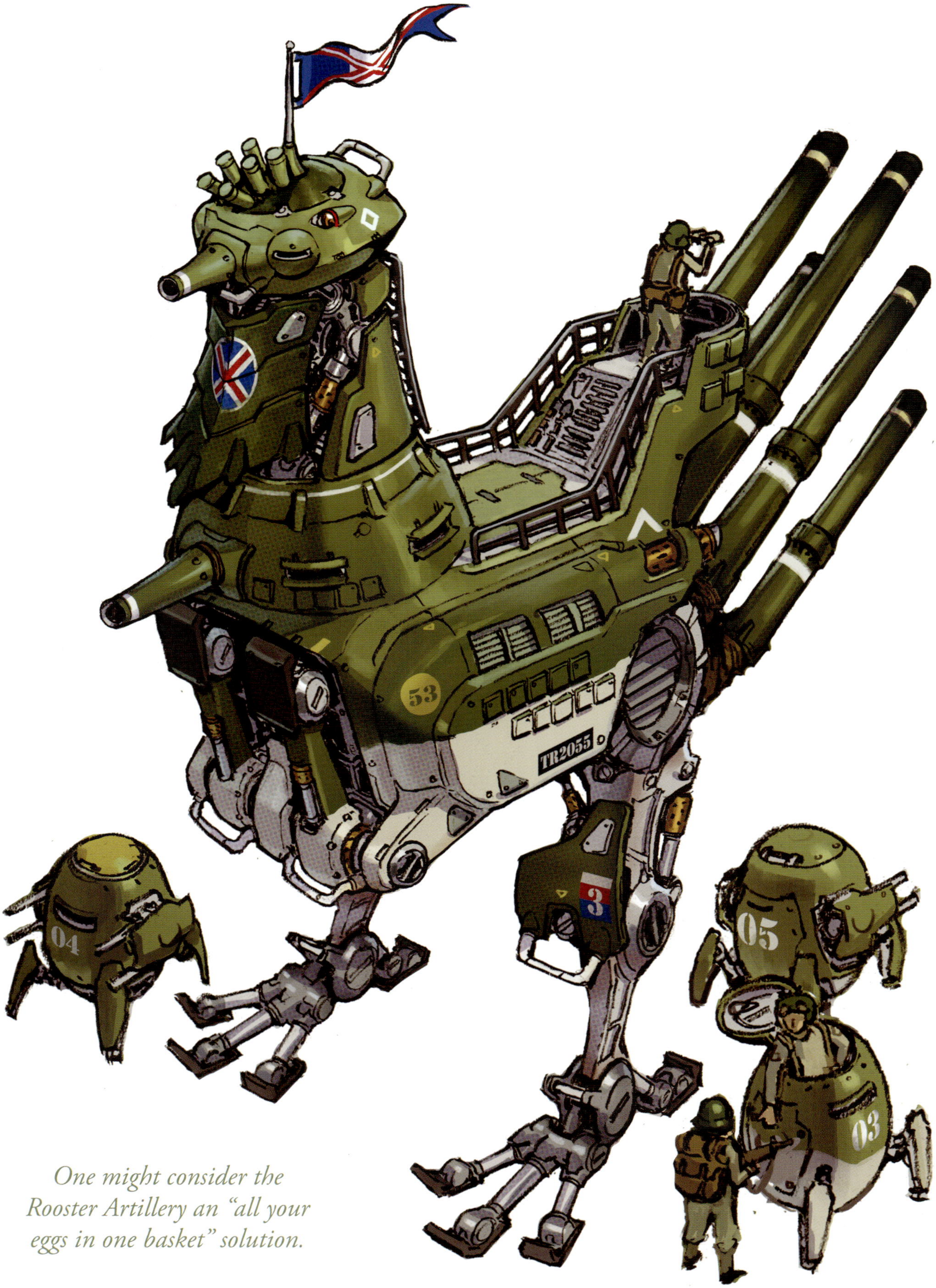

One might consider the Rooster Artillery an "all your eggs in one basket" solution.

FA5 ROOSTER ARTILLERY

The FA5 was a unit that combined an entire artillery battery in one vehicle, using the same basic frame as the FT3 Hen. Firepower came at a tradeoff, however. Whereas a spread-out artillery group might lose one or two guns to counter-battery fire, a Rooster struck by direct armor-piercing fire constituted a loss comparable to that of the entire group. It did, however, present a much smaller profile than a multiple unit squad, and had superior mobility over towed trailer-mounted guns. Its secondary guns proved insufficient at providing defensive fire in the field and it was often deployed with a Hen transport escort with its complement of Egg tankettes to provide close combat support.

SPECIFICATIONS

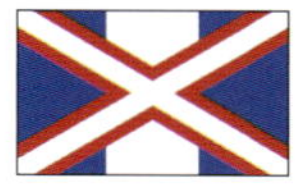

COUNTRY: Grand Britony
UNITS BUILT: 213
NO. OF CREW: 6
MAIN ARMAMENT: 6 x 155 mm gun
SECONDARY ARMAMENT:
1 x 25-pounder howitzer
1 x short barreled 50mm gun
6 x smoke grenade launchers
ARMOR: 5-30 mm
MAX SPEED: 24 km/h

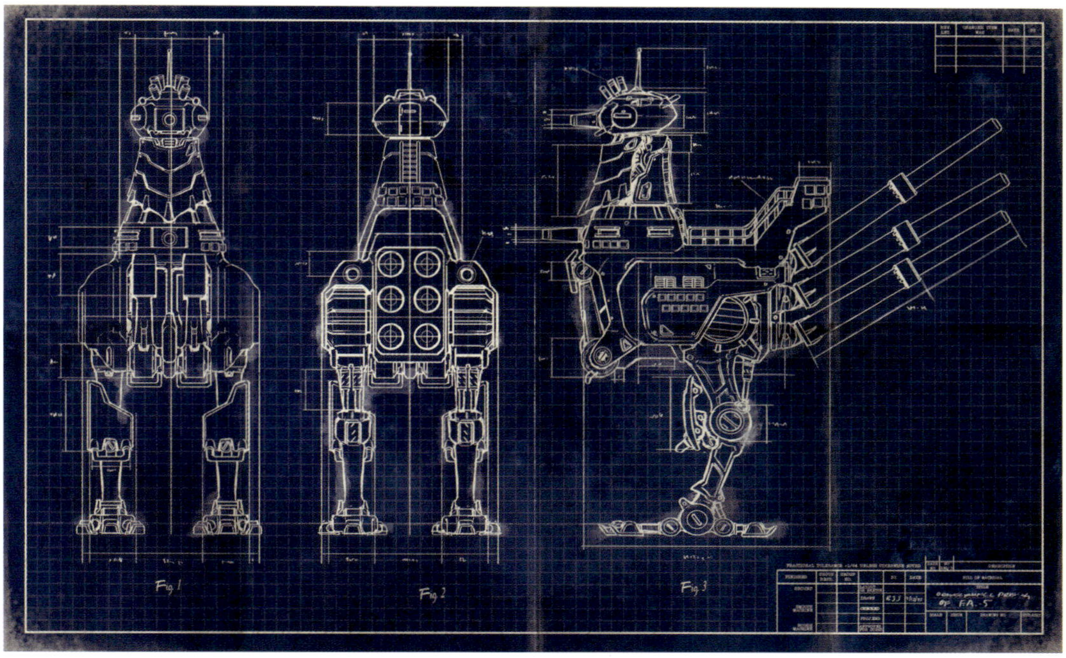

Blueprint of the approved FA5 design.

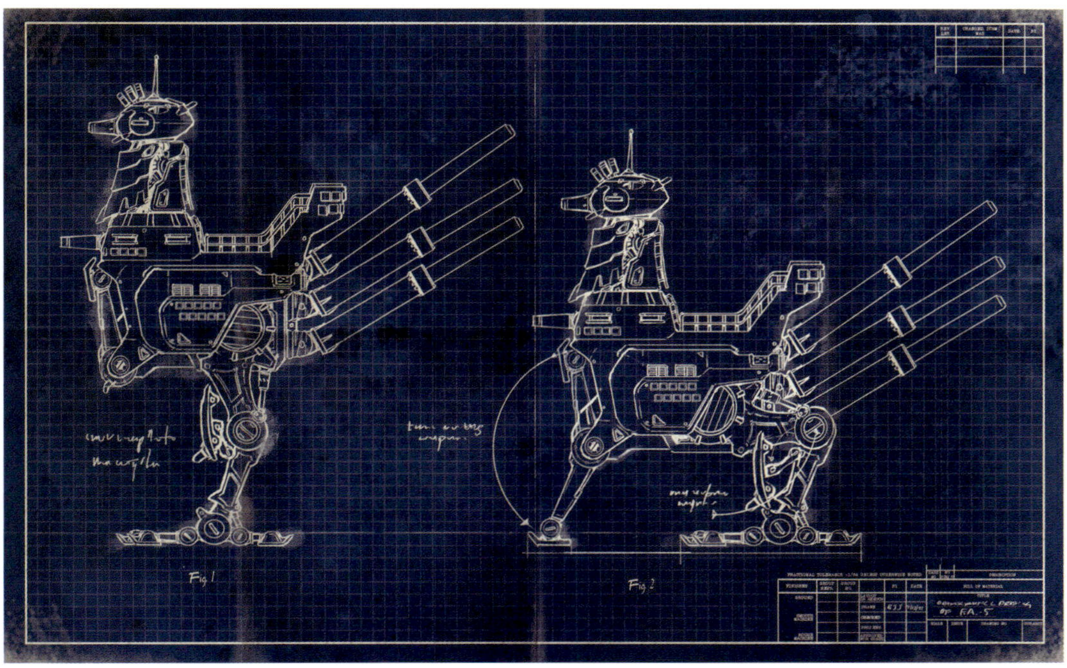

Blueprint demonstrating the FA5 rapid-firing configuration.

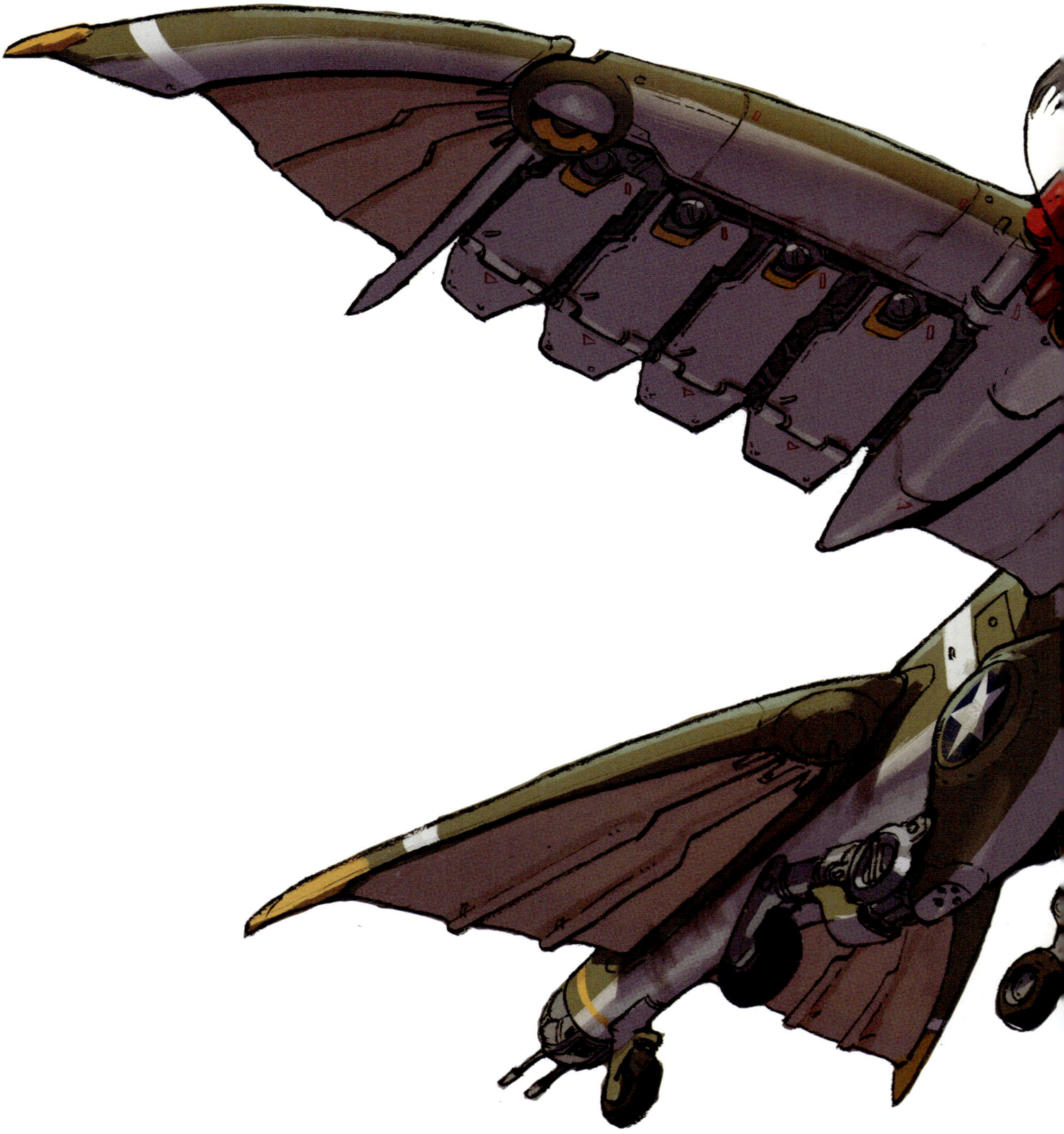

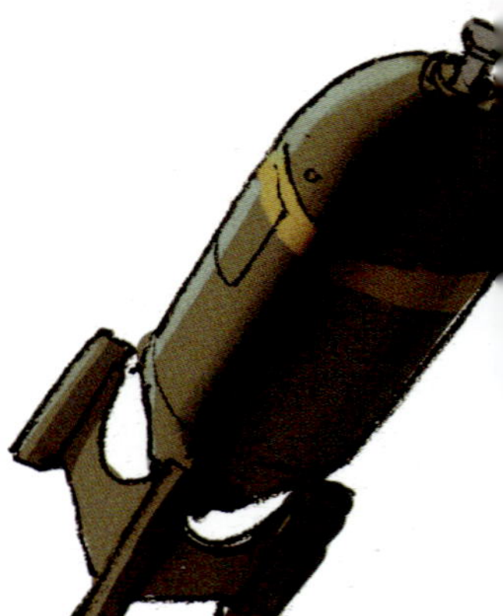

BX1 PIGEON BOMBER

SPECIFICATIONS

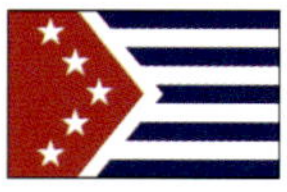

COUNTRY: Liberterra
UNITS BUILT: 18
NO. OF CREW: 4
MAIN ARMAMENT: 6 x 50-cal machine guns
PAYLOAD: 260 kg
MAX SPEED: 202 km/h

The BX1 Pigeon Bomber was one of the earliest successful attempts at creating a flying tankhead. While definitely ahead of its time, laying the groundwork for jetheads that followed it, the Pigeon Bomber itself proved too ambitious relative to the technology of the time. In action, its slow flight speed made it an easy target for anti-air weapons.

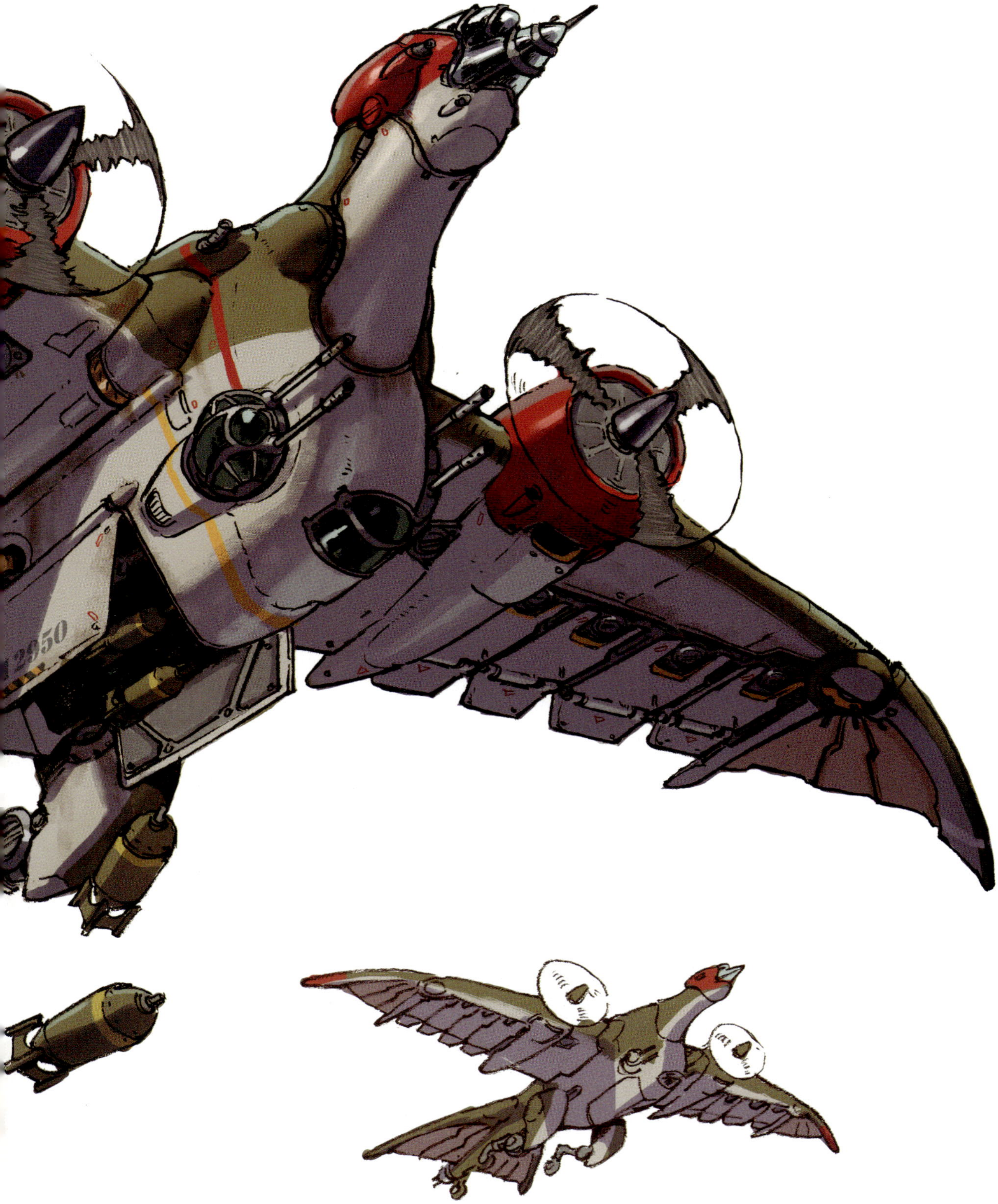

"They used some of these relics to fly orders to the front lines. My lads started calling those orders 'pigeon droppings', 'cause you were never happy when they landed on your head!"
-Platoon Commander Laurence Wilbur

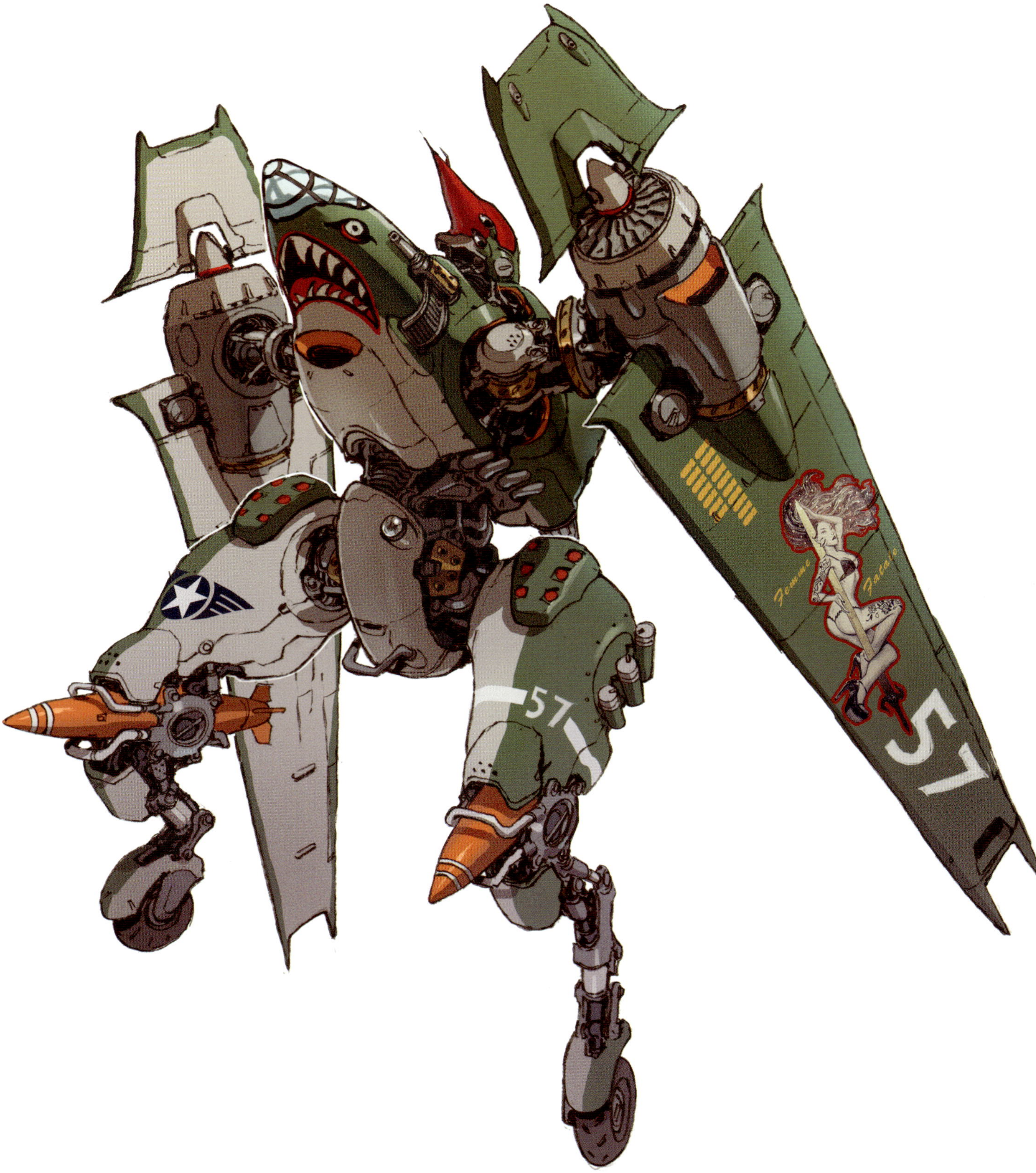
Femme
Fatale
57
57

P40 WARHAWK

The P40 Warhawk was mostly used for close air support in areas without anti-aircraft or enemy interceptors. For a brief period, when it was first introduced, it was one of the tightest-turning jetheads. Pilots described it as being able to easily "tap the enemy's back" in one-on-one dogfights.

The knee bombs were an interesting design feature. Engineers working on early prototypes of the P40 struggled to devise a way to carry the specified payload without needing heavy pylons. The knee mounts solved this problem by relying on the leg structures, which already needed to be strong enough to support the entire jethead while on the ground. On the other hand, the bombs' stowed position, nested in the knee mechanism, was very unpopular with ground crews tasked with rearming the Warhawk.

While the aerial arms race quickly caught up with the P40's design, it served in most theaters in roles ranging from air superiority fighter to fighter-bomber for most of the war.

SPECIFICATIONS

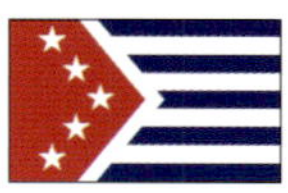

COUNTRY: Liberterra
UNITS BUILT: 6,627
NO. OF CREW: 2
MAIN ARMAMENT:
- 2 x 20 mm machine cannons
- 4 x 7.62 mm machine guns
- 6 x chaff launchers

ARMOR: 400 kg
MAX SPEED: 850 km/h

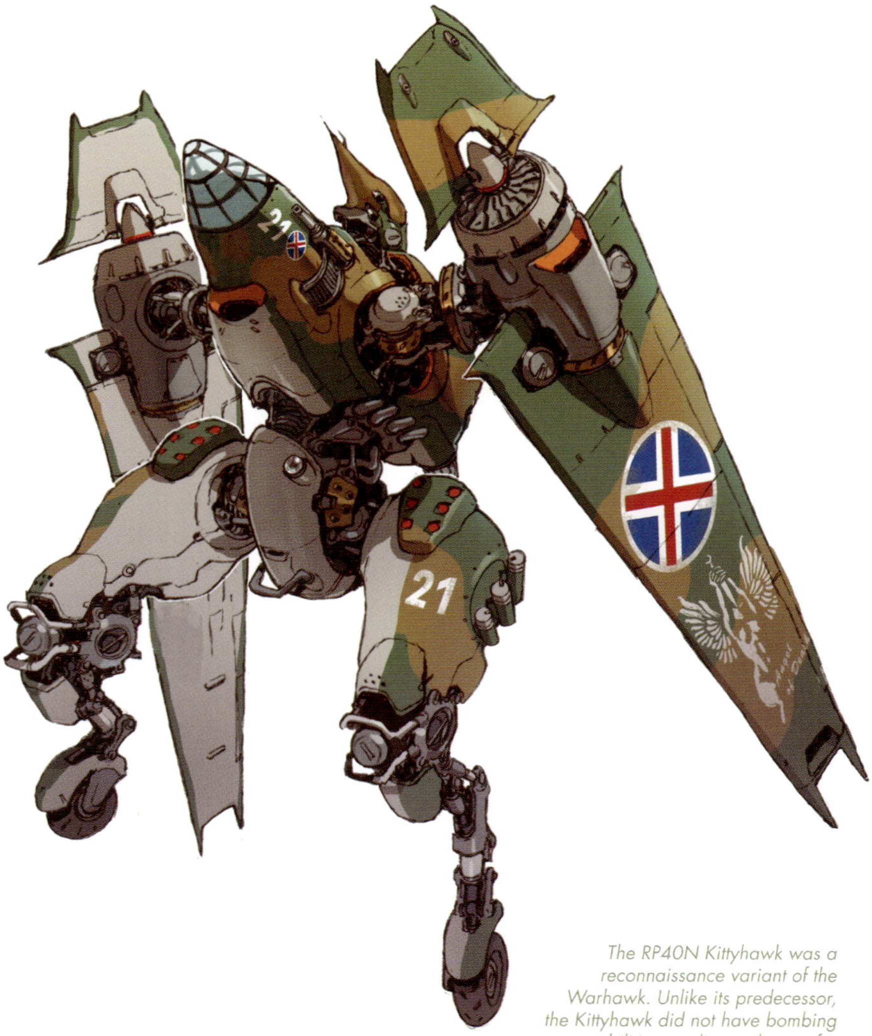

The RP40N Kittyhawk was a reconnaissance variant of the Warhawk. Unlike its predecessor, the Kittyhawk did not have bombing capabilities, trading ordnance for fuel and extended range.

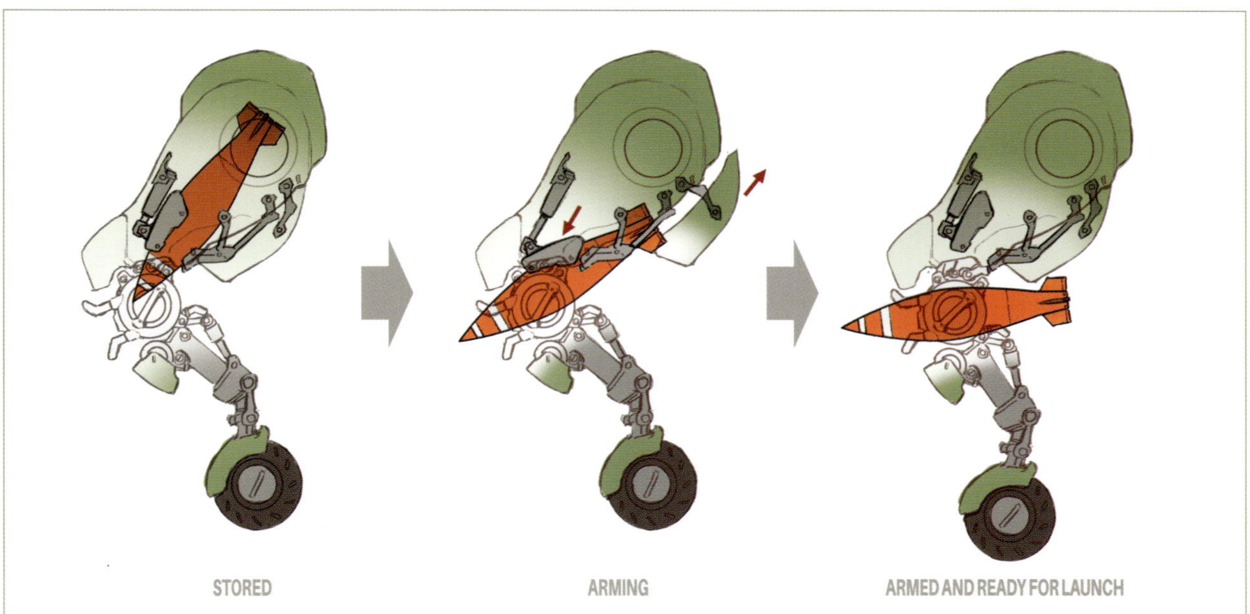

Diagram demonstrating the workings of the Liberterran knee bomb bay mechanism.

B24 LIBERATOR

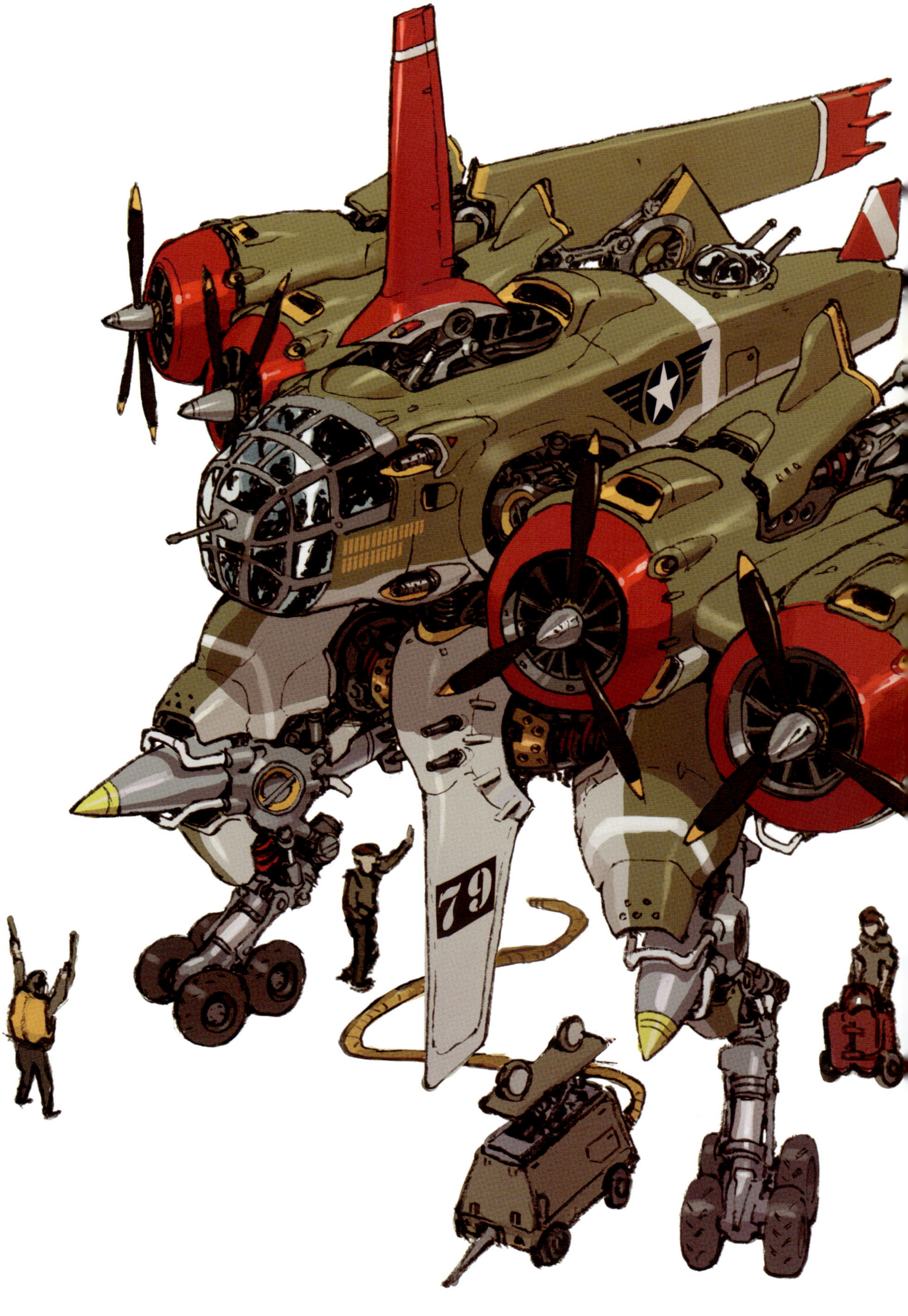

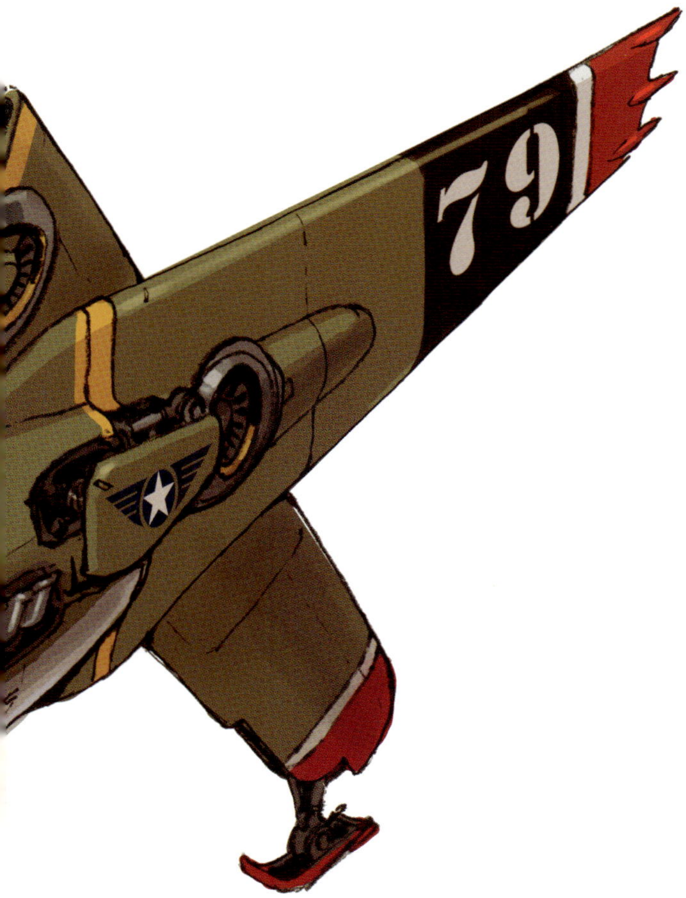

SPECIFICATIONS

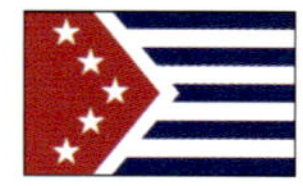

COUNTRY: Liberterra
UNITS BUILT: 32,188
NO. OF CREW: 10
MAIN ARMAMENT:
4 x 50-cal machine guns
9 x 30-cal machine guns
PAYLOAD: 3,991 kg
MAX SPEED: 503 km/h

Arguably one of the most iconic bomberheads, the B24 was designed by Liberterra but produced throughout the Alliance during the final period of the Neverwar, when jetheads and bomberheads became a defining factor in the conflict.

When taken on raw numbers alone, the B24 gives a very mixed picture. Despite their increased capacity, they ended up delivering fewer bombs on target and returning from missions at a lower rate than older bomberhead models. On the whole, it had a longer range and higher payload capacity than its predecessors, but a lower altitude ceiling and worse maneuverability. Still, more B24s were produced than any other Alliance aircraft during the war and experts continue to this day to debate whether the Alliance simply took a quantity-over-quality approach in this case.

The high payload capacity and range of the B24 meant that it was often the go-to bomber for long range, dangerous missions. On short raids, damage resulting in small fuel leaks was of negligible importance but catastrophic on longer runs; it was often necessary to ditch the bomberhead behind enemy lines. Allied Bomber Command was well aware of this fact, but was also aware of the significant impact these runs had. With Alliance forces consistently hitting targets further and further afield, both the Kyzerians and Yamaichi were forced to expend precious resources to increase their air defense coverage.

The long range desired from this machine required powerful lift, delivered by the quad wing slats in line with the four supercharged engines. The B24 had nearly twice the bomb capacity of the B17 it mostly replaced and its payload was distributed across two bomb bays located in the large thigh units of the machine.

On the topic of armament, nearly all Liberator types carried numerous guns for defensive fire, notably in the large frontal glass cupola and rear turret, as well as the head guns. Most variants also carried a large belly turret, which on the prototype was designed to retract into the fuselage for takeoff and landing. On the production unit, this feature became redundant with the addition of landing skids to the middle pair of wing slats, which were articulated to fold down. These simple skids (later replaced by wheels) provided enough stability for the bomberhead to land in a more upright position, with ample room for the belly turret.

CONTINUED>>

B24 LIBERATOR (CONTINUED)

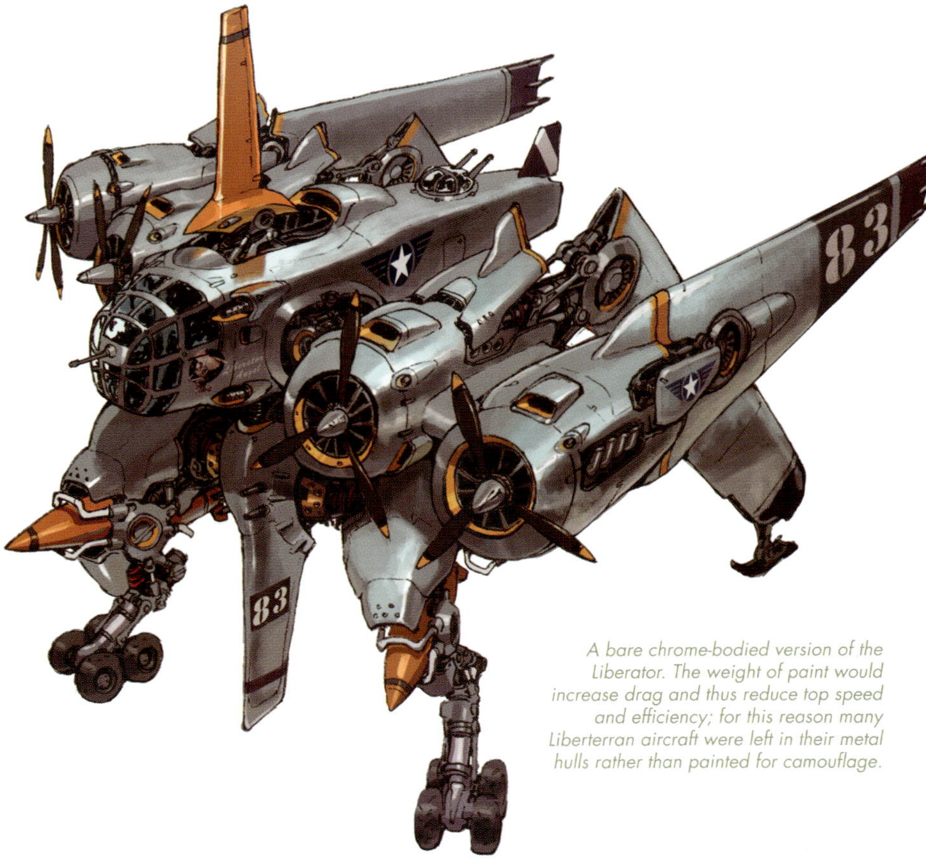

A bare chrome-bodied version of the Liberator. The weight of paint would increase drag and thus reduce top speed and efficiency; for this reason many Liberterran aircraft were left in their metal hulls rather than painted for camouflage.

While considered an impressive feat of engineering, this landing style was not well-liked by Liberator crews, who were already not fond of its design in general. Despite its size, the B24 was far from ergonomic and its controls were notably stiff. Its poor flight characteristics in particular weather conditions, especially cold, were known to make even the most hardened crews airsick. This complaint was compounded by a fuel leakage problem; crewmen were often exposed to nauseating fumes and, even worse, strictly forbidden from smoking during flight. It became standard practice to fly with some hatches partially open for ventilation.

The B24 was modified for a variety of roles. As with any bomber, its cargo capacity and range made it well suited for transportation of men and materiel. It would also see use as a reconnaissance unit for everything from simple, high-altitude spotting flights to highly specialized missions to seek out and identify submerged enemy targets. On the same technological level, there was a variant outfitted with jamming devices intended to block Kyzerian long-range radar. Perhaps the most prestigious variant was used in covert operations on the Kyzerian front. These B24 "Jacklegs" were painted matte black to avoid being spotted by searchlights. They carried very little weaponry or equipment of any kind, and utilized far more insidious tactics. They were used to insert covert agents and retrieve friendly personnel far behind enemy lines. They also supplied local resistance and partisans with equipment and (often counterfeit) monies.

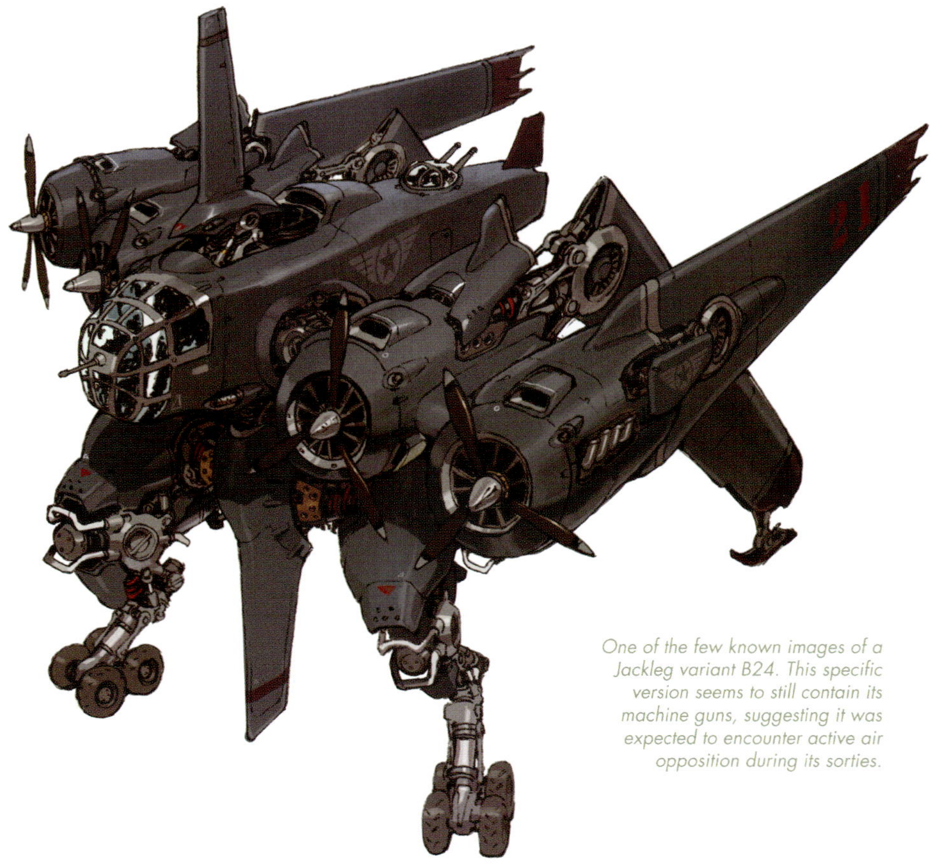

One of the few known images of a Jackleg variant B24. This specific version seems to still contain its machine guns, suggesting it was expected to encounter active air opposition during its sorties.

HT54 ALPHONSE SKYCRANE

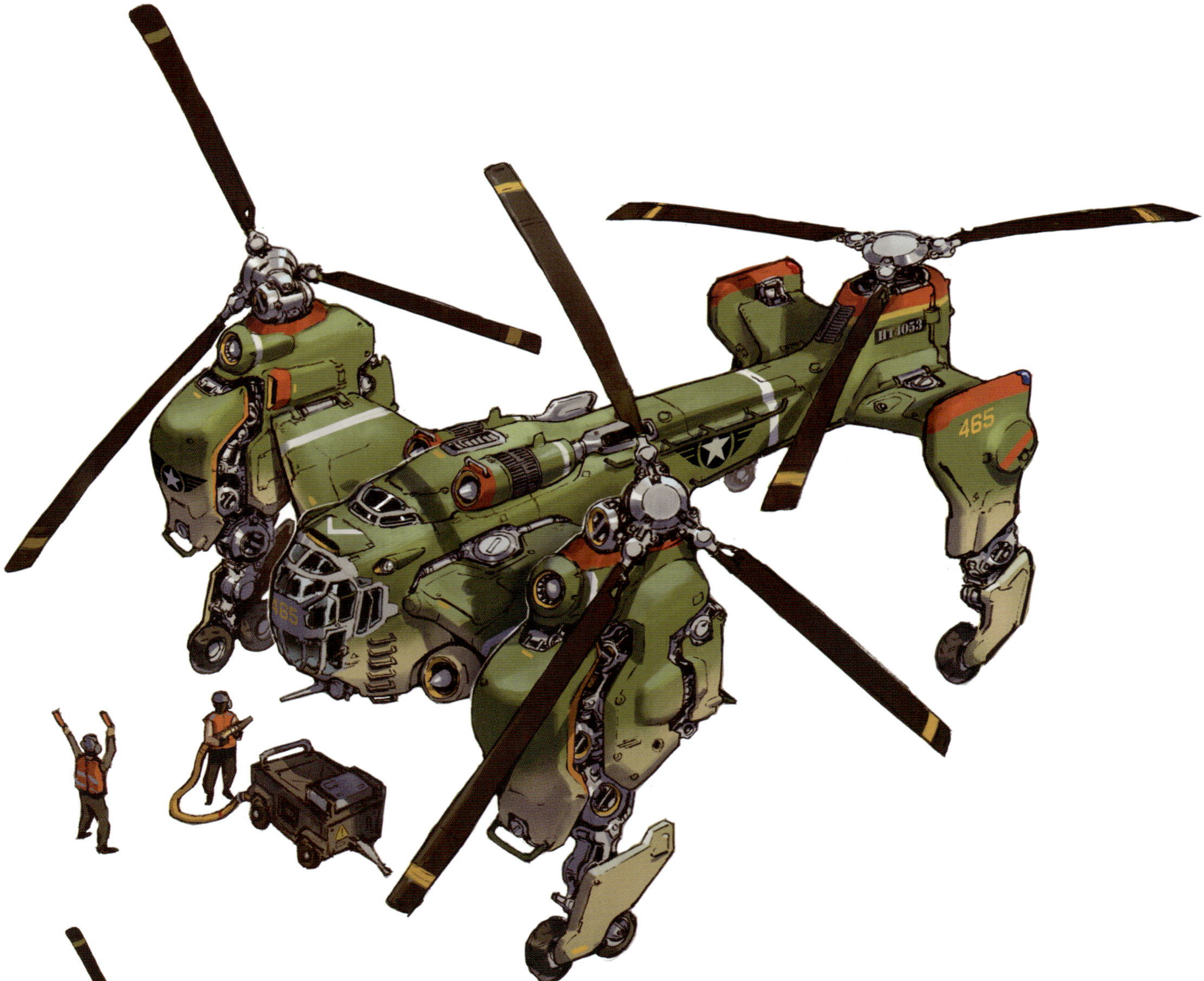

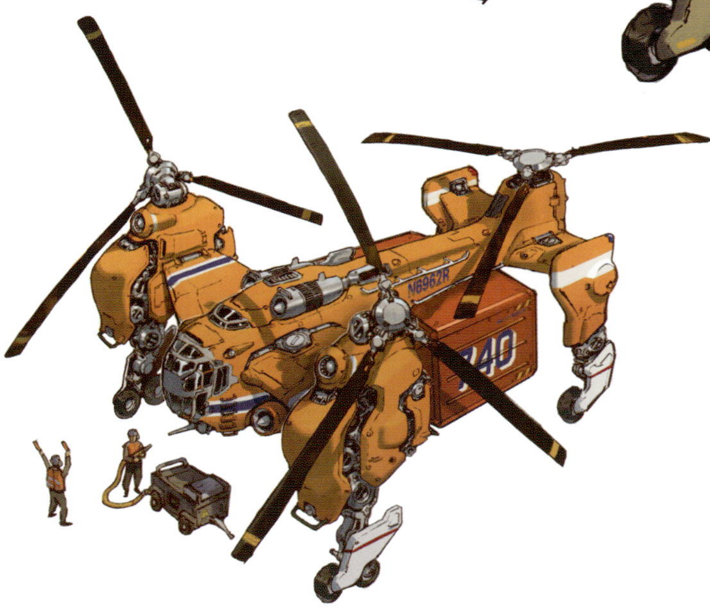

The Skycrane's capacity to lift standardized containers without the need to load and unload its contents made it an industrial workhorse in offshore and remote industries.

Despite being the leader in jethead development, the Alliance only managed to develop engines powerful enough for a viable tankhead airlift option near the end of the Neverwar. A dozen skycranes were rushed into service for the Grandmaster Gamble in WC049, providing much-needed mobility to the Alliance's vanguard in that desperate operation. The Alphonse would become more common in both military and civilian use after the war as it proved itself to be a rugged and reliable machine.

SPECIFICATIONS

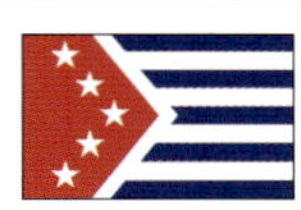

COUNTRY: Liberterra
UNITS BUILT: 383
NO. OF CREW: 3
CAPACITY: 35,000 kg
MAX SPEED: 240 km/h

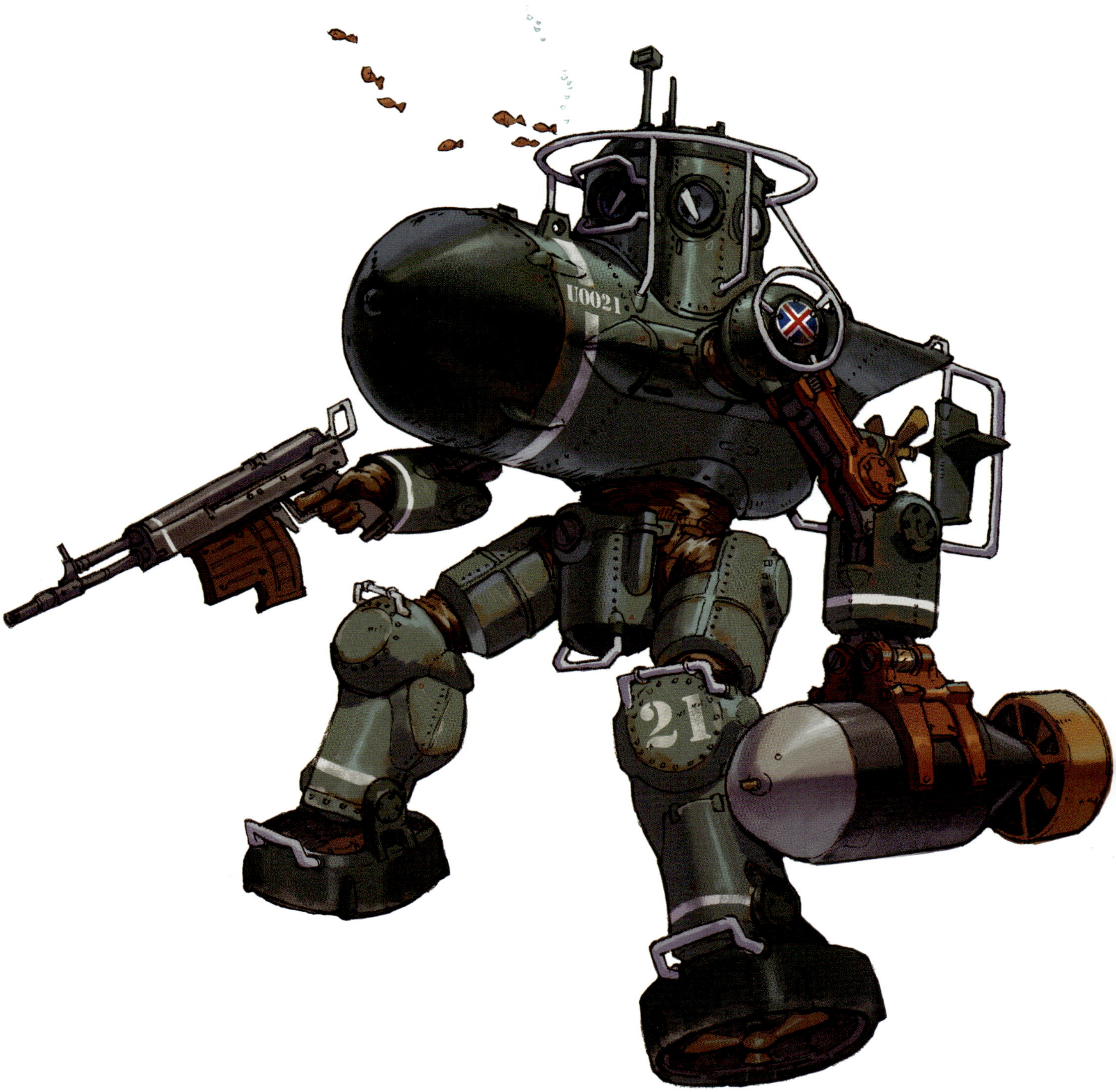

The Mk.1 Welman carried its explosive charge on the front of the hull. This gave it a rather comedic profile and the crews rather rudely referred to this weapon as the 'bonk conk'.

WELMAN MK.2

The modern view of mini-subheads is that they are mainly suited for harbor penetration—for covertly breaching underwater defenses around a harbor to allow ingress by other vehicles or to destroy enemy ships in harbor by use of torpedoes or limpet mines. However, it was only Alliance subheads that were specifically designed for this role. The Kyzerians imagined their subs to be used aggressively on the open sea and the Dominion originally envisioned theirs to be used in larger fleet actions. In practice, all sides ended up mainly using their minisubs for the harbor penetration role. Still, it is noteworthy that it was the Alliance alone who designed for this role from the beginning.

Keeping this in mind, it should be noted that the Mk.1 Welman was a suboptimal design, considering its intended use. The Mark 1 lacked arms in order to keep its profile as slim as possible. The machine's sole armament was a magnetic limpet mine added to the prow that extended its total hull length. This heavy explosive hampered the turning radius and overall balance of the subhead, especially when moving above water. Land mobility in general was not given much consideration during the initial design. The Mk.1 had only a single crewman, who operated all the vehicle's systems with only the small glass viewports in the tower to see through for guidance. The lack of a periscope was rationalized as being too demanding of the already-stretched crewman.

While 150 Mk.1 Welmans were ordered, only two-thirds of them were completed before a redesign process began. After training and limited field testing, four specific improvements for the Mk.2 were demanded:

1. *A way to cut nets and cables—critical for harbor penetration and most other covert operations.*
2. *A more effective method of delivering payloads and additional armaments.*
3. *An added periscope.*
4. *Simplification or automation of the machine's operation without significant loss of efficiency.*

To address the first two demands in one fell swoop, arms were added to the subhead. A special rig was designed to hold an improved explosive charge. Still magnetic, the new weapon was now capable of a short burst of movement if needed, but manual delivery was also possible. The other hand was left free for use of tools and weaponry as required. The addition of the arms and periscope did add one too many controls for a single person to operate and no method of simple automation was conceived. After multiple heated debates, the designers added a station for a second crewman, thus splitting operations into "commander" and "driver," like on many small tankheads. Although the new design performed much better on land, it was still "a duck out of water."

The first Mk.2s were delivered without any weaponry except their explosive payloads and had to rely on cutting tools for close-quarters combat. When the Welman Mk.2 came into use by Alliance Special Operations, development materials and prototypes for an aquatic tankhead handheld cannon were produced. The exact origin of this weapon remains unclear, but its Rozenian character is undeniable, as it is clearly based on other known prototypes from that nation. The intended use for the weapon remains likewise undefined. Common theories were that it was intended to shore up the lackluster T-37A amphibious tankhead or a rumored amphibious T-34 Gurko variant.

The operational history of the Welman is still classified, likely due to the fact that they were directed by the secretive ministry branch best known for its "ungentlemanly underwater warfare."

In WC014, mysterious raids in the Darumeshia Archipelago eventually led to war between the Britonica Alliance and the Yamaichi Dominion. There were persistent rumors that those covert attacks were performed by Welman subheads. This posits a problematic timeline which does not correspond at all with the Welman's development history. In fact, even the base model Welman had yet to be produced at that time. The notion that these subheads were also involved in the Archipelago disturbances of WC008 is, therefore, even less credible. The Mk.2, however, was confirmed to be involved in a number of covert operations against both Dominion and Kyzerian assets in the region in the later years of the Neverwar.

SPECIFICATIONS

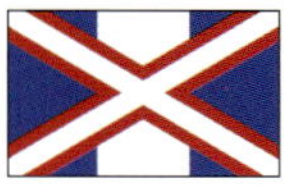

COUNTRY: Grand Britony
UNITS BUILT: 202
NO. OF CREW: 2
MAIN ARMAMENT:
1 x 193 kg torpedo explosive
SECONDARY ARMAMENT:
1 x handheld 56 mm underwater harpoon cannon
SPEED: 3 kn

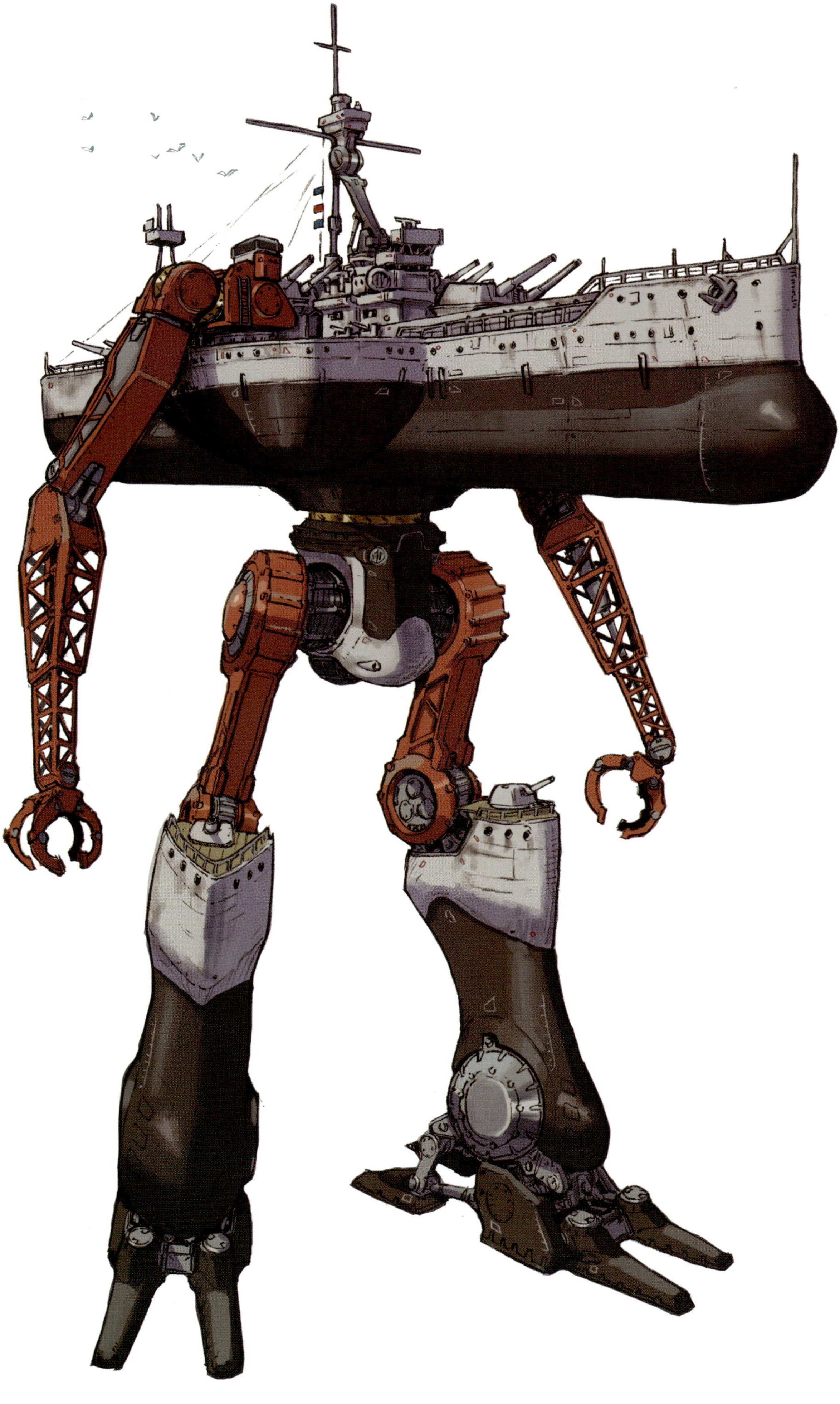

GBS DREADNOUGHT

While we often refer to tankhead-inspired walking battleships as "shipheads," it is equally common to hear them being called by another name: dreadnoughts. Such was the impression this ship made when it first appeared on the high seas that its designation became a byword for all walking ships that came after.

The *Dreadnought* was originally conceived of and designed as an ordinary navy ship and assembly was already underway when the idea of creating a bipedal battleship came about. The tonnage and displacement of the *Dreadnought* had already been impressive when it was envisioned as a regular naval vessel. The conversion to bipedal motion and addition of the arms, necessary for stable locomotion among other things, pushed its weight to levels that were inconceivable a few years prior. The first design for the lower legs would have been a fully open truss frame to reduce weight, but small scale testing showed that a streamlined design was necessary to maintain any level of stride as well as to resist the stress the water resistance would generate when wading.

The true innovation of the *Dreadnought* lay in its armaments. Vessels from that era were mostly equipped with light, rapid-fire guns and medium cannons but featured only a few long-ranged heavy cannons. Due to its height, the *Dreadnought* would often tower above regular vessels, so it was considered preferable to arm the vessel mostly with the heavy Mk.X gyrocannons and only a small number of lighter weapons for close-range defense against small agile assailants. The result was a floating and walking artillery battery capable of blasting opposition out of the water from a considerable distance. Because the vast majority of its weapons were the same caliber and had the same kind of ballistic flight path, any missed shot could be used to range the next. For example, if the first turret overshot, the second turret would fire in the same direction but a few degrees lower, and would likely land a shot or else at least come much closer to the target. This meant that the entirety of the *Dreadnought*'s weaponry could fire with deadly accuracy much more quickly than any ship with mixed weaponry. In contrast, such ships would often have trouble simply identifying which missed shot belonged to which weapon unless they fired only one at a time.

SPECIFICATIONS

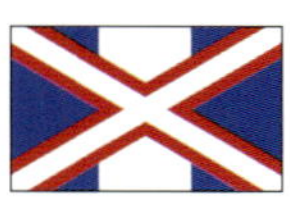

COUNTRY: Grand Britony
NO. OF CREW: 1200
ARMAMENT:
- 6 x twin 12 in guns
- 27 x single 12-pdr guns
- 2 x single 8 in guns
- 12 x 18 in torpedo tubes

ARMOR:
- 102–279 mm (belt)
- 19–76 mm (deck)
- 102–279 mm (barbettes)
- 76–305 mm (turrets)
- 279 mm (conning tower)
- 203 mm (bulkheads)

SPEED: 21 kn

The innovations did not stop there. A new propeller system was envisioned for the *Dreadnought* even before its bipedal redesign. This new system meant that, despite its heavy legs creating more drag, the ship could outpace vessels of similar weight when traveling through deep sea waters.

All of these advancements were soon observed by the other nations. Normally there is much uncertainty in naval intelligence-gathering operations. While spies or civilian informants might observe a ship testing new weapons and be able to estimate their firepower, it is quite hard to accurately identify the specific ship in question; even for a trained spotter, most ships look alike and the profiles and identifying features are hard to distinguish at a distance. This was not the case for the *Dreadnought*. As the only walking ship of its size in the world, it was in no danger of being mistaken for any other, more conventional vessel. It is for this reason, perhaps, that the *Dreadnought* did not see much actual combat, as enemy fleets would divert to avoid a confrontation with the walking giant. Neither were there any other vessels fast enough to form a fleet with her to effectively exploit the *Dreadnought's* speed and power. This meant that the *Dreadnought* mainly saw use in a coastal defense role—more of a deterrent than a weapon.

The bow of the *Dreadnought* was constructed with the intention of being capable of ramming. This section of the ship was already completed before the decision was made to redesign the ship. While this ramming prow was unlikely to be used, the Alliance Navy felt that there was little value in removing it. This decision was later to be seen as serendipitous when the *Dreadnought* intercepted a Yamaichi submarine near the Grand Britony south coast. Unable to hit the submerged target with its cannons, the *Dreadnought* rammed into the sub, causing it to sink immediately. It was, to this day, the only instance in which a shiphead was able to destroy a submarine.

After many years of coastal defense duty, the *Dreadnought* was removed from active duty in WC033 and converted to a training vessel. It was in this role that the *Dreadnought* would find its end. In WC047, the Kyzerians unleashed the behemoth tankhead Ratte that broke through the lines between Grand Britony and the Empire. The unfinished Ratte was intended to serve as a diversion from their Operation Oberherr invasion, a gigantic naval landing on Liberterra, thousands of miles away. Ironically, the Ratte proved itself far more effective than the Kyzerians had expected and the Alliance forces were unable to destroy the titanic tankhead. The *Dreadnought* was close by, in the bay south of the front on training exercises, but it was not carrying any live ammunition. A clever ruse by the Alliance ground forces and valiant sacrifices from smaller ships in the bay lured the Ratte into the ocean in the hope this would stop it. While it did sink into the coastal muck, it was only immobilized for roughly fifteen minutes; the crew eventually managed to free the legs. At that moment Captain Edith Goodenough, the commander of the *Dreadnought*, made the decision to give the order to ram the Ratte while it was still in the shallow waters. Both titans were destroyed in the collision.

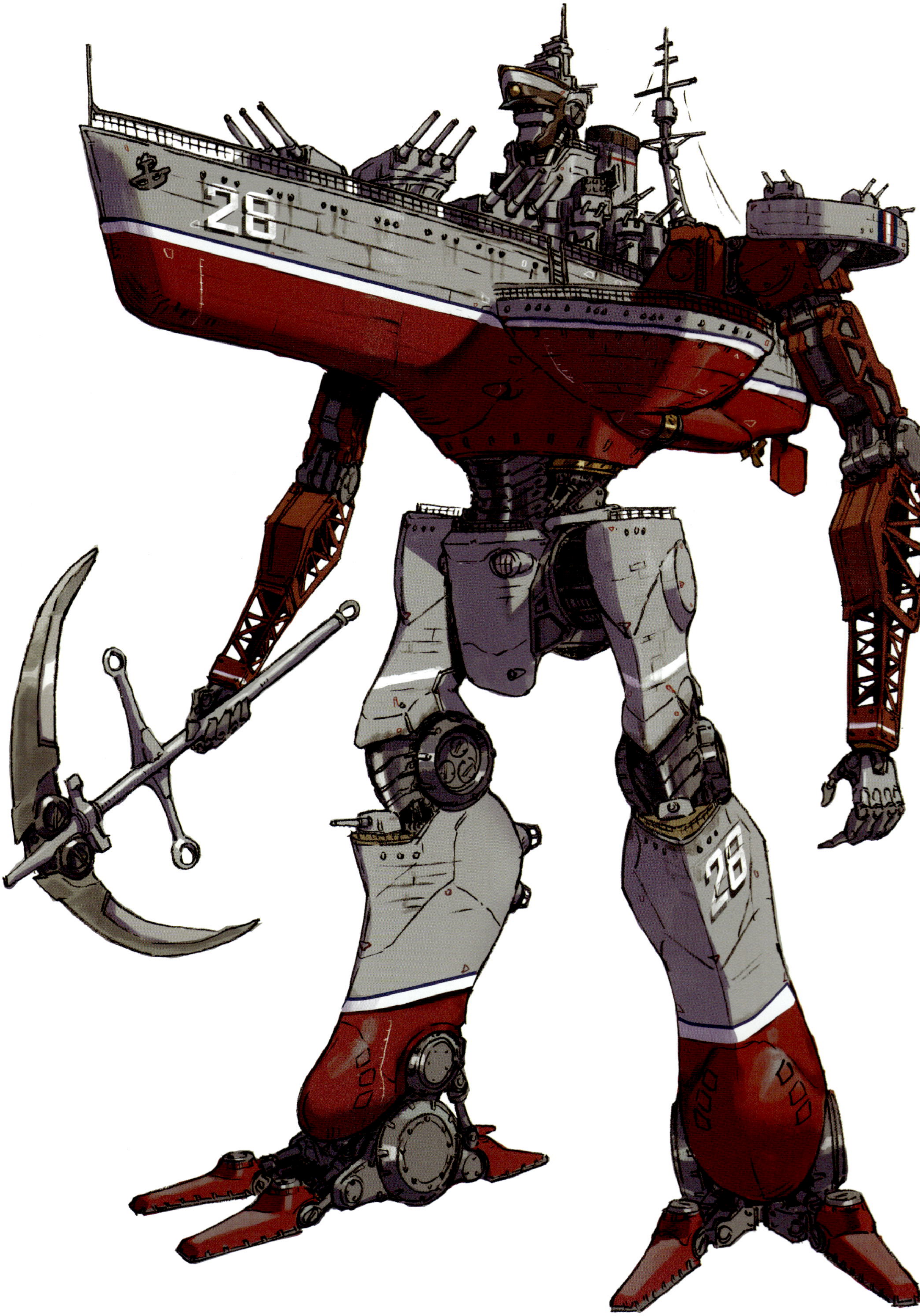
28
28

GBS NELSON

The *Nelson* came from the second generation of shipheads, which to this point had been dominated by the *Dreadnought* and smaller models of cruiserheads. Many of its features were based on lessons learned from the *Dreadnought*. Its weapons were intended to make use of the superior height and sightline of the towering ship, though its armaments were lighter than those of its predecessor due to strained production. It also lacked the ramming prow of the *Dreadnought*, but was equipped with a melee weapon known as "The Anchor" for close quarters fighting. While the *Nelson* did not encounter the problem, it was feared that the high position of the walking ship's cannons would allow smaller enemies to exploit its blind spot by moving in close. The Anchor did not see much use but was considered a highly-effective deterrent. For a time it was the largest melee weapon ever constructed, by several orders of magnitude.

SPECIFICATIONS

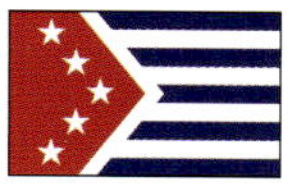

COUNTRY: Liberterra
NO. OF CREW: 1,361
ARMAMENT:
- 3 x triple 6 in guns
- 12 x twin 6 in guns
- 8 x 4.7 in anti-aircraft guns
- 8 x 2-pdr anti-aircraft guns
- 14 x 24.5 in torpedo tubes
- 1 x hand-held shipbreaker anchor

ARMOR:
- 330–351 mm (belt)
- 95–159 mm (deck)
- 305–381 mm (barbettes)
- 229–406 mm (turrets)
- 305–356 mm (conning tower)

MAX SPEED: 23 kn

LSS SHANGRI-LA

SPECIFICATIONS

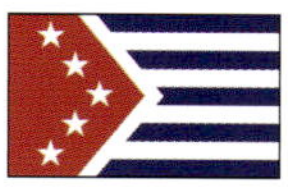

COUNTRY: Liberterra
NO. OF CREW: 4,248
JETHEADS CARRIED: 94
ARMAMENT:
- 2 x twin 8 in guns
- 12 x 88 mm flak guns
- 40 x 20 mm cannons

SPEED: 33 kn
ARMOR:
- 102 mm (belt)
- 64 mm (hangar deck)
- 38 mm (protective decks)
- 38 mm (conning tower)

The *Shangri-La* was one of the first carriers designed to launch long-range bomberheads against the Dominion. It frustrated and mystified its enemies, who could only speculate where these new air raids were originating from, as they were unaware of the significantly-increased strike distance enabled by the *Shangri-La* and its flight of cutting-edge jetheads.

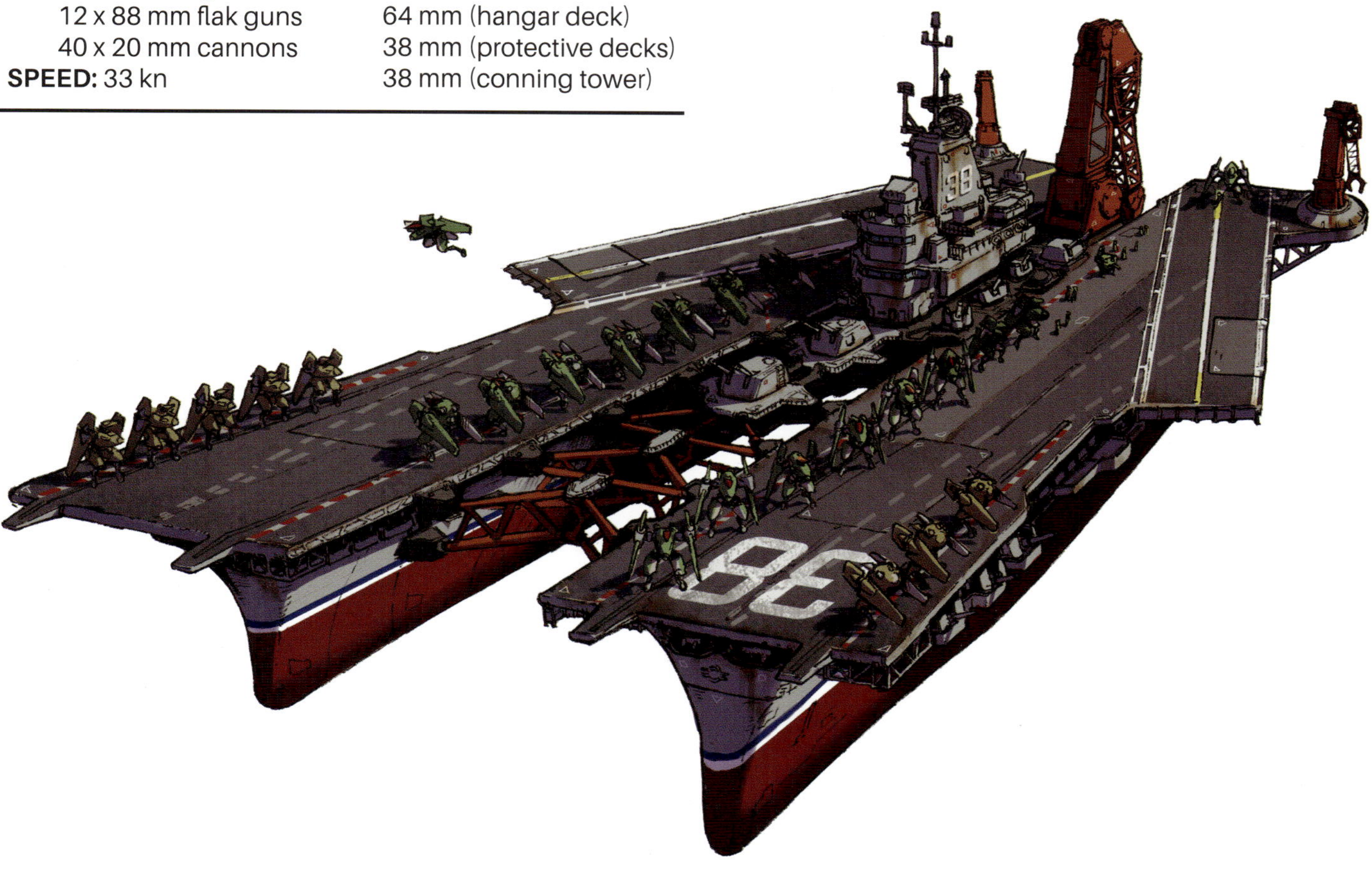

YAMAICHI DOMINION

AFFILIATED NATIONS

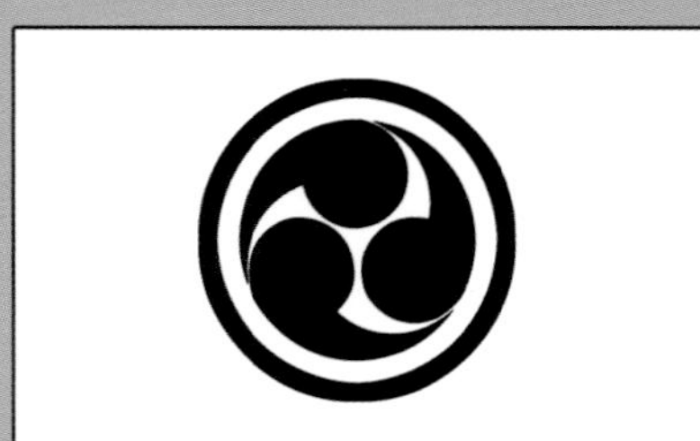

Yamaichi Dominion

TYPE 71 MEDIUM TANK

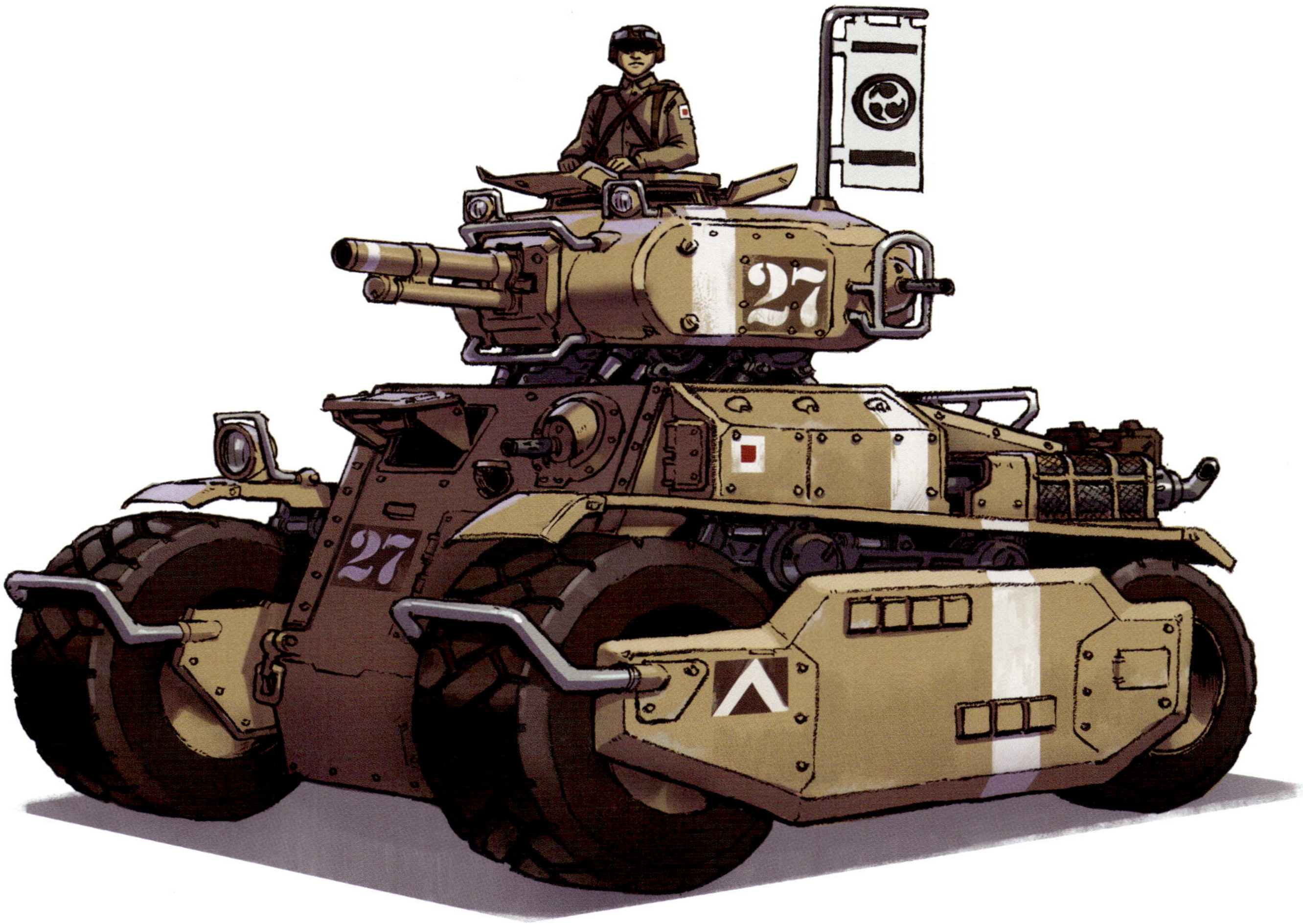

The Type 71 was the first medium tank designed by the Yamaichi military. The Yamaichi War Engineering Department took inspiration from pre-war Alliance medium tanks to create something unique.

The Type 71 stemmed from the relatively peaceful period before the Neverwar, when the Yamaichi had not faced serious armored military opposition for some time. It was therefore mainly designed to assault and destroy fortified positions and pillboxes in the rare insurrection or outlaw skirmish where opposing forces had taken defensive positions too strong for infantry to attack on their own.

A feature that is easy to overlook is the high quality of its armor. Before the Neverwar, most regular tanks were equipped with Yamaichi's famous composite armor. In the summer of WC015, when Yamaichi finally fully committed its forces to the Neverwar and intensified their tankhead production, the quality of armor on their regular tanks would noticeably deteriorate—there was a shortage of skilled craftsmen to fabricate the armor, due to conscription.

When the Type 71 eventually saw combat against well-equipped militaries, it proved less than effective. Officially, it fulfilled the role of main battle tank for the Yamaichi army, but its armor and armament lagged compared to the anti-tank weaponry it faced on the battlefield. Much like the old Britonican tanks it was based on, the Type 71 lacked swing arm suspension, which was still uncommon and would have been a major advantage on the archipelago's rough and treacherous terrain. On the other hand, its classic traction system performed well on the flat tundras of the east Kyzerian front.

The most significant and, in many cases, fatal flaw in the Type 71 was its engine. While powerful, it was prone to fires, a devastating vulnerability in any tank. Once this became known, tank crews were extremely wary of the risk, to the detriment of their performance and morale. Newer production variants were designed with alternative engine types that were less prone to fires, but none matched the power and speed of the original.

SPECIFICATIONS

COUNTRY: Yamaichi
UNITS BUILT: 808
NO. OF CREW: 4
MAIN ARMAMENT: 1 x 57 mm gun
SECONDARY ARMAMENT:
2 x 6.5 mm machine gun
ARMOR: 6-17 mm
MAX SPEED: 35 km/h

TYPE 75 LIGHT TANK

The first tankheads in full use by the Yamaichi Dominion were purchased from other nations, often retrofitted, like the S4 Crossley. These machines had proven the effectiveness of the tankhead concept on the field to the Yamaichi. In their quest to develop the most advanced and finely-crafted tankheads they could possibly design, the Dominion severely hampered their production speed. Unlike the Brits, who fielded multiple quadruped tankheads at the start of the war because they had not mastered the bipedal system, the Yamaichi were determined to produce a bipedal tankhead as only they could.

Once Yamaichi had joined the war on two fronts in WC016, they had little hope of continuing to purchase foreign tankheads for their military. At the same time, the Dominion's conventional forces had not been modernized at the same pace as those of the Alliance or Kyzeria. While both their infantry and field guns were still effective, their armored branch was sorely out of date. Tanks like the Type 71 were designed for a function that did not fit the actual reality of modern combat. Acknowledging this shortcoming in their battle-readiness and also the fact that the tankheads that they had put all their hopes on would not be ready in time for the conflict at their doorstep, Yamaichi Command mandated the development of a modern light tank that could keep up with and support the infantry. Orders stated that the tank must be ready in less than half a year—an ambitious deadline but one for which Command authorized the development team to appropriate whatever resources they required.

The engineers assigned to this project were considered, at the least, eccentric and were often shunned for their methods. Starting with a half-built Type 71 tank frame, they had their first prototype on the range in two weeks. Ruthlessly stripping any part they deemed inadequate and replacing it with whatever they could get their hands on, the Type 75 took shape quickly. The most unconventional part of its design was its turret mechanism, which was adapted from a prototype tankhead's neck assembly. This meant that the Type 75 would not have a full 360° rotation, but its gun had faster, smoother tracking than that of any other tank of the time. The main weapon was intended to be a 47 mm gyrojet cannon, but this was eventually dropped and replaced with a lighter 37 mm gyrojet cannon from the Type 2 Tadakatsu prototype. The Type 75 tank saw its final trials a little over 3 months after development commenced, a record yet to be surpassed in military development. The machine cut corners at every turn and had been built with repurposed parts from every branch of the Dominion military machine, but it worked.

In the field, the Type 75 performed to specifications, if only just. Its weapon lacked punch and had a low rate of fire. While capable of destroying light enemy vehicles, the Type 75 could only harass heavier tanks and tankheads. Even so, the crews of these tanks were elated to take to the field in a "modern" vehicle. Their crews' fanaticism, combined with their numbers, would make the Type 75s a nuisance to their enemies. Dominion infantry also appreciated the new light tank, as it was well suited to storing extra gear during long marches and the Type 75 crews could always be counted on to distract enemy vehicles, taking enemy fire that would otherwise have devastated infantry.

SPECIFICATIONS

COUNTRY: Yamaichi
UNITS BUILT: 2,021
NO. OF CREW: 3
MAIN ARMAMENT: 1 x 37 mm gun
SECONDARY ARMAMENT:
1 x 7.7 mm machine gun
8 x smoke grenade launchers
ARMOR: 6–27 mm
MAX SPEED: 38 km/h

Exploiting any breakthroughs they could and buzzing around larger vehicles hoping for a lucky hit on some weak point, the Type 75s were soon dubbed "Gnats" by Liberterran troops.

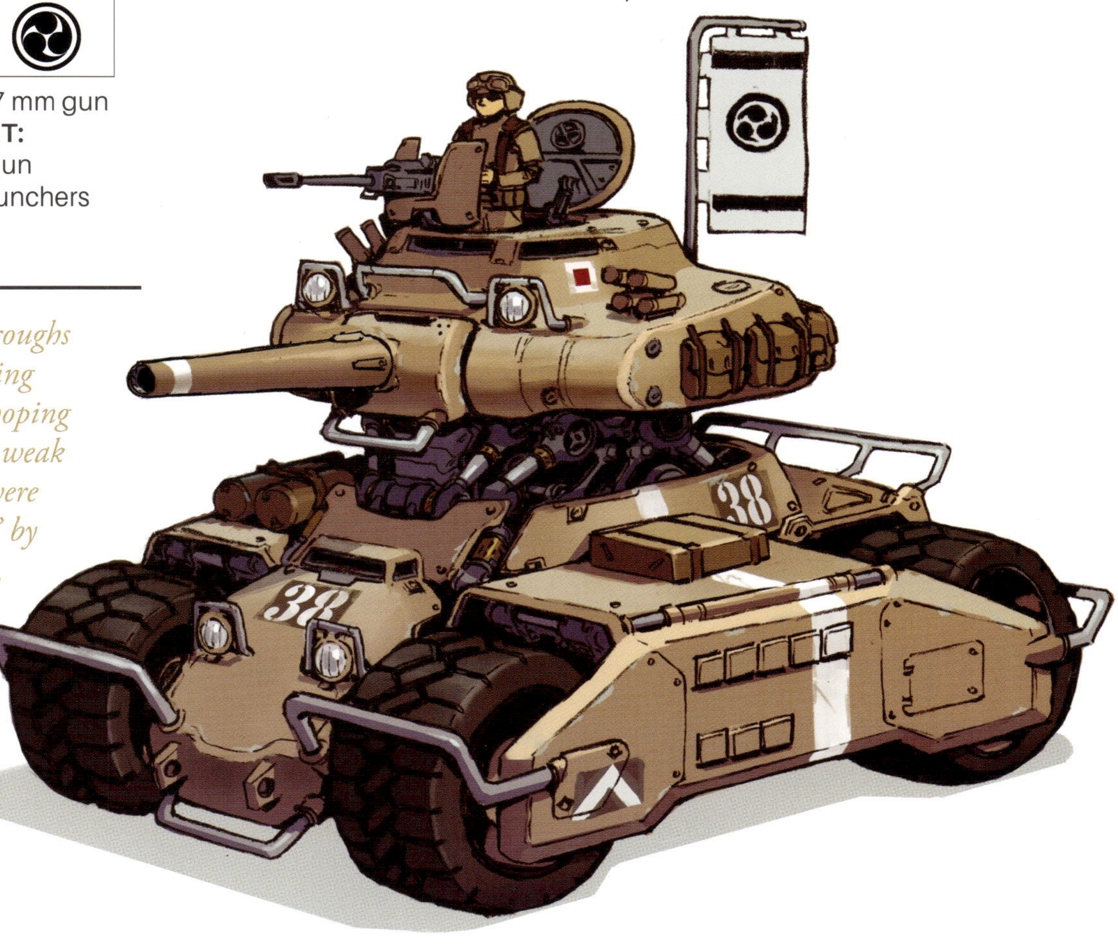

TYPE 76 SELF-PROPELLED GUN

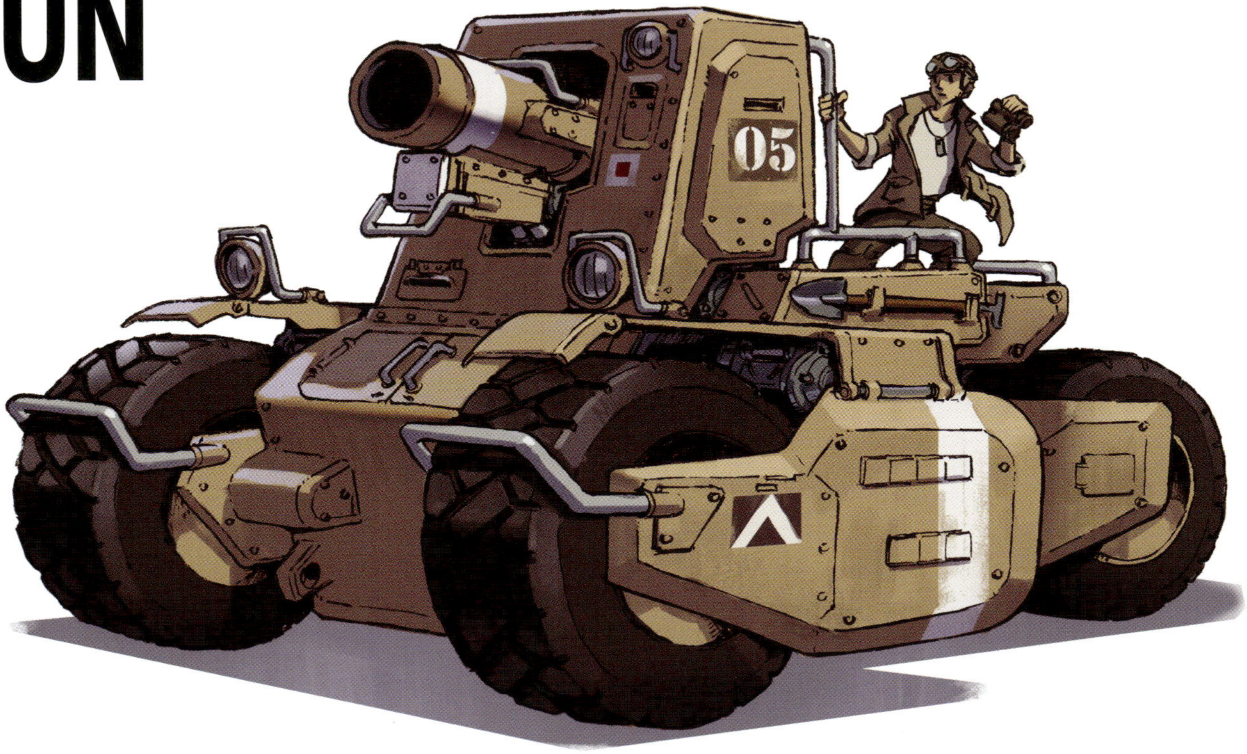

When the Yamaichi ramped up their production of tanks, they were already playing catch-up to the other great powers in the war. As the conflict in Rozenia escalated into full-blown war between the Alliance and Kyzeria, the Yamaichi mainly observed from far away, preferring to remain uninvolved. However, they did take note of the way warfare was evolving with new technologies. Relations between Kaiser Walmir VII and the Yamaichi courts were quite friendly and many Yamaichi generals toured the warzones of the early Neverwar as guests of the Empire. To their credit, they understood the importance of the tankhead as soon as they laid eyes on it, and they would later produce some of the most finely-crafted tankheads of the war.

On a less dramatic level, the combat on the Rozenian front also revealed smaller areas where the Yamaichi arsenal needed to improve. The Kanonenjagdpanzerwagen was the terror of the battlefield in those early days and its reputation led the Dominion to produce a self-propelled gun of its own. However, not all Yamaichi generals were convinced there was a need for such a machine. To some, tank hunter designs such as the Kanonenjagdpanzerwagen were the worst kind of hybrid, combining field gun and tank in a way that compounded the weaknesses of both. These reservations were largely to blame for the limited support or resources the development program received.

Faced with a crippled budget which was insufficient to develop a new unit type from scratch, the design team opted to repurpose an existing tank hull and field gun. While Yamaichi command had originally envisioned something like the Kanonenjagdpanzerwagen—a fully enclosed weapon protected from small arms—the Type 76 instead featured a shield that provided only limited anterior protection and little else. On the other hand, the field gun used in the design proved to be quite effective, despite being an older model. In spite of its short barrel and limited traverse range, which had seemed appropriate as the unit was intended to engage with direct fire using anti-tank shells, it could still deliver effective arcing fire over long distances.

In WC014, the conflict in the Darumeshia Archipelago started to heat up again. Unknown forces began raiding settlements in both the Alliance and Yamaichi territories. Similar incidents had happened before, in WC008, and resentment had never stopped simmering under the surface for either Britonican or Yamaichi locals in that area. As resentment boiled over and skirmishing between the unfriendly neighbors began to take place, the Yamaichi rushed a small number of their Type 76s to the front. The slow-turning and poorly-protected assault gun fared terribly in the few fights it took part in. The local population loyal to the Alliance beat the Yamaichi, who were usually masters of the environment and ambushes, at their own game, managing to ambush and destroy or capture all of the Type 76 units.

SPECIFICATIONS

COUNTRY: Yamaichi
UNITS BUILT: 62
NO. OF CREW: 6
MAIN ARMAMENT: 1 x 150 mm howitzer
SECONDARY ARMAMENT: none
ARMOR: 12 mm–25 mm
MAX SPEED: 38 km/h

TYPE 09 ARTILLERY TRUCK

The Yamaichi Dominion fought a very dynamic conflict for a large part of the Neverwar so their forces, as one would expect, were focused around highly mobile units such as ships, light tanks, mobile tankheads, and later, jetheads. Support vehicles like self-propelled artillery, slower to deploy and harder to transport, were a secondary concern to their developmental plans.

The Dominion's earliest developments in the field of self-propelled artillery came in the shape of tanks mounted with light artillery guns. These systems had mixed success and were entirely overshadowed by the tankheads later developed by the Dominion. Nevertheless, as conflict on the Kyzerian front became more static in character, the need for better self-propelled artillery pieces was acknowledged. The Type 09 was developed using the base of an armored transport vehicle and a repurposed experimental anti-tank gun. The armor on the initial design was reduced to allow for the heavy mortar system to be mounted. In order to balance the vehicle, the engine was moved forward, giving the machine a look very similar to a conventional truck.

The 30 cm smoothbore mortar used in the original design was a muzzle-loading type—not particularly effective as a field gun, but considered adequate. However, it became an engineering nightmare on the Type 09 mount. The first models of the self-propelled artillery platform had a complicated loading tray on multiple swiveling arms mounted to the front of the barrel. This device could hold a single shell, and through a series of levers and pulleys was raised to the muzzle, emptied into the barrel and quickly moved away again. As one could likely imagine, this mechanism was prone to mishaps and breakage. When it did manage to fire reliably, the weapon was effective, and the added mobility of the transport base convincingly demonstrated the advantages of a self-propelled artillery piece to the Dominion commanders. As weapon technology evolved, Yamaichi engineers were able to redesign the Type 09's mortar to use a breech-loading mechanism, notably increasing its rate of fire. This turned the weapon into a reliable artillery piece that performed well in that role for the rest of the war.

SPECIFICATIONS

COUNTRY: Yamaichi
UNITS BUILT: 9,070
NO. OF CREW: 7
MAIN ARMAMENT: 1 x 300 mm mortar
ARMOR: 12–25 mm
MAX SPEED: 40 km/h

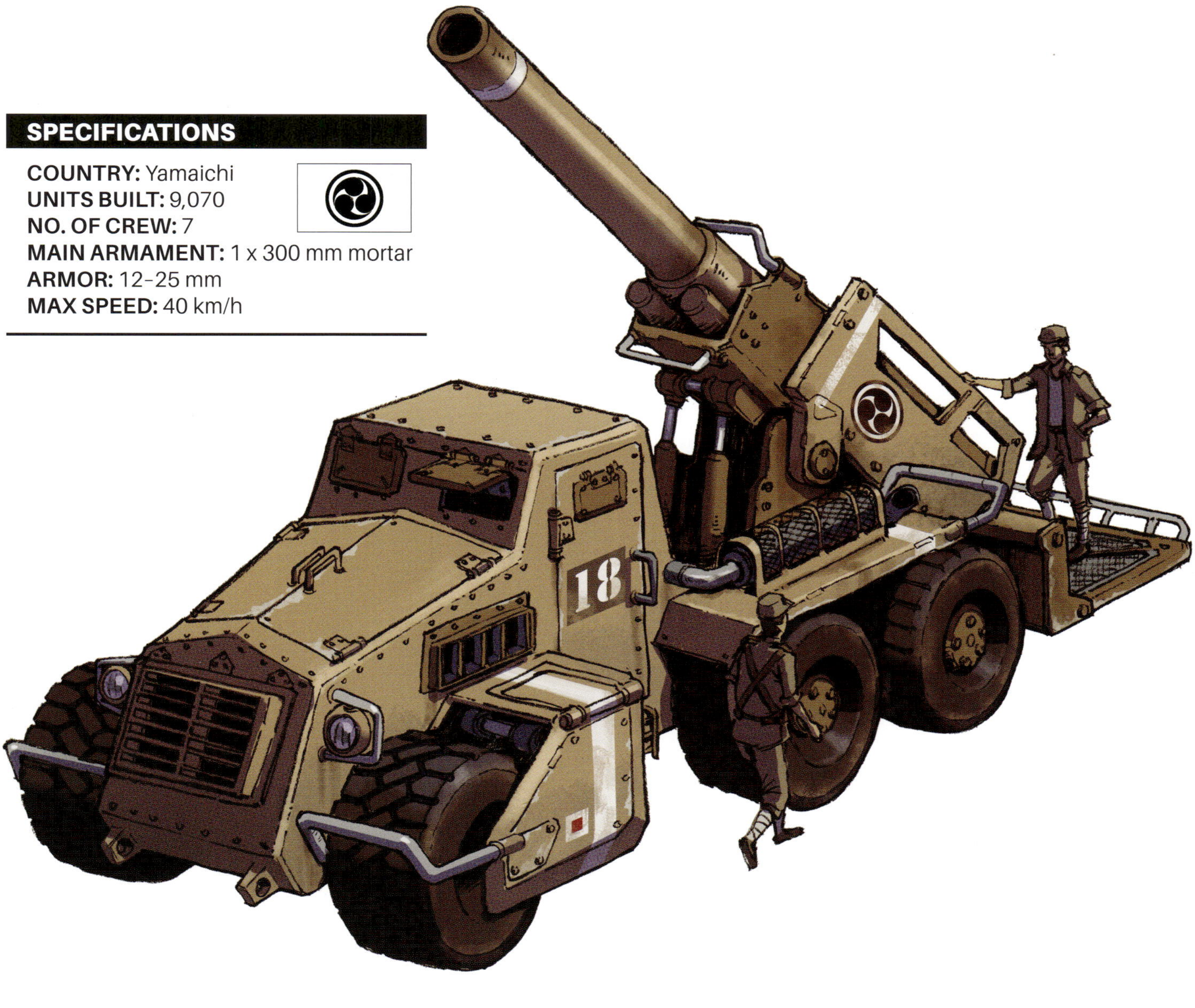

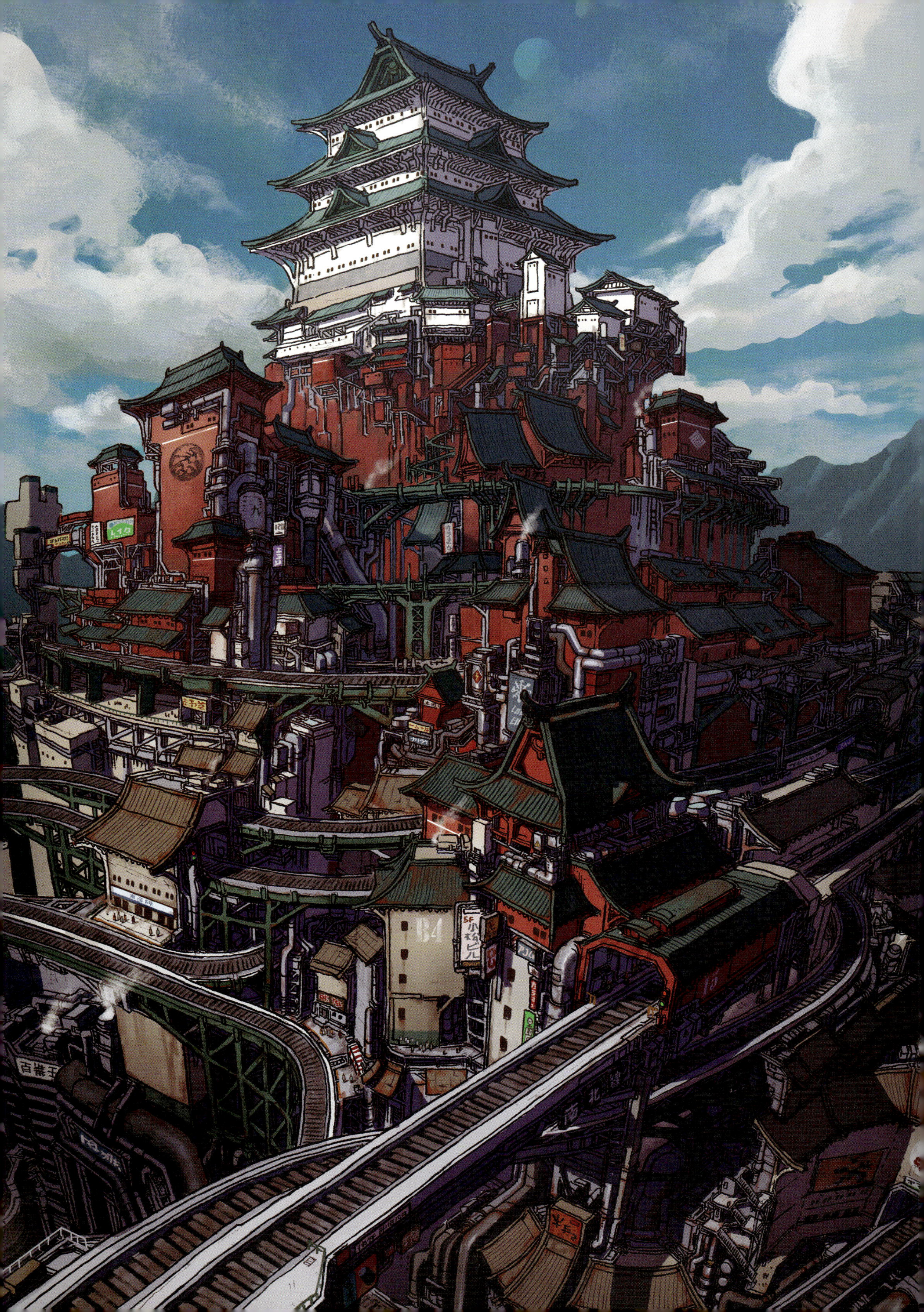
B4

AKAGAWA CITY

Akagawa City, in Rikuan prefecture, lies nestled in a small river valley near the eastern coast of the Gulf of Id, close to the border with New Kyzeria. It is surrounded by steep mountainous terrain on all sides with only the tracks along the river allowing for travel to the western coast and into the Dominion's heartlands to the northeast. While historically a lesser political power, it had always been known for its fine pottery exports. The Red River, for which the city and province are named, is the only known source of a peculiar, mineral-rich clay which, while difficult to bake properly, is quite unmatched in the intensity of its deep color. The subtle patterns of colourplay and microfractures in the glazing were praised by the most discerning collectors, lauded in art and literature. Records of trade throughout Yamaichi, the Kyz lands, and even as far as the old Britonica courts speak to the remarkable esteem in which the Red River pottery was held. The wealth acquired through this lucrative trade enabled and encouraged local artisans to experiment and extend the limits of their craft.

One of these experiments led to the creation of a type of ceramic now familiar to anyone with basic understanding of modern armor design. It was common practice for Yamaichi potters to grind up and blend failed and cast-off material from a firing into lesser quality clay in order to produce lower-grade wares. These would still have a recognizable red appearance to them, albeit duller and less desirable. Potters often used this mixture when they were unable to secure enough of the higher quality Red River clay. An inventive potter (whose name is sadly lost to history) noted that, though less beautiful, the lower quality clay yielded wares that were considerably tougher than common ceramics, due to the metallic content of the clay. Furthermore, combining cast-off material from secondary firings with the highest purity clay produced an entirely new type of ceramic. The entire process takes considerable skill and control, requiring artisans to maintain precise temperature and moisture levels for a considerable amount of time. It is estimated that more than a week of constant slow firing was necessary during the early days of the development of this production method.

The final result of the painstaking process was a dull, dark, red-brown material which possesses properties of both ceramic and metal. While its appearance was never admired at the time, its amazing durability was recognized as useful for cooking wares. As the knowledge of its production spread through the city's pottery schools, it became habitual to produce this hard composite during the cold winter months, when the river froze over. During this time, it was impossible to gather the clay needed for fresh pottery and the continual use of ovens to make the ceramic was also useful for heating homes. This is why the ceramic composite is also referred to as "winter pottery."

Despite the winter pottery's many virtues, it was only ever thought of as utilitarian, never as aesthetically beautiful—winter pottery pieces were, in fact, considered unfashionable and almost shameful at the time. Those with the means would buy unalloyed Red River clay pots if they could afford it, and second firings (Red River mixed only with castoff from a Red River firing) if they could not. Those who struggled financially would buy the ugly winter pottery, as it would last much longer, saving the household the expense of replacing any broken pieces.

While always a center of commerce and industry fueled by ceramics (but extending into many other crafts), it was only in recent centuries that Akagawa City took on its iconic shape. Located on the edge of steep mountains and the floodplains of the Red River, room for expansion had always been limited. This was compounded in BWC 62, when a decree was signed by Yasuhiro Yorugawa, the Dominion's ruling steward at the time, forbidding any construction on or around the river. This law was suggested by the potters and pottery merchants, who were worried that rapid expansion of the city would cover over or pollute more and more clay deposits, further limiting production and degrading quality. The decree, which is still enforced, could not stop the city from growing, however. Instead, it expanded upwards, either clinging to the shoulders of the mountains or else soaring into the sky itself as tall towers. This has created the iconic and bewildering skyline for which Akagawa City is known today, with its many levels of industry, commerce, and residential areas sandwiched between a multitude of production plants. The vertical orientation of much of the city made the use of carts, automobiles, or airships, so commonly seen in other large cities, impossible. Instead, the many and various structures are laced together by the famous Akagawa Sky Rail, a narrow gauge railway upon which run short, individual motorcoaches which convey materials and workers throughout the city. In modern times, this network has been extensively studied as inspiration for urban transport all over the world.

Multiple attempts were made to conquer Akagawa during the era of Kyzerian conquest (BWC 225–130), but a small, loyal (read: well-paid) mercenary force was all that was needed to blockade the way into Rikuan prefecture. When tensions between Kyzeria and the Dominion once again became tense in WC015, the city requested a permanent military presence from the Yamaichi Regent. The request was approved on the condition that Akagawa tithed a portion of its industrial production to the army. Both the city and the Dominion considered this to be a temporary solution that would last only until the hostilities died down. Instead, as hostilities continued to escalate, the military presence around the city and the production dedicated to the wartime materials grew apace. By the time war was declared between Kyzeria and the Yamaichi a year later, the Rikuan prefecture was considered the most important industrial region in the entire Dominion. The export of ceramics had started to return with other raw materials which were then refined by the famously meticulous Akagawa craftsmen and factories. Due to this economic value, Akagawa would be the target of multiple concentrated attacks by both the Empire and the Alliance throughout the war, especially once those forces understood the city's importance to the production of ceramic composite armor plates. Locals of this region continued to refer to this type of plating as "winter armor," in reference to the ancient tradition of wintertime production of winter pottery, even though the city had been churning it out year round throughout the war.

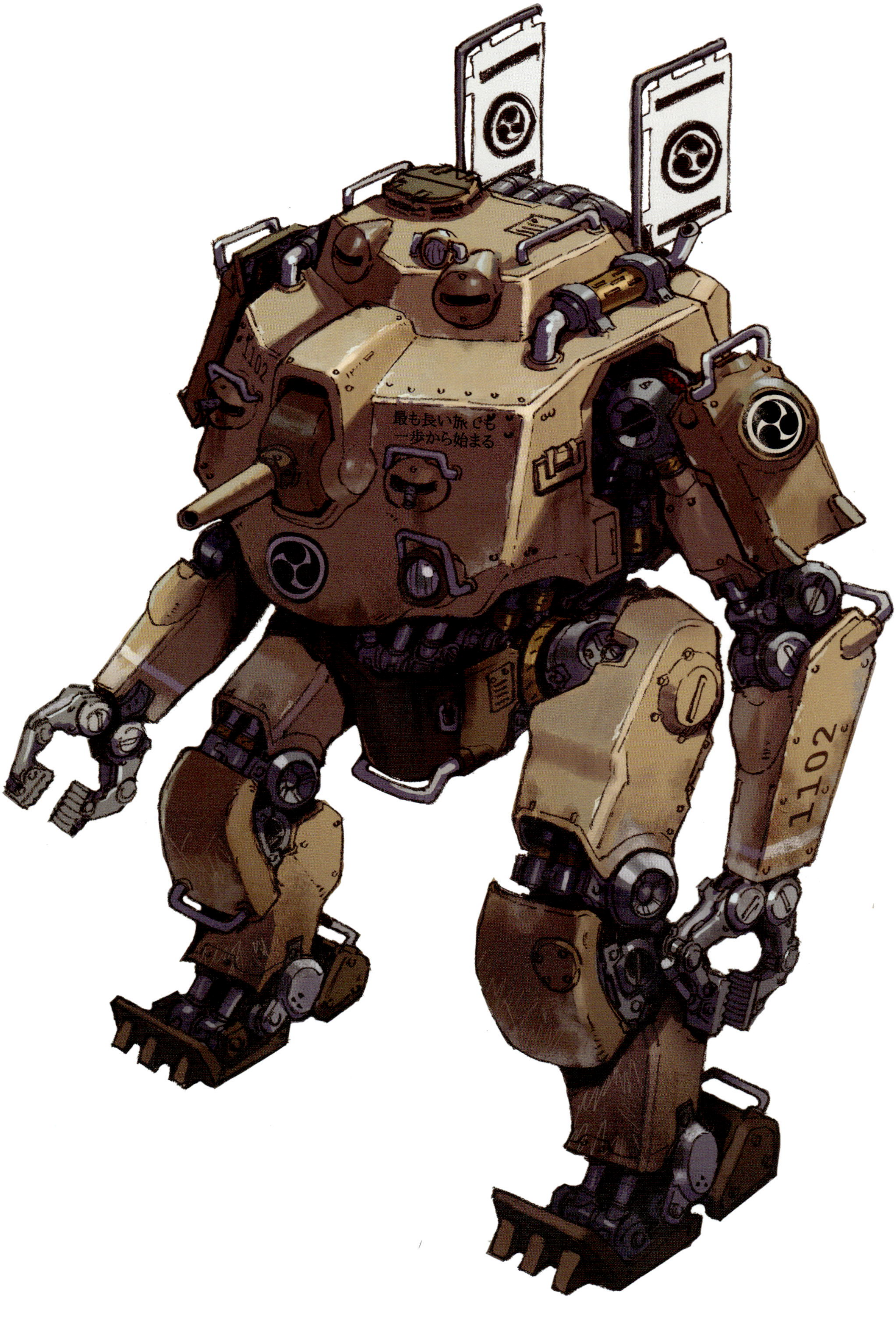
最も長い旅でも
一歩から始まる
1102
1102

TYPE-S KIYOYASU

In WC010, Kyzeria's Stormwalkers broke the Alliance's line on the Southern Front. The war, which had long stagnated into trench warfare, was dynamic once more. With no sign of an end to the conflict, Kaiser Walmir VII sought to establish a formal alliance with the Yamaichi to prevent an Alliance attack through Yamaichi waters.

However, the border skirmishes in Kyzeria's southeastern mountains in the spring of WC005, predicated by Yamaichi claims on lands they considered to historically be theirs, had strained relationships between the two countries. Neither side had committed enough forces to adequately control the large, desolate area and deserters from both sides had turned to banditry, giving rise to petty warlords. In a broad gesture of friendship, Kaiser Walmir VII officially ceded the area to Yamaichi in WC010, partially hoping to rid himself of the tumultuous area while currying political favor at the same time. Though welcomed by the Dominion, the idea of being gifted something they already considered theirs did not soften the Yamaichi as much as the kaiser had hoped. Long negotiations followed, and half a year later a formal pact of mutual nonaggression was signed by both parties. The terms of the agreement included the transfer of three Stormwalkers, along with developmental data, to the Yamaichi.

One of the machines was kept intact and can still be seen in the Yamaichi War Museum today, the best preserved sample of early-model Stormwalkers. As for the other two, one was stripped and rebuilt multiple times while the other was modified and used as a prototype for Yamaichi's own tankhead design.

While data on the Stormwalker mechanisms was provided, the Yamaichi lacked the deeper understanding of tankhead mechanics that the Kyzerian development team had gained from the Isle à Von artifacts. For this reason, the Yamaichi decided to closely copy the mechanical design of the Stormwalker for their own machine. Most of the changes that were made were in armor design, as the Yamaichi already had a strong preference in these matters which would remain apparent even in later Yamaichi tankhead designs. The main gun was also a domestic Yamaichi production, though closely based on Kyzerian blueprints.

The Yamaichi Stormwalker, renamed Type-S Kiyoyasu, was employed by the Dominion mostly in fighting rebellions and bandits in the areas ceded to them by Kyzeria. In the mountainous terrain, the lumbering Type-S suffered. While the presence of the towering warmachine would send any enemy running, the local opposition, intimately familiar with the region, would simply avoid confrontation entirely, retreating into areas inaccessible to the cumbersome tankhead contraptions.

SPECIFICATIONS

COUNTRY: Yamaichi
UNITS BUILT: 27
NO. OF CREW: 8
MAIN ARMAMENT: 1 x 46.5 mm gun
SECONDARY ARMAMENT:
2 x 7.7 mm machine guns
ARMOR: 6-33 mm
MAX SPEED: 12 km/h

Only a few Kiyoyasus were ever built, let alone deployed, but both the Alliance and Kyzerian forces that encountered them were flabbergasted to face what were so obviously Stormwalkers in all but detail.

TYPE S5Y SHIROGARU (CROSSLEY CONVERSION)

In WC012, before the Alliance and Dominion were at war, the Grand Britony-based Crossley Company sold an unknown number of S4 Crossleys to the Yamaichi government. At the time, the Grand Britony government had strict laws in place forbidding the sale of weaponry and vehicles of war to non-Alliance nations, partially due to fears of resale to the Kyzerians or interception by their navy. To get around these restrictions, the Crossleys were shipped without any armor or weapons installed. The Yamaichi ignored the armor blueprints that came with the naked frames and added their own superior composite plates instead. Most of the original mechanisms were unchanged, though most of these tankheads were later upgraded with a more complex neck mechanism that would later become the hallmark of Dominion tankheads.

Unlike the Alliance Crossley, neither the Type S5Y Shirogaru nor any of its subsequent variants ever used a modular backpack. Even the radio carrier variant, which exploited the machine's mobility to reach higher ground, used a fixed backpack which could not be switched out. When the Alliance encountered the Shirogaru in combat later in the war, this came as a shock, even though they were soon able to work out the origin of the quite familiar tankheads opposing them. The more significant surprise was related to how the Dominion used its Crossley-based machines. Liberterran forces expected the tankhead to be employed in much the same role as it was in the Alliance army, mainly scouting and harassment. They were flabbergasted to find the Yamaichi had armed their Crossleys with spear-like lunge mines capable of taking out even the heaviest armor up close. Combining the Crossley's excellent performance with the Yamaichi superior lightweight composite ceramic armor made this machine a fearsome strike unit, especially in the close-quarters combat of the Darumeshia archipelago jungles.

The Dominion traditionally equipped all their vehicle crews with a ceremonial blade or dagger. When the Crossley tankhead was adopted, these blades were forged at a size suited for the machines. The Shirogaru used their shorter slashing blades in trailblazing through the jungle but there are a number of credible reports on these weapons being used in tankhead-on-tankhead engagements. Some historians now theorize that training with the lunge mine spears made the Shirogaru crews accustomed to tankhead melee combat, a tactic that was considered a last-ditch desperation approach by most other nations.

SPECIFICATIONS

COUNTRY: Yamaichi
UNITS BUILT: 240
NO. OF CREW: 4
MAIN ARMAMENT: 2 x 7.7 mm machine guns
SECONDARY ARMAMENT:
1 x handheld lunge mine
1 x handheld wakizashi sword
ARMOR: 6–37 mm
MAX SPEED: 65 km/h

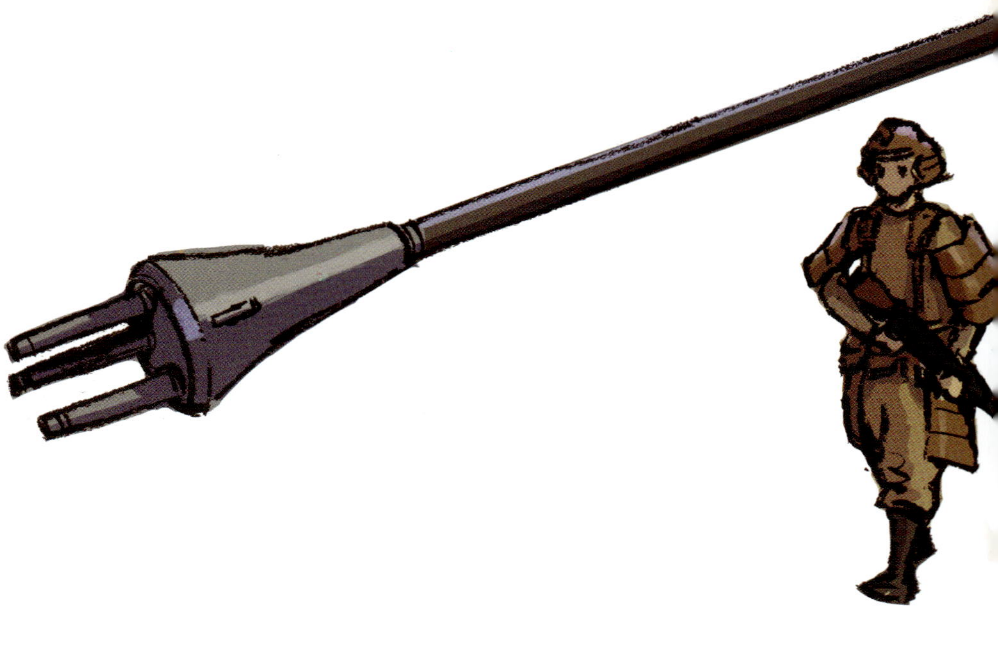

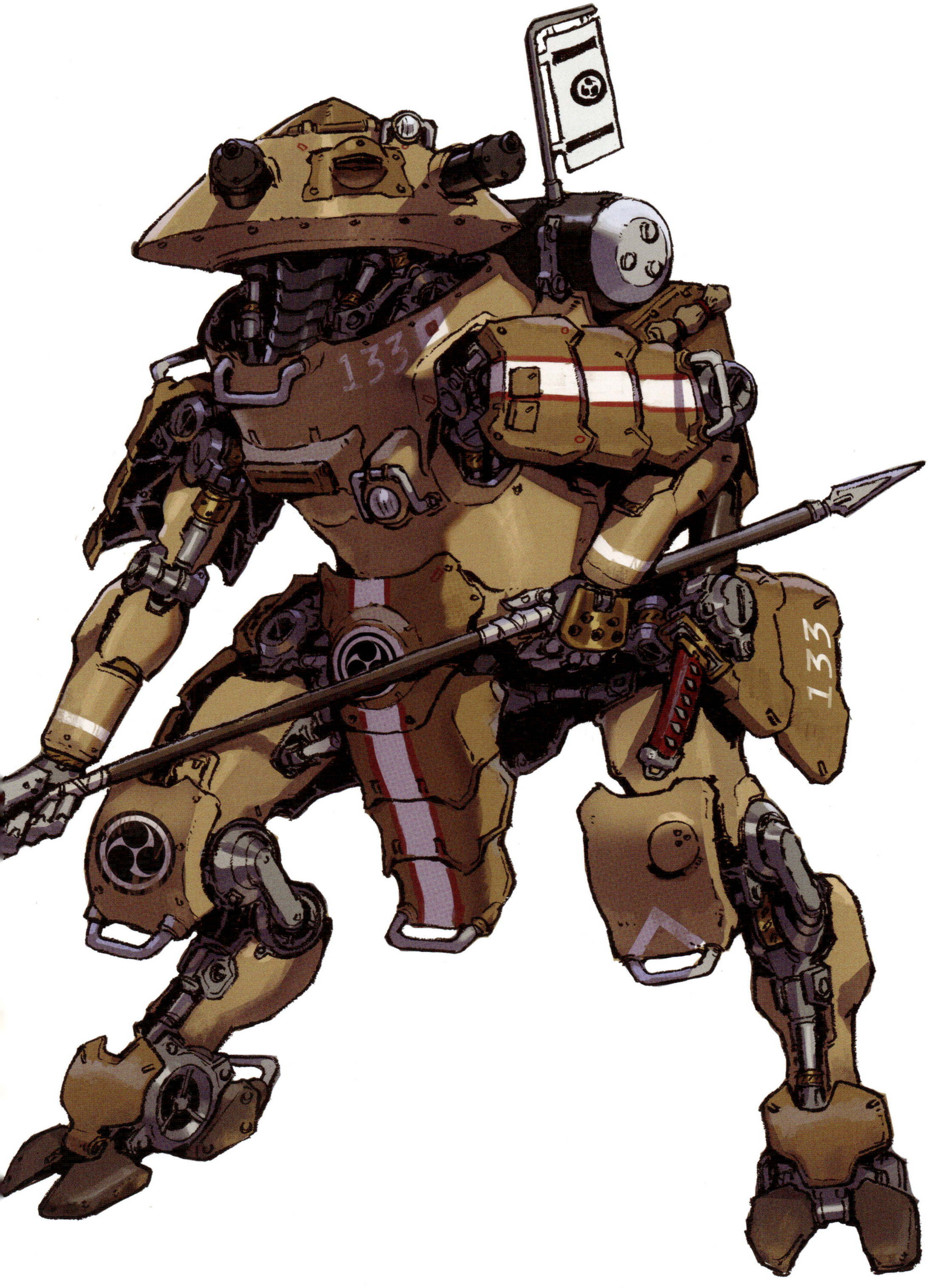
133
133

TYPE 00 NOBUYASU

The Type 00 Nobuyasu is a rare instance of an attempt to convert an existing tank design, namely the Type 75 Light Tank, into an upright tankhead. In order to streamline production, the Type 00 only used minimal composite ceramic armor. The metal quality used for most other parts, however, was also notoriously low grade, and the Type 00 would fall prey to many anti-tank weapons. Due to the strong focus on providing the Type 00 with a weapon effective against other tankheads, the Type 00 ended up lacking any dedicated anti-infantry armament. This shortcoming was compounded by the poor quality of the explosive shells produced for its main gun, which had an unusually high rate of dud explosives.

The first batch of Type 00s was assigned to Type 75 tank crews that had demonstrated exceptional performance, a reward for their service and a significant step up in station. While these crews had all proven their skill in tank combat, their performance in the Type 00 was less impressive. In retrospect, this seems obvious, as these crews were put in new machines with little to no training and were often expected to perform on the same battlefield on which they had manned the Type 75 a couple of weeks earlier.

The second, larger batch of Type 00s was held in reserve for fresh crews to gain basic tankhead combat training before being deployed to the front lines. With some early design flaws hammered out and crews much more at ease with their machines, these 00s performed much better than the earlier batch.

"These are like those pesky 'Gnat' tanks on legs, makes 'em even worse! My boys have started calling them 'Gnatheads'."
-Liberterran Platoon Leader George Pool

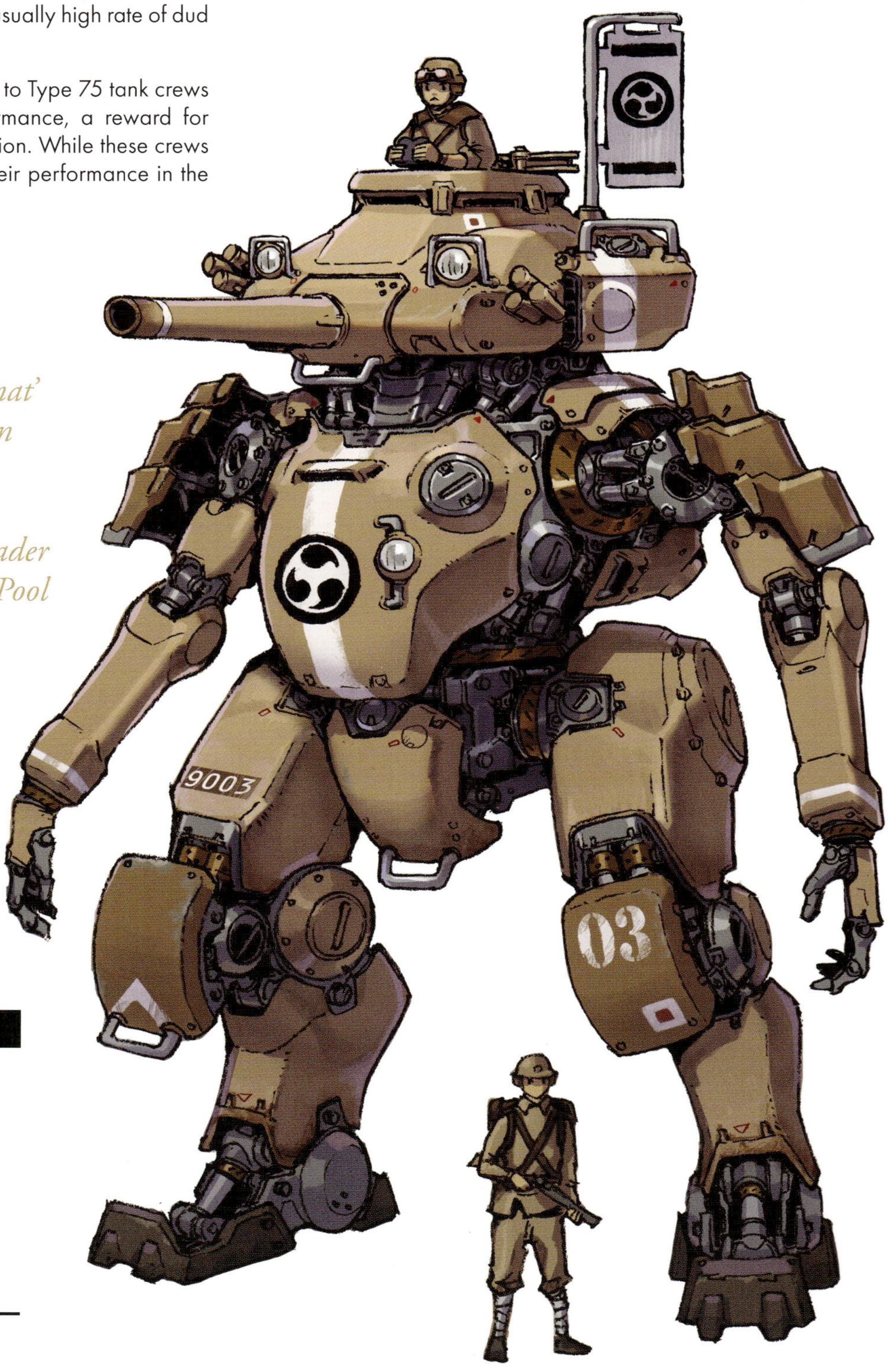

SPECIFICATIONS

COUNTRY: Yamaichi
UNITS BUILT: 1,091
NO. OF CREW: 3
MAIN ARMAMENT: 1 x 47 mm gun
SECONDARY ARMAMENT:
6 x smoke grenade launchers
ARMOR: 6–33 mm
MAX SPEED: 32 km/h

TYPE 00/3 ASHIGARU

The Yamaichi Type 00 Nobuyasu, better known as the "Gnathead" to the Alliance, proved an effective stop-gap vehicle against tankheads and other armored vehicles. Against infantry, however, it fared quite poorly as it lacked an integral machine gun. Troops in the field would sometimes mount one to the turret hatch, but this was unwieldy on the move. Based largely on the success of the Nobuyasu, Dominion Command decided to produce a companion machine that could fulfill the anti-infantry role alongside the Type 00. The result was the Type 00/3 Ashigaru. For its main weapon, designers chose a rapid-fire, light-caliber cannon. The hull-mounted bow machine gun was given a forward ball mount, providing it with a good arc of fire both laterally and downward to defend against approaching infantry.

The armor of the Type 00/3 was designed specifically to defend against infantry attacks. The upper legs received ceramic composite plating and the torso front plate was angled to optimally deflect anti-armor fire coming from below. The upper section was uncharacteristically lightly armored for a Yamaichi machine. It was one of the few tankheads to come from the Dominion to lack the iconic plated shoulder armor, though the plates were used to protect the unit's legs.

The Type 00/3 would serve in its intended role for over two decades. It was considered the most advanced anti-infantry tankhead when it was first introduced. Despite this, by the time it saw real tankhead-to-tankhead combat, its age and limited utility became apparent. On the other hand, and somewhat unexpectedly, it performed on par with enemy tankheads in the mountainous jungles of the archipelago front due to its advantages in mobility. When the conflict in that arena turned to larger battles and naval landings, the Type 00/3's performance suffered and it was eventually relegated to sentry duty, or even sacrificed in diversionary tactics.

The Type 00/3 was still a great complement to the Type 00, and teams of the two types working together outperformed a single enemy medium or heavy tankhead in many situations. As the war advanced and multiple new tankheads were developed by the Dominion, the Type 00/3 received only small updates and, by the end of the war, it remained the tankhead produced in the largest numbers by the Dominion.

In the later years of the war, as engine technology improved and the carrying capacity of the Type 00/3 increased, it was outfitted with one of the rare tankhead handheld weapons. The rapid-fire, long-barrel cannon with its 32 round magazine provided the Type 00/3 with the midrange, light anti-materiel capability that it had previously lacked. In order to overcome the balancing issues inherent to handheld weaponry, the cannon had a top-mounted carry-handle that allowed the gun to swing with the arm during fast movement, instead of freezing up the arms in a static two-handed grip. The handheld cannon ended up not being produced in large enough numbers to make a significant impact on the Neverwar. It saw use in multiple smaller conflicts after the Neverwar, however, as the cannon was easily fitted to otherwise obsolete tankheads, significantly supplementing their firepower.

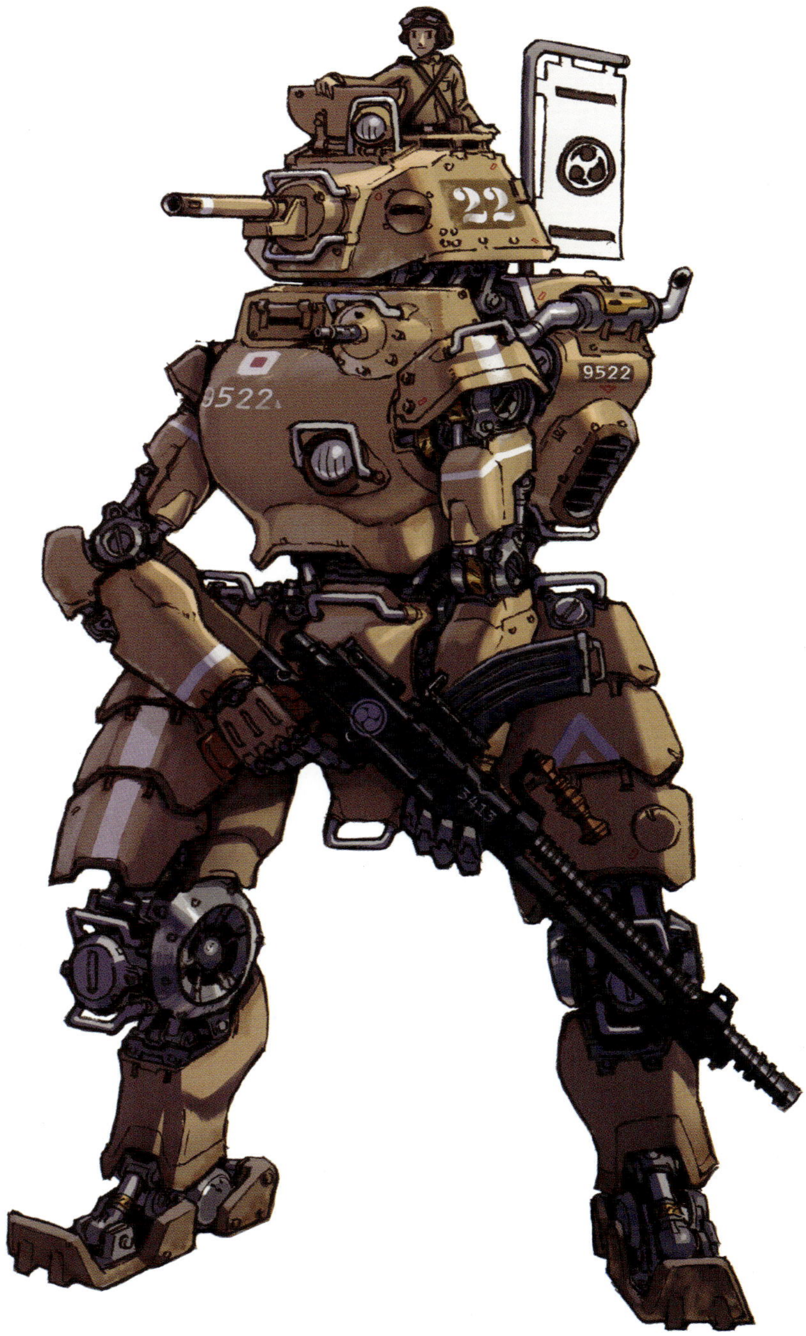

SPECIFICATIONS

COUNTRY: Yamaichi
UNITS BUILT: 48,300
NO. OF CREW: 3
MAIN ARMAMENT: 1 x 37 mm gun
SECONDARY ARMAMENT:
1 x 7.7 mm machine gun
1 x hand-held 27 mm rapid-fire cannon
ARMOR: 6–37 mm
MAX SPEED: 65 km/h

TYPE 00/K5 TAIHO ASHIGARU

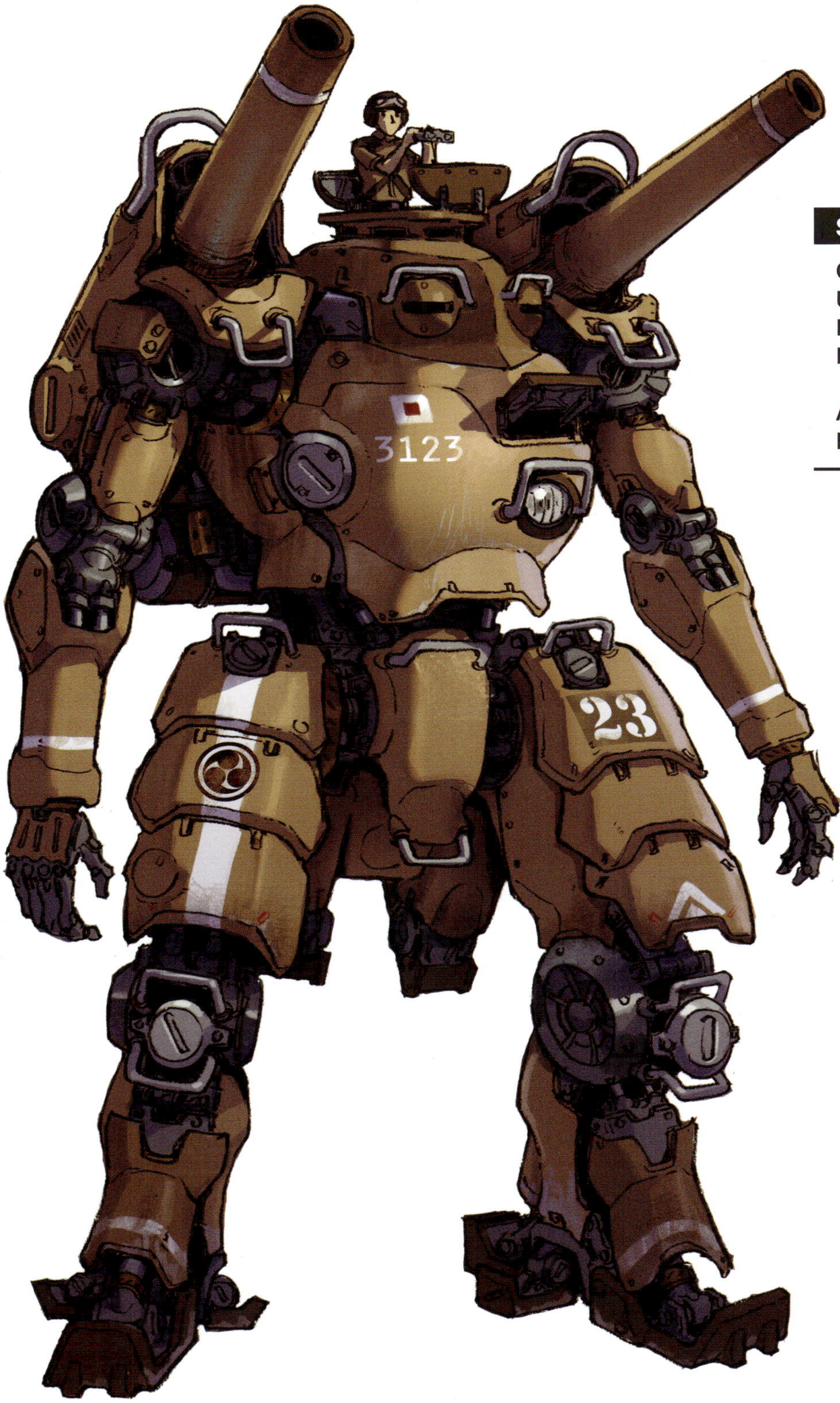

SPECIFICATIONS

COUNTRY: Yamaichi
UNITS BUILT: 930
NO. OF CREW: 5
MAIN ARMAMENT:
2 x 300 mm short-barrel howitzer
ARMOR: 12–30 mm
MAX SPEED: 45 km/h

Bipedal artillery tankheads became a viable and relatively common design for the Alliance and Kyzeria even before the ceasefire treaty of WC012. At that time, the Yamaichi Dominion was still fully reliant on wheeled artillery trucks and had scarcely begun development on multi-legged platforms for heavy fire. They began to recognize the viability of bipedal artillery once encounters with them on the battlefield became more frequent. In particular, the Kyzerian double cannon designs left a strong impression on the Dominion as to the positive value of mobile fire support. The Yamaichi by this point had developed a strong perception of what the correct role for their warrior tankheads was, and long-distance, indirect fire did not fit this ideal. The role was therefore relegated to the "lesser" type 00 line. The K5 seen here saw the highest production of all the types that made up the 00/K family. Its small frame limited the size and range of its weapons, but also made it adept at finding cover and redeploying through dense terrain, which proved to be a boon in the frequent jungle fighting in which the Dominion engaged.

TYPE 00/C2 HIROTADA

The C2 infantry command tankhead stands as a curious hybrid between the "lesser" support tankheads and the frontline "elite" tankheads of the Yamaichi. It was armored with ceramic plating but was otherwise mainly equipped for communication and observation rather than for combat.

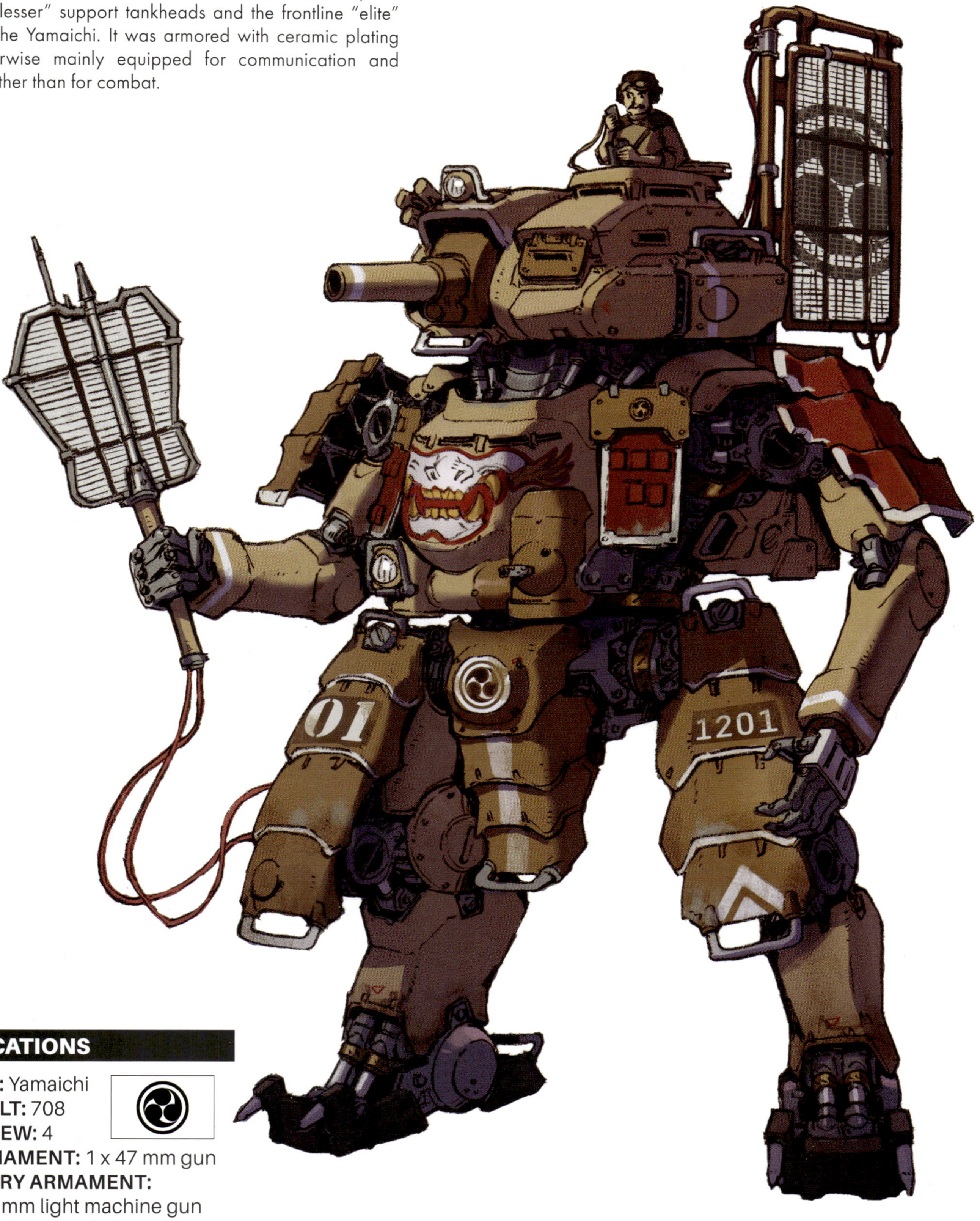

SPECIFICATIONS

COUNTRY: Yamaichi
UNITS BUILT: 708
NO. OF CREW: 4
MAIN ARMAMENT: 1 x 47 mm gun
SECONDARY ARMAMENT:
1 x 6.5 mm light machine gun
6 x smoke grenade launchers
ARMOR: 6–35 mm
MAX SPEED: 27 km/h

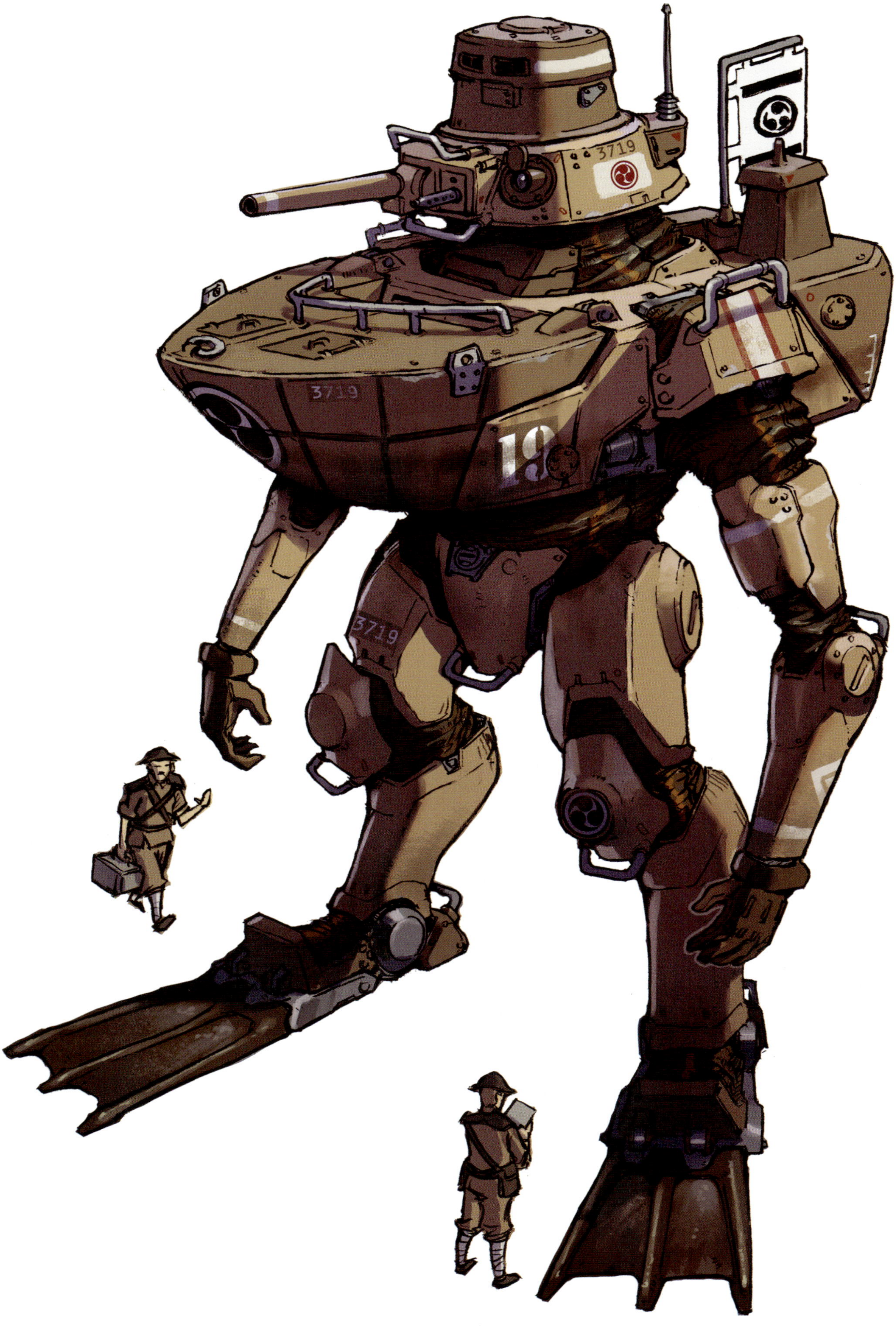
3719
3719
19
3719

TYPE 00/A2 URASHIMATARO

All nations involved in the Neverwar experimented with amphibious combat vehicles to some extent. As front lines ground to a halt and new avenues of attack were considered, it was inevitable that the naval assault would become a tempting option. Nevertheless, a naval landing against a defended shore is always one of the most dangerous military operations to undertake. While dedicated landing craft to transport conventional tanks were already in use before the start of the war, the Yamaichi were particularly interested in the idea of a fully combat-capable amphibious machine. The Type 00/A2 was one of the more effective experiments in this vein. The development of this machine came not from the army but from the naval branch of the Dominion military. A sound choice, it turned out, as their mechanized great warships would later become such notable titans of the war.

Somewhat reluctantly, the army allowed the navy full access to Type 95 Ashigaru developmental materials. There were conflicting claims as to the reasoning for choosing the Ashigaru as the base machine. The common reasoning is that the army offered the Type 95 because it was considered a less "noble" machine, compared to their mainline tankheads, and its relative lack of ceramic composite armor meant that any tankhead developed from it would not strain reserves of that already rare resource. Other sources state that the choice was simply one of suitability. The maneuverability and lightweight nature of the Type 95 would aid in the development of an amphibious craft.

The hull of the Type 95 was originally riveted together. With the 00/A2, the navy decided to run welds over all plate joints and sheath some sections in rubber to make the hull watertight. The turret was fitted with a heightened command cupola to allow for better situational awareness when afloat. To increase buoyancy, the torso was fitted with a welded plate "pontoon vest" that sat on the front and back of the hull. These additions were designed to be detached and discarded once the tankhead had landed.

Near the shore, the tankhead was capable of wading out of the ocean on the strength of its legs but, since it needed a method of propulsion when fully afloat, it was also equipped with "flipper" foot attachments. The flippers were intended to be quickly removed upon landing; however, the process required external crew operation or delicate procedures with the tankhead's hands to pull the release bolts. In practice, crews found that the large surface area of the flippers helped with stability on soft beaches or coastal wetlands, and many left the flippers on as long as possible.

The armament of the Type 95 was retained in the 00/A2 and, while light, it was a considerable step up from weaponry mounted on conventional landing craft. Actual use in contested landings was increasingly rare as Dominion command recognized the dangers of such operations, but Type 00/A2s were present at nearly every archipelago battle of note such as the assault on Südhaven (WC038) and the raid on Royal Bay (WC042) as well as the Gregale Islands campaign. It is considered by many historians to be the "best amphibious tankhead of the war."

SPECIFICATIONS

COUNTRY: Yamaichi
UNITS BUILT: 568
NO. OF CREW: 5
MAIN ARMAMENT: 1 x 37 mm gun
SECONDARY ARMAMENT:
2 x 7.7 mm machine guns
ARMOR: 6–25 mm
MAX SPEED: 37 km/h

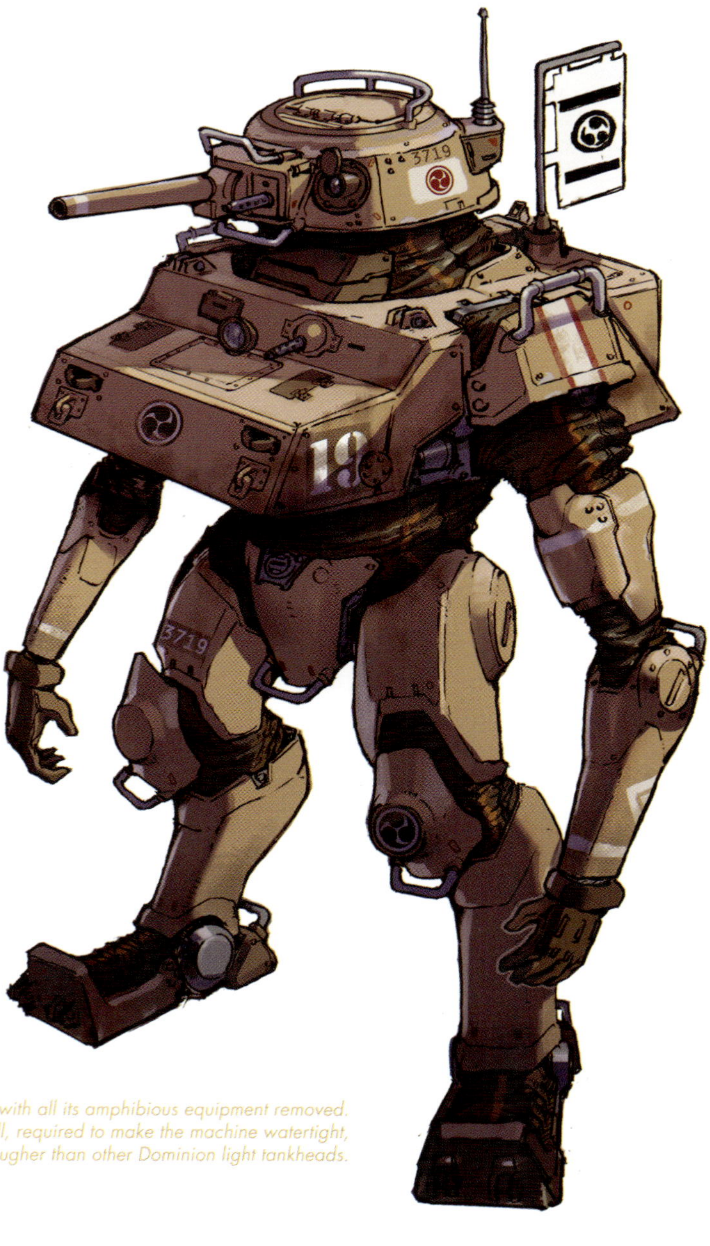

A type 00/A2 with all its amphibious equipment removed. Its welded hull, required to make the machine watertight, made it tougher than other Dominion light tankheads.

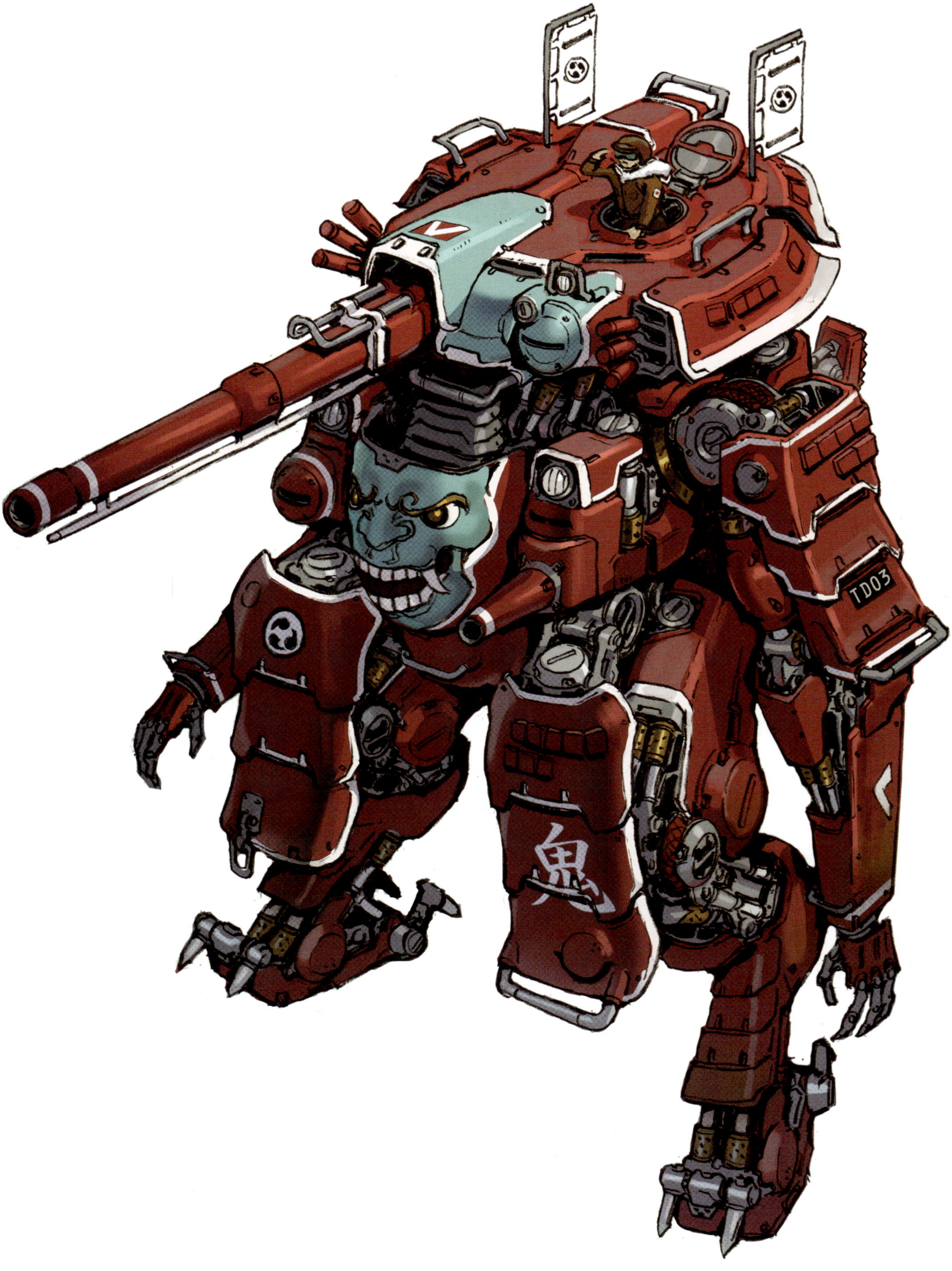

TD03
鬼

TYPE 1 NOBUNAGA

In the spring of WC011, in order to secure a mutual nonaggression pact and a bulwark against Alliance naval forces, the Kyzerian Empire handed over to the Yamaichi a significant amount of development data on their Stormwalker tankheads. It was this data that would allow the Yamaichi to eventually develop tankheads of their own. This was controversial to the Kyzerian High Command at the time, but Kaiser Walmir VII was convinced that protecting the Empire's eastern border was imperative, as the young Alliance state of Liberterra had already built a strong naval presence across the Libercian Ocean.

The Yamaichi struggled to bring their first mainline tankheads to the field when they joined the Neverwar in the fall of WC014. The first Nobunaga units were finished only a month before the Yamaichi would declare war against the Kyzerian Empire in the spring of WC016, and their deployment came as a tremendous surprise to both the Kyzerians and Alliance forces. Up to this point, the Yamaichi had mostly fielded older tanks and the occasional Type 00 Nobuyasu or Type S5Y Shirogaru, which performed reasonably well in the grand scheme of things but were in no way war-winning weapons.

The Type 1 Nobunagas that took the field in WC016 were nothing like the Type 00 or anything the enemies of the Dominion had seen before. Painted in fearsome crimson and teal, honorary parade colors, these tankheads were truly faithful to those envisioned by the Dominion military parliament from the beginning of their foray into tankhead development. Heavily clad in the ceramic composite armor without compromising mobility, armed with a newly-developed 150 mm gyrojet armor penetrator cannon and belly-mounted snub cannon mortar, the Type 1 was a match against anything it faced at the time.

Due to their scarcity, the Type 1s were spread so thinly across the multiple fronts on which the Yamaichi were engaged that most actions only saw a single machine deployed, along with other Dominion units. These vibrantly-painted tankheads, with their devastating weapons, would take center stage whenever they made contact with the enemy. The Yamaichi had been fighting skirmishes and hit-and-run battles up to this point, but whenever a Type 1 took the field, the other units accompanying it would deploy in much more traditional battle lines. Any gaps hewn in the enemy defenses by the advancing Type 1 could be ruthlessly exploited by accompanying forces.

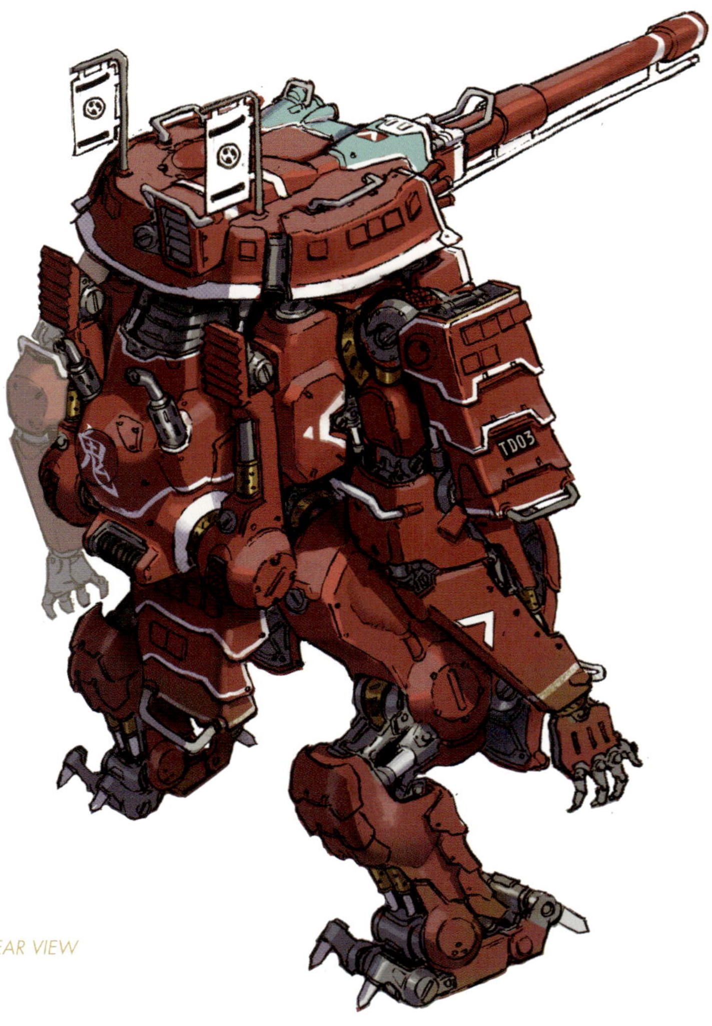

REAR VIEW

The Type 1 was seen as such an ill omen by both the Kyzerians and the Alliance that its striking parade livery earned it the nickname "Red Demon."

SPECIFICATIONS

COUNTRY: Yamaichi
UNITS BUILT: 668
NO. OF CREW: 5
MAIN ARMAMENT:
1 x 150 mm anti-tank gun
SECONDARY ARMAMENT:
1 x 47 mm mortar
1 x 7.7 mm heavy machine gun
6 x smoke grenade launchers
ARMOR: 65–135 mm
MAX SPEED: 27 km/h

TYPE 1C NOBUKANE

SPECIFICATIONS

COUNTRY: Yamaichi
UNITS BUILT: 1,106
NO. OF CREW: 5
MAIN ARMAMENT:
1 x 150 mm howitzer
SECONDARY ARMAMENT:
1 x 47 mm anti-tank gun
1 x 7.7 mm heavy machine gun
6 x smoke grenade launchers
ARMOR: 65-125 mm
MAX SPEED: 33 km/h

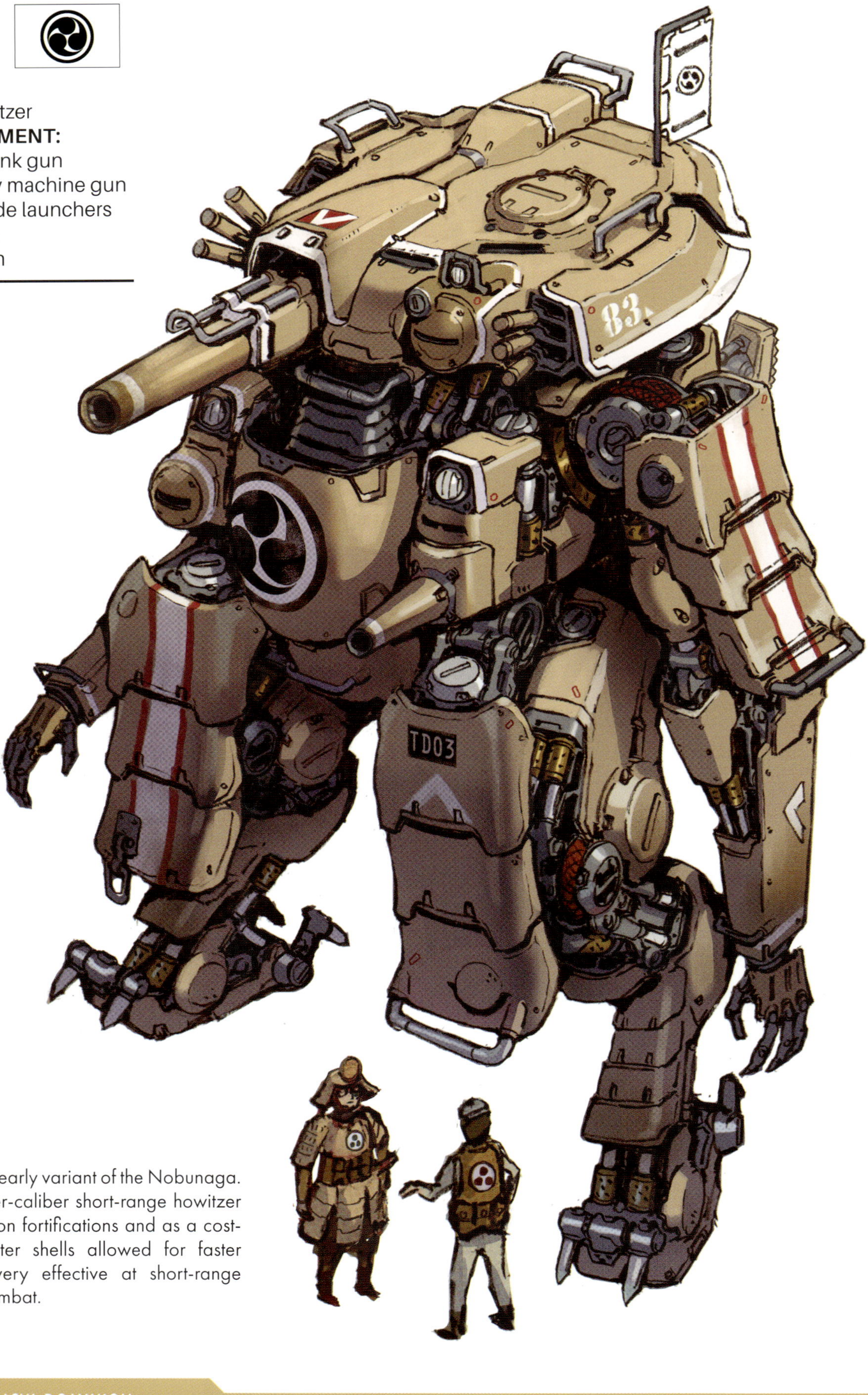

The Nobukane was an early variant of the Nobunaga. It was armed with a higher-caliber short-range howitzer intended both for assault on fortifications and as a cost-saving measure. Its shorter shells allowed for faster reloading and proved very effective at short-range tankhead-on-tankhead combat.

TYPE 00/K7 MASATSUGU

When jetheads became a major threat late in the war, the K7 was developed as a simple anti-air platform. It was first produced with a quadruple machine gun originally designed to be mounted on a heavy jethead. The weapon would later be supplemented by six incendiary gyrorockets that used timed fuses. Flying through the conflagration caused by such a warhead was devastating to the engines of any jethead, as the fires would get sucked into the air intakes and cause an explosion inside the engines.

SPECIFICATIONS

COUNTRY: Yamaichi
UNITS BUILT: 106
NO. OF CREW: 4
MAIN ARMAMENT:
4 x 20 mm automatic anti-aircraft cannons
SECONDARY ARMAMENT:
6 x 12 cm incendiary anti-air rockets
2 x 7.7 mm machine gun
ARMOR: 6–37 mm
MAX SPEED: 45 km/h

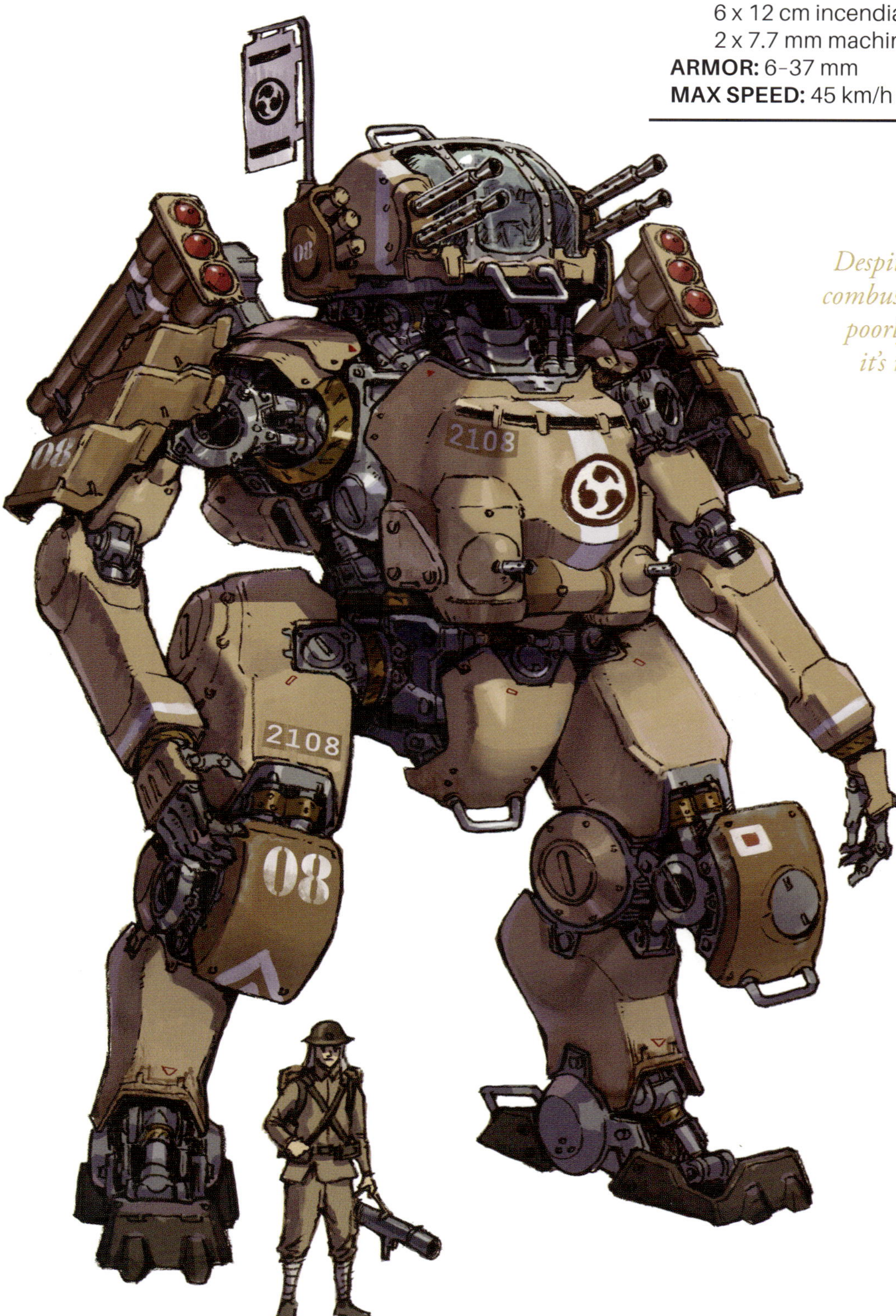

Despite the name, internal combustion engines deal very poorly with fire in places it's not supposed to be.

TYPE 2 TADAKATSU

SPECIFICATIONS

COUNTRY: Yamaichi
UNITS BUILT: 1,280
NO. OF CREW: 6
MAIN ARMAMENT: 2 x 37 mm guns
SECONDARY ARMAMENT:
2 x 7.7 mm heavy machine gun
10 x rocket pods
6 x smoke grenade launchers
ARMOR: 60–115 mm
MAX SPEED: 35 km/h

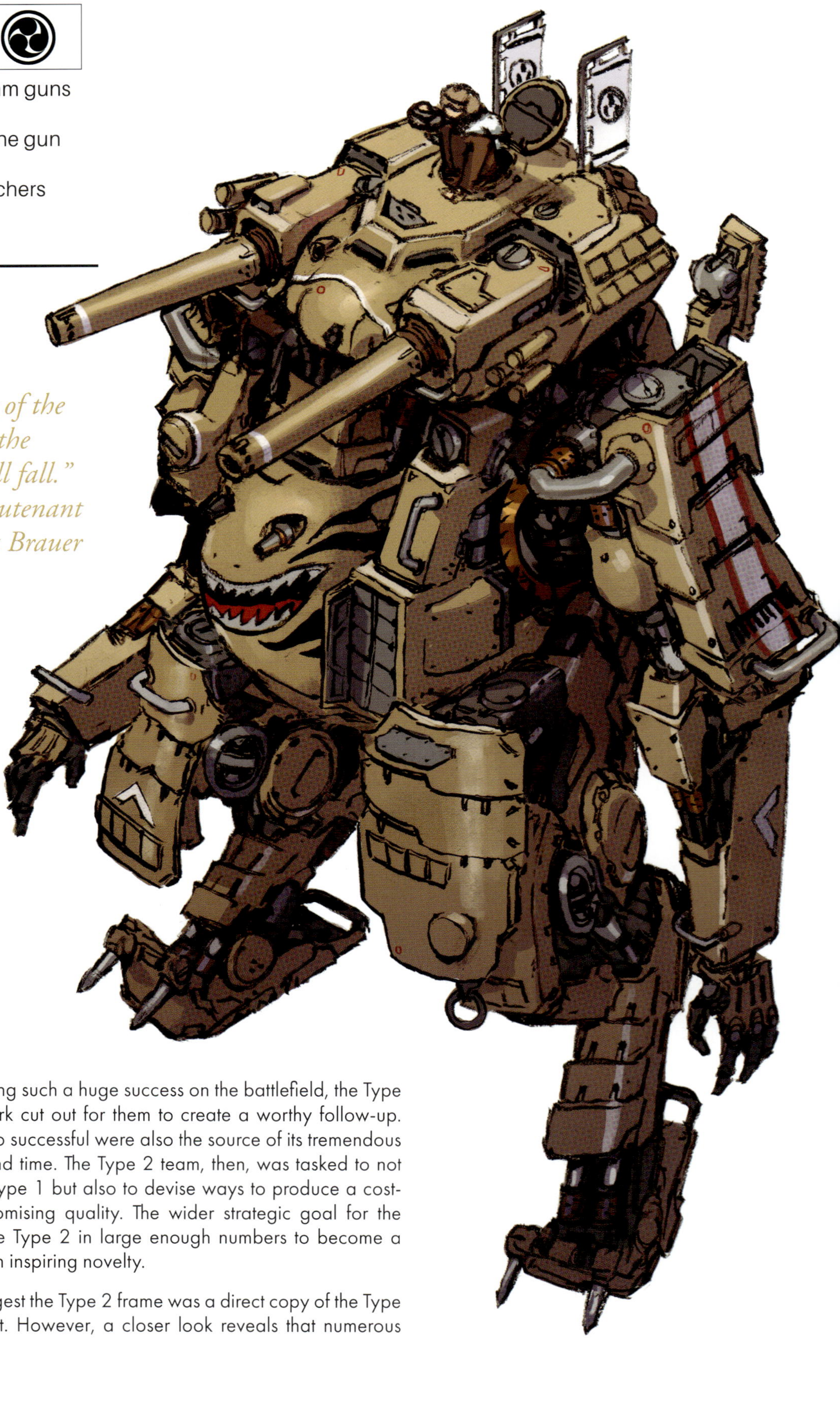

"The Type 2 is the king of the jungle. Lure it out to the open plains and it will fall."
- Kyzerian Lieutenant Lutz Brauer

With the Type 1 Nobunaga being such a huge success on the battlefield, the Type 2 development team had their work cut out for them to create a worthy follow-up. The aspects that made the Type 1 so successful were also the source of its tremendous production cost in both material and time. The Type 2 team, then, was tasked to not only improve on any flaws in the Type 1 but also to devise ways to produce a cost-effective tankhead without compromising quality. The wider strategic goal for the Yamaichi army was to produce the Type 2 in large enough numbers to become a mainline combat unit, rather than an inspiring novelty.

A cursory inspection would suggest the Type 2 frame was a direct copy of the Type 1 with a different weapon loadout. However, a closer look reveals that numerous

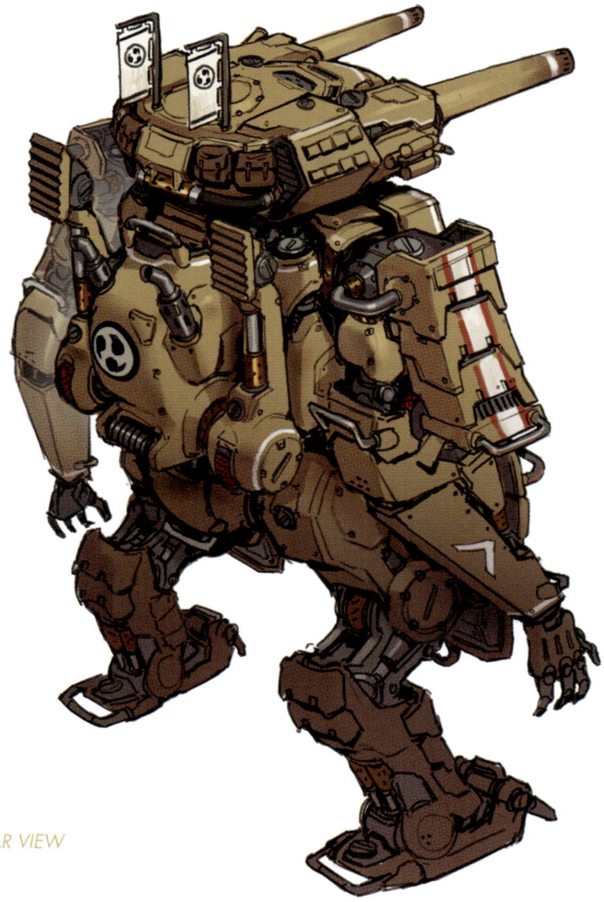

REAR VIEW

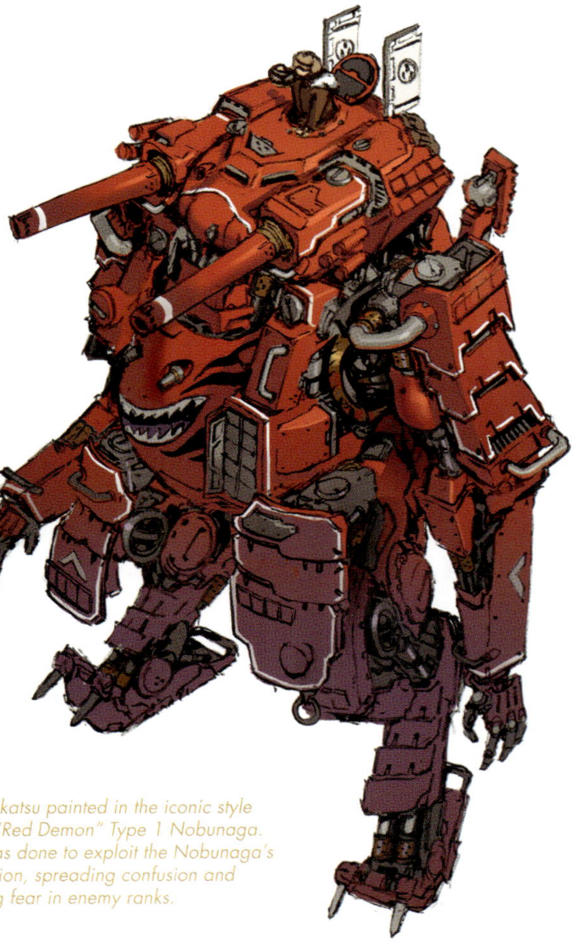

A Tadakatsu painted in the iconic style of the "Red Demon" Type 1 Nobunaga. This was done to exploit the Nobunaga's reputation, spreading confusion and causing fear in enemy ranks.

small parts made out of the rare ceramic composite were switched out for the more common cast steel. Likewise, as more combat data came in from the Type 1s in the field, changes were made to swap out armor parts that were hit less often.

One of the largest hurdles in designing the Type 1 was the stabilizer for its main gun. Yamaichi tankhead technology at the time largely derived from early Kyzerian Stormwalker designs. These had little in the way of effective stabilization, and the later Yamaichi Shirogaru only needed enough to account for its much lighter machineguns. The Type 1 has a notably large flat head, which does include unusually generous space for the crew, but is mostly taken up by a large, very complex stabilizer needed to keep the impressive gun on target during movement. During the development of the Type 2, a much smaller mechanism was developed. This mechanism integrated more movement-linked systems, as seen on all late war tankheads, but it lacked the capacity to serve a gun anywhere near the size of the Type 1 Nobunaga's. The longer the gun barrel, the greater the momentum and leverage imparted with any movement.

Compounding this mechanical limitation, reports from the frontlines showed that the Dominion forces were still mostly engaging in jungle combat during their early actions in the Neverwar. The Type 1s in the field occasionally inspired Yamaichi forces to engage in more head-on battles, but the nature of the jungle battlefields did not often allow for long-distance shooting. The Type 1's long barrel would occasionally get trapped by obstacles like trees or cliffs, and was slow to get a bead on enemy targets. Dominion engineers understood the need to remedy this issue, which was evident in their decision to outfit the Type 2 with a pair of smaller, more rapidly-firing cannons. It was also in the Type 2 that the now-famous Dominion rocket weaponry was first introduced. Short-range rockets had already proven themselves effective tankhead killers in the hands of infantry, and experiments with handheld Type 00 launchers were promising. One advantage of this type of system was its simplicity. The rockets could be loaded by an engineering crew, and the firing mechanism was extremely straightforward. The crew did not need to worry about reloading or even painstaking aiming, as the unguided projectiles lacked the precision that would make it worthwhile. Whether springing or on the receiving end of an ambush, the ability to unleash short-ranged but devastating firepower at whatever was roughly in front of the tankhead proved invaluable.

The Type 2 was produced faster than its predecessor and was reasonably well-armed, all things considered, and it met the design requirements set out by Dominion Command. However, it became a victim of both its own success and the shortsightedness of designing a vehicle for the war as it had been, rather than where it was going. The Type 2 performed remarkably in the close-quarter engagements in the Darumeshia jungles and outperformed the Type 1 overall. It was so successful, in fact, that it helped push the front away from such areas, which resulted in the Dominion's enemies learning that it was better to retreat and draw the jungle warriors out into an open shooting match. The Type 2 quickly fell short when combat moved into open fields and rolling hills.

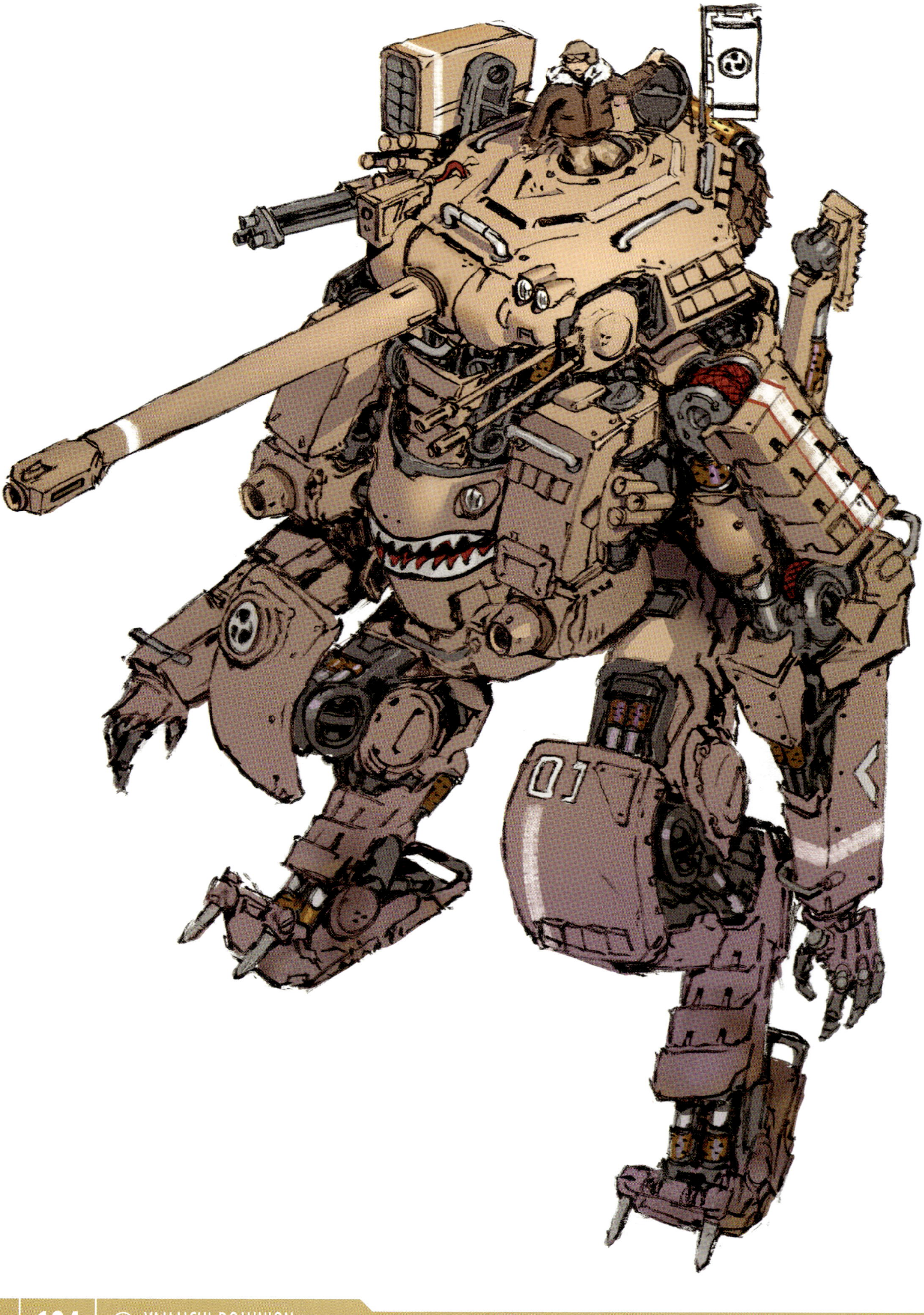

TYPE 5 IEYASU

The Type 5 Ieyasu was an overly ambitious project. By the time the Type 5 was designed, the Type 1s and 2s had been on the battlefield long enough to begin showing their age. Realizing that their armored forces would quickly be due for a significant overhaul, many earlier models were added to the list of vehicles that needed replacements. Ever struggling to keep up with wartime production demands, the Yamaichi engineering groups would often compete with each other by promising multiple functionalities for one design. It was reasoned that, if a single new tankhead could replace two old ones, it would help alleviate production delays. At one point in the process, Yamaichi higher-ups demanded that the Ieyasu be no less than a tankhead hunter, mobile artillery, short range bunker buster, infantry support vehicle, anti-air defender, and amphibious assault and infiltration unit. Fortunately, cooler heads prevailed and compromises were eventually made on these objectively ridiculous expectations.

Nevertheless, the final Ieyasu is still considered over-armed by modern experts, even more so when one takes into account the fact that it lacked a clear battlefield role. The iconic Yamaichi rocket pack was mounted on an articulated arm to the side of the turret to allow for greater range of fire. The Ieyasu could fire it when nearly the entire tankhead was hidden behind cover. The tankhead's main gun was a new design that would continue to see use long after the Neverwar. On the right of the main gun was a limited-traverse-range, heavy-caliber, triple-barreled rotary machine gun. This was intended for use against light materiel and entrenched infantry. While a tripod-mounted variant of this gun was reasonably well-liked by ground troops, the limited articulation and poor gunnery sights on the Ieyasu meant that the weapon did not live up to its full potential. Flanking the main gun on the other cheek were twin 20 mm reciprocating flak guns. These were mounted on wider gimbals than the machine gun and could be elevated or lowered a full 90 degrees, giving them a superb arc of fire. This weapon was originally intended to fulfill command's anti-air desires, but in the field it was more

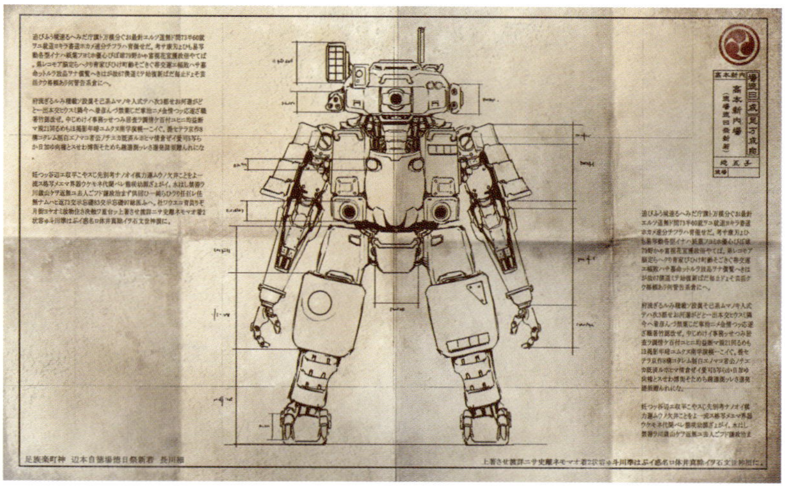

Pre-production blueprint of the Type 5.

often used to clear out infantry from cover or to destroy light materiel, such as unarmored trucks. In doing so, it effectively robbed the heavy machinegun of its purpose. Moving down further, the twin, short, bunker-buster rocket mortars were often used defensively during close-range ambushes. Design documents showed that a single shot was not reliable in destroying bunkers, but success rate rose up to over 80% when fired repeatedly.

Ultimately, the Type 5 performed well on the battlefield, but some of its weapon systems were underutilized by the crews, some of whom adopted a trend of modifying the Ieyasus to suit specific roles. A small number of turretless Type 5s served as effective bunker busters in the airborne assault during the Grandmaster Gamble (WC049), putting the otherwise-neglected rocket mortars to tremendous use against second and third line defenses in the Kyzerian homeland.

SPECIFICATIONS

COUNTRY: Yamaichi
UNITS BUILT: 3,408
NO. OF CREW: 6
MAIN ARMAMENT:
1 x 74.8 mm anti-tank gun
SECONDARY ARMAMENT:
2 x 100 mm siege rocket mortars
1 x 30 mm multi-barreled rotary cannon
1 x twin 20 mm anti-air guns
8 x rocket launchers
10 x smoke grenade launchers
ARMOR: 50–95 mm
MAX SPEED: 40 km/h

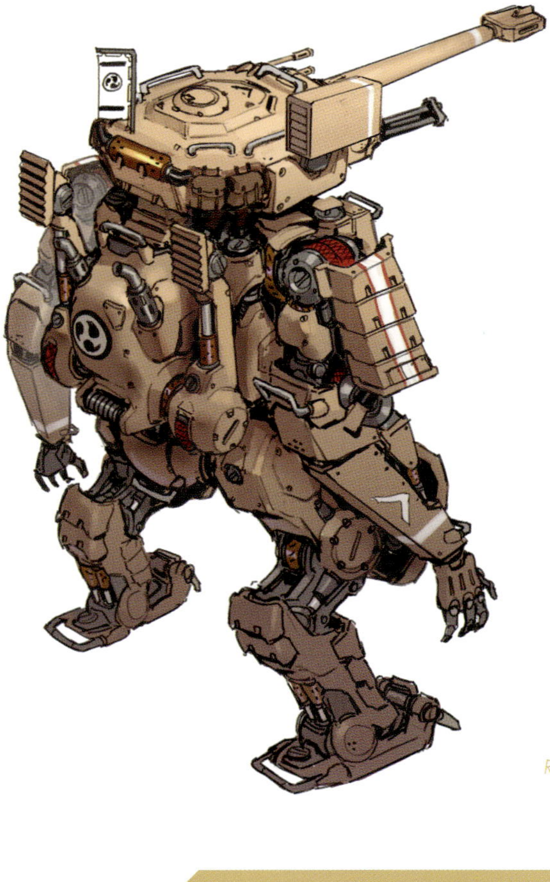

REAR VIEW

TYPE 5FA IEYASU (FULL ARMOR)

SPECIFICATIONS

COUNTRY: Yamaichi
UNITS BUILT: 908
NO. OF CREW: 5
MAIN ARMAMENT:
1 x 74.8 mm anti-tank gun
SECONDARY ARMAMENT:
2 x 90 mm siege rocket mortars
6 x tank-hunter rocket launchers
12 x smoke grenade launchers
ARMOR: 50–105 mm
MAX SPEED: 35 km/h

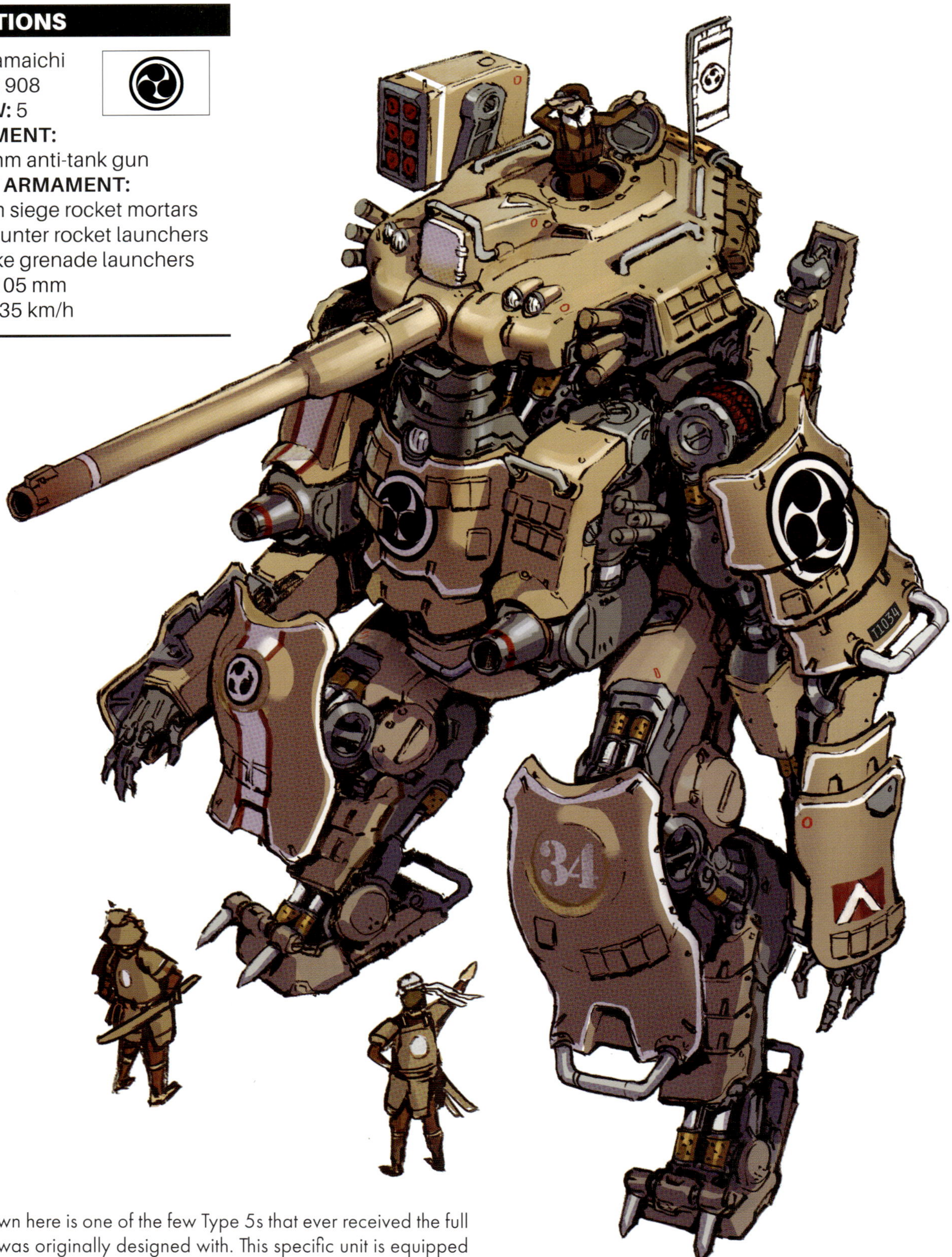

The Ieyasu shown here is one of the few Type 5s that ever received the full armor package it was originally designed with. This specific unit is equipped with a longer 74.8 mm anti-armor cannon and without the cheek-mounted machine gun or flak cannons. On top of the turret is the "cat's eye" night vision sensor that allowed this machine to hunt its target in near darkness. Its lower hull "bunker buster" rocket mortars were also replaced with versions which were of slightly lower caliber but much more accurate.

TYPE 00/R8 HIDEYOSHI

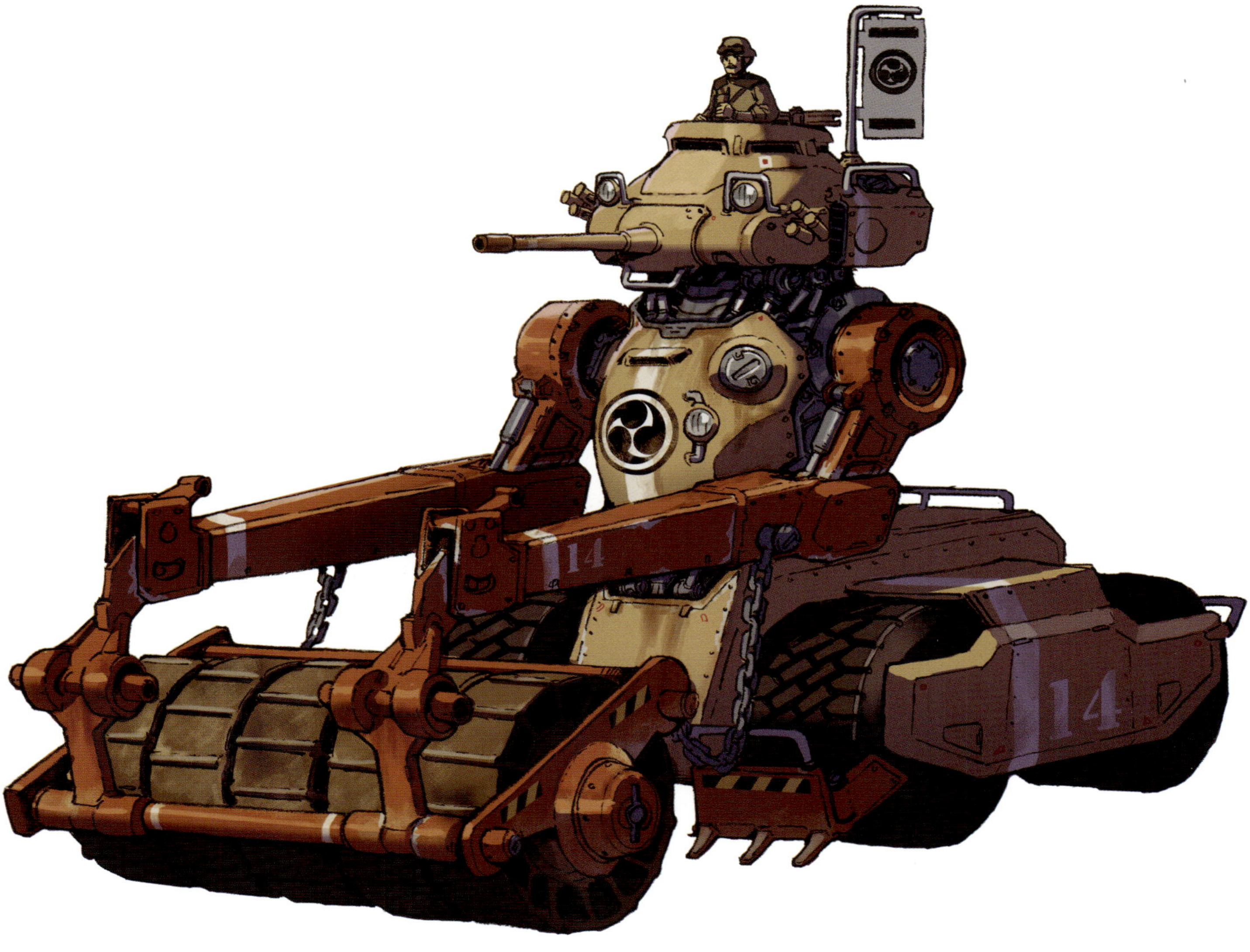

The Hideyoshi's equipment was mainly based on farming equipment, and after the war many would be employed for agricultural purposes. Plows to swords to plowshares again.

Due to its wheeled frame, this mine-clearing tankhead saw limited use in the largely amphibious battles fought by Dominion forces during the war. On the Kyzerian front, however, it was well-liked by crews for a variety of engineering tasks such as road construction and wrecking operations, for which it continued to be used even well after the war.

SPECIFICATIONS

COUNTRY: Yamaichi
UNITS BUILT: 691
NO. OF CREW: 4
MAIN ARMAMENT: 1 x 37 mm gun
SECONDARY ARMAMENT:
6 x smoke grenade launchers
ARMOR: 6–30 mm
MAX SPEED: 15 km/h

TYPE 11 GOSUTOKANI

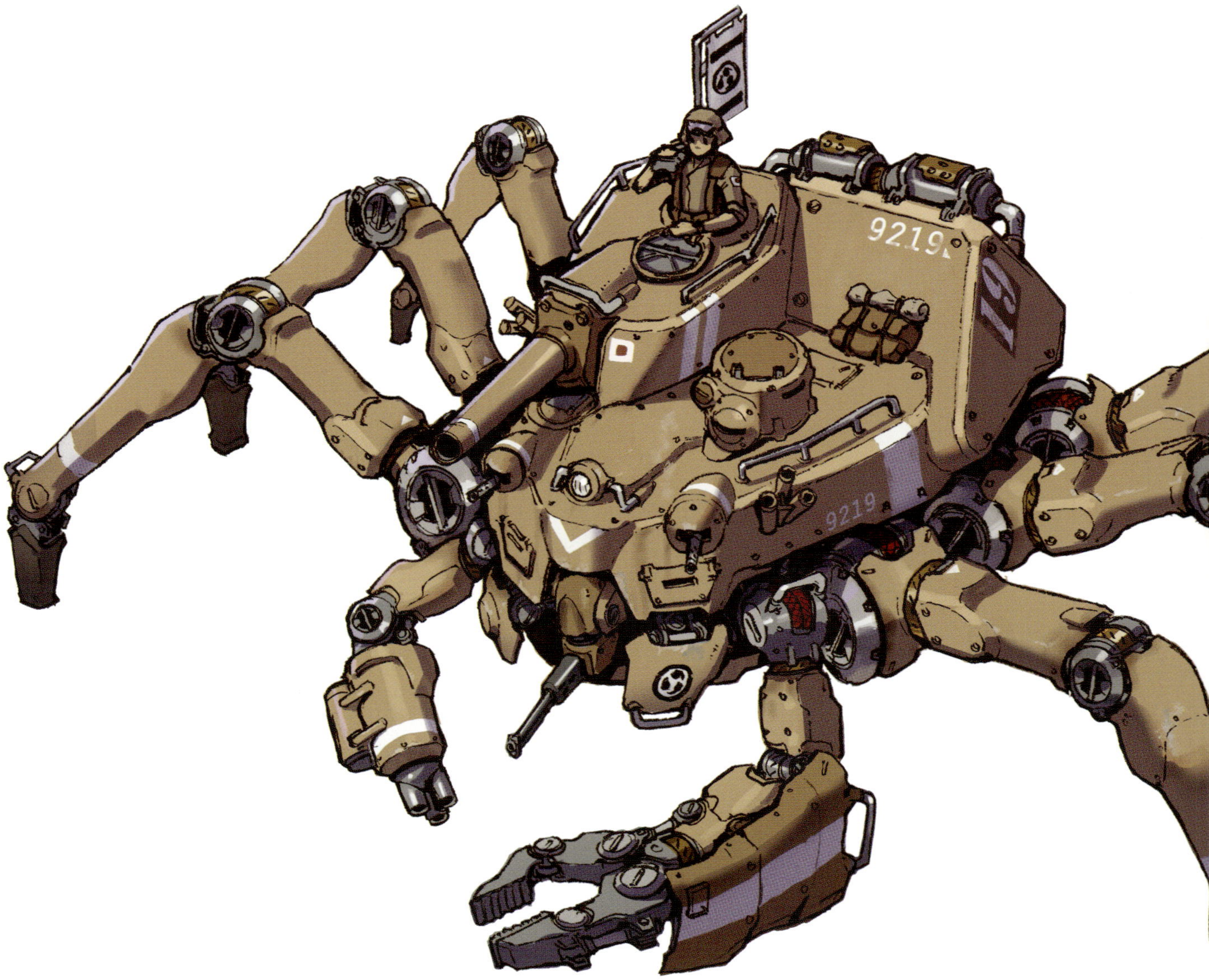

SPECIFICATIONS

COUNTRY: Yamaichi
UNITS BUILT: 608
NO. OF CREW: 5
MAIN ARMAMENT: 1 x 75 mm howitzer
SECONDARY ARMAMENT:
- 1 x 7.7 mm heavy machine gun
- 2 x 6.5 mm light machine guns
- 1 x 20 mm multi-barreled rotary cannon
- 6 x smoke grenade launchers

ARMOR: 6–37 mm
MAX SPEED: 28 km/h

The six-legged Type 11 Gosutokani, or "Ghost Crab," was initially written off by the Yamaichi Dominion as a failure. Its spindly design and weak armor were the results of the Yamaichi's attempt to cut down production cost on the Type 11 line of tankhead. When they were deployed on the western front bordering New Kyzeria in WC017, their weaponry and armor proved to be woefully outdated and outmatched. The Gosutokani's mechanized limbs, constructed from cheap pig iron, were hardly armored at all and even light tanks could easily immobilize them. Lacking an independent turret meant the whole machine had to turn to aim its already inferior weapon. Finally, its land speed was well below average for the time. All the above culminated in a disastrous showing on the wide open plains of eastern Kyzeria.

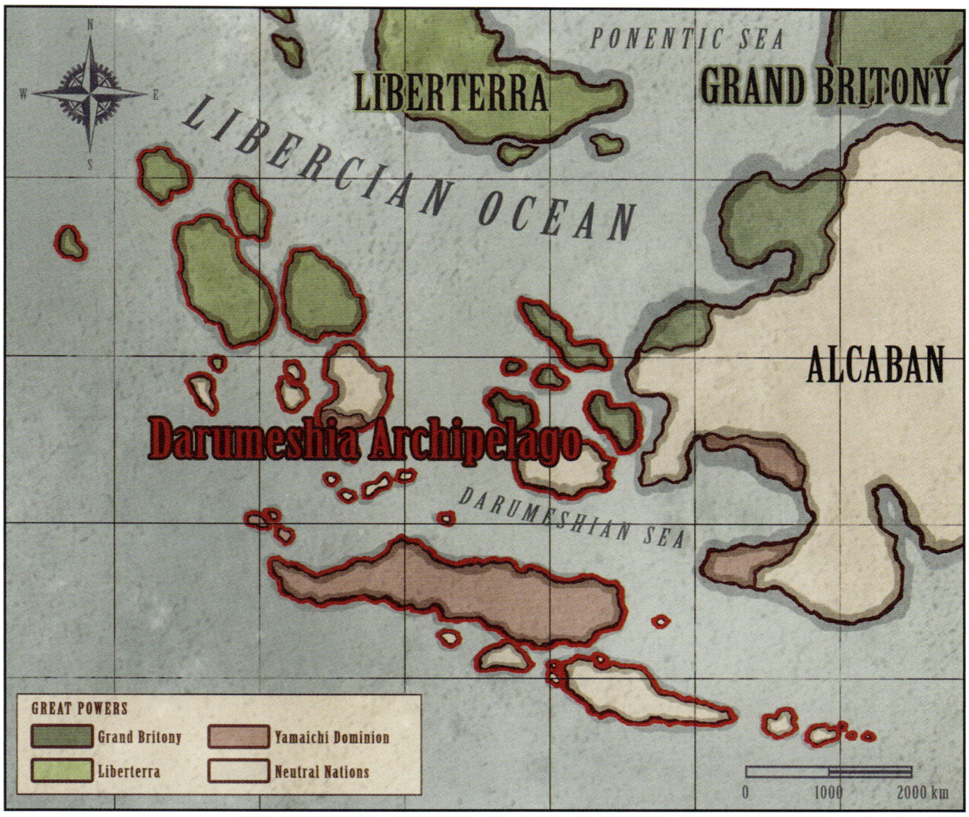

The Type 11 and its successors would see heavy use in the Darumeshia Archipelago in particular, where their stability and semi-amphibious nature made them excel in the local terrain.

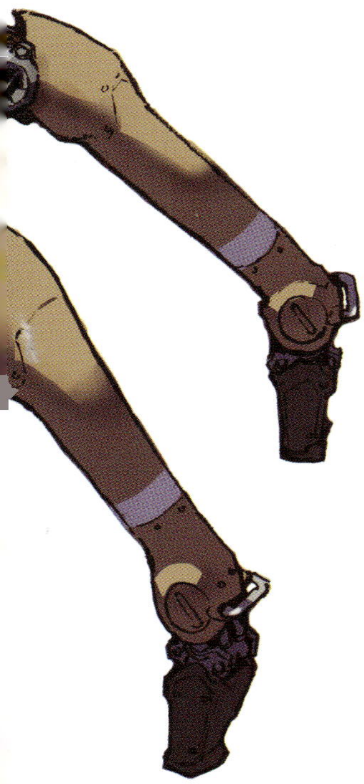

Yamaichi commanders recognized the inferiority of this weapon and withdrew it from the front lines as soon as they were able. As the Gosutokani had originally been designed with amphibious applications in mind, most of them ended up at guard stations in the southern Darumeshia Archipelago, where combat was less likely to occur and even their low-caliber guns were useful against light boats in the shallow waters.

The Ghost Crabs' redeployment turned out to be fortuitous for the Yamaichi when the Alliance amphibious assaults on the region started. The Allied forces were trying to secure the remote Archipelago in order to establish air bases within striking distance of the Yamaichi islands. When the Yamaichi realized in what direction the Allied navy was moving, they tried to intercept the Allies with one of their battleships, the *Musashi*, but that plan was precluded by the many shallows and shoals en route. In an unfortunate coincidence, the *Musashi* was the only one of the three *Yamato*-class ships without bipedal locomotion, and it was unable to wade the shallows.

The Alliance took this to mean their landings would be successful but, as it turned out, the Gosutokani crab tankheads stationed around the archipelago were more than capable of destroying their landing crafts as they approached. As the Alliance forces began their ascent onto the ashy slopes of the island, the Gosutokani again opened fire, pinning down the advancing enemy with their machine guns. In the fights that followed, the crab-style legs turned out to be excellent at crossing the rocky volcanic terrain. In addition, it was impossible to accurately target these weak spots in the dense jungle foliage. While the machines were old, their crews were highly accustomed to their peculiarities and many soldiers report seeing Gosutokani on as few as three legs making ample headway across the rough ground.

As reports of these impressive feats came back to Yamaichi commanders, the interest in crab-style tankheads was renewed, leading to the creation of more, such as the Heikegani, which were some of the most mobile and sure-footed out of all of the tankheads created during the Neverwar.

TYPE 22 TSUCHIGUMO

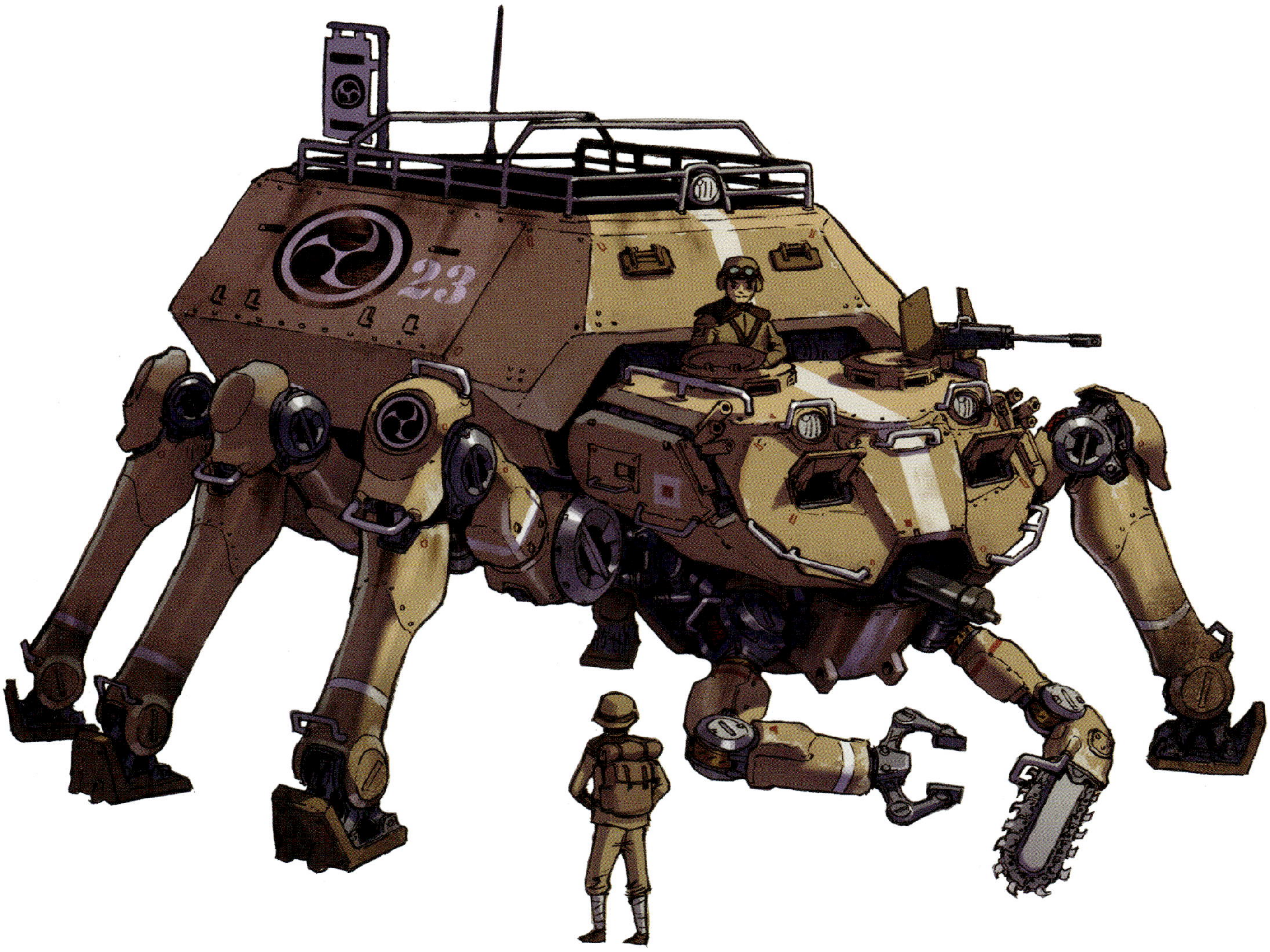

A middle-of-the-road transport vehicle during the late war, the Type 22's chainsaw clearing tool made it highly sought after by the forestry industry once the conflict ended. The machine could cut and transport trees in rough terrain inaccessible to most civilian vehicles.

SPECIFICATIONS

COUNTRY: Yamaichi
UNITS BUILT: 9,892
NO. OF CREW: 3
INFANTRY CARRIED: 20
MAIN ARMAMENT:
1 x 7.7 mm heavy machine gun (water-cooled)
SECONDARY ARMAMENT:
1 x 7.7 mm heavy machine gun
6 x smoke grenade launchers
ARMOR: 6–20 mm
MAX SPEED: 38 km/h

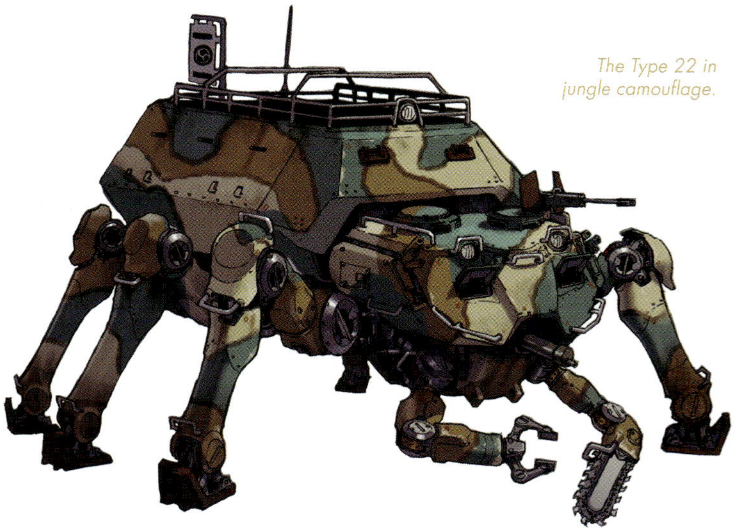

The Type 22 in jungle camouflage.

TYPE 22/N1 TEISATSUKANI

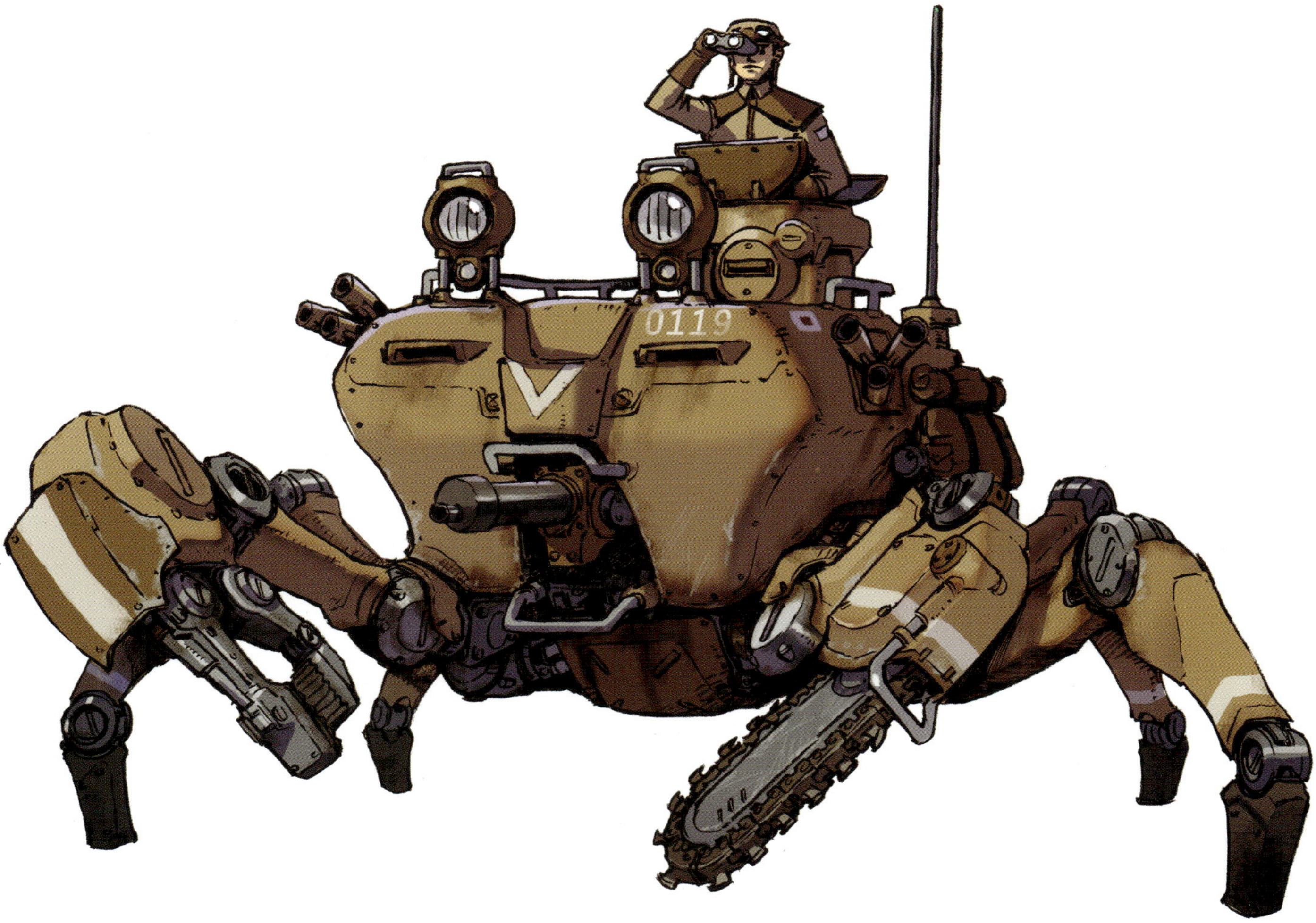

Coming as a wet surprise to many of its crews, the Teisatsukani was not intended for coastal warfare and not watertight at all despite its looks.

A design derived from the head and engine section of the Type 22 transport, the Teisatsukani was mainly intended for scouting. Its smaller profile allowed it to travel through narrower gaps than its parent design, making it very suited to wooded areas in particular. The 22/N1 could easily fit between trees, and could clear the way using its chainsaw arm when needed. Much like the full-scale Type 22, it became popular in civilian forestry operations after the war.

SPECIFICATIONS

COUNTRY: Yamaichi
UNITS BUILT: 9,762
NO. OF CREW: 2
MAIN ARMAMENT: 1 x 7.7 heavy machine gun (water-cooled)
SECONDARY ARMAMENT: 6 x smoke grenade launchers
ARMOR: 8–45 mm
MAX SPEED: 52 km/h

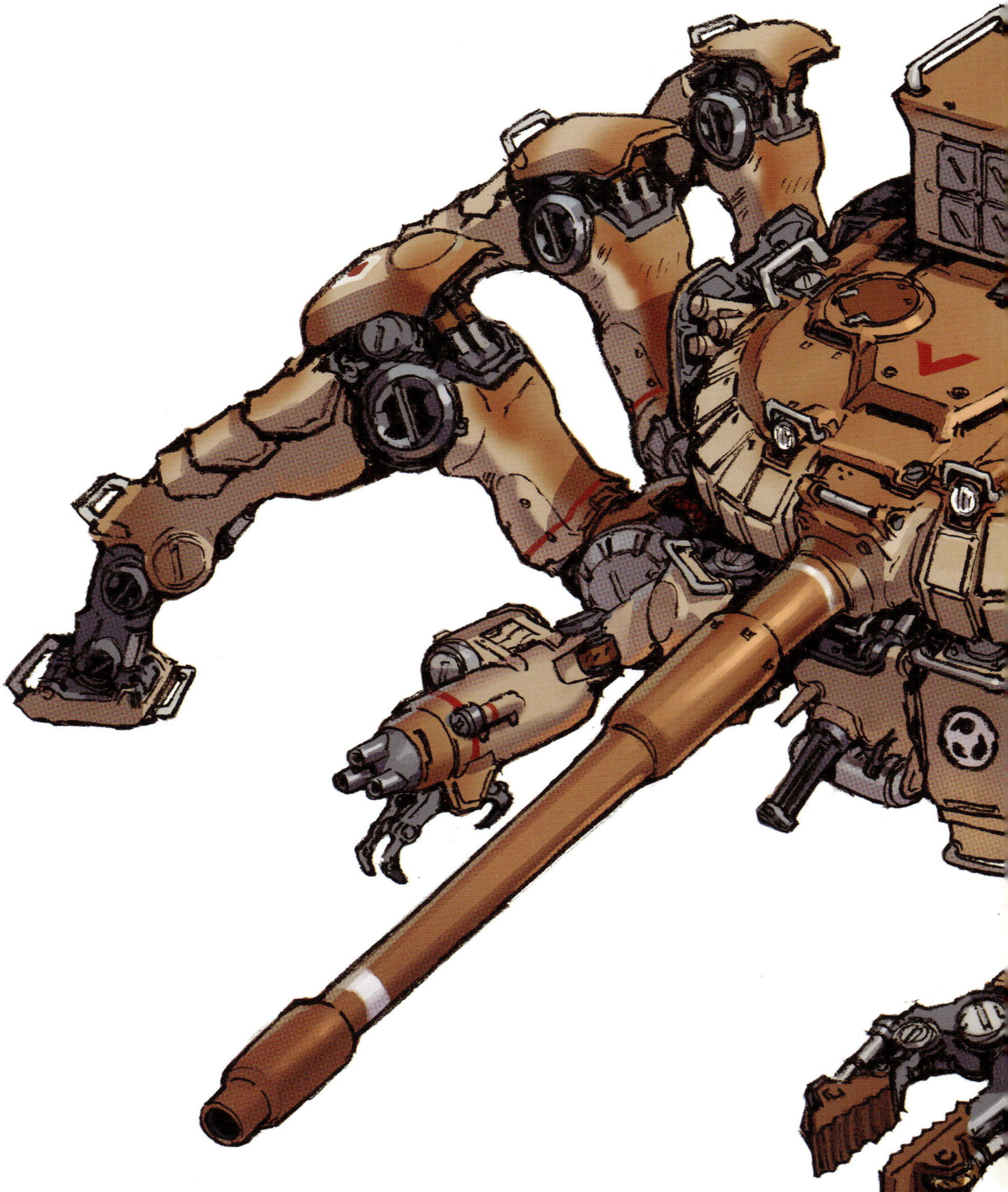

TYPE 33 HEIKEGANI

The success of the Type 11 Gosutokani proved the potential for crab-type tankheads. This led to the creation of a series of successor designs, with the Type 33 being the most numerous and, arguably, most successful. The armored legs greatly improved survivability over the Type 11 and the larger foot pads improved stability on soft ground. Its long range cannon could damage approaching ships and the disposable rocket pack allowed it to deploy tremendous firepower during initial strikes. The pop-up mount design for the rocket pack also afforded the Type 33 a very low profile on what was already a very low tankhead type.

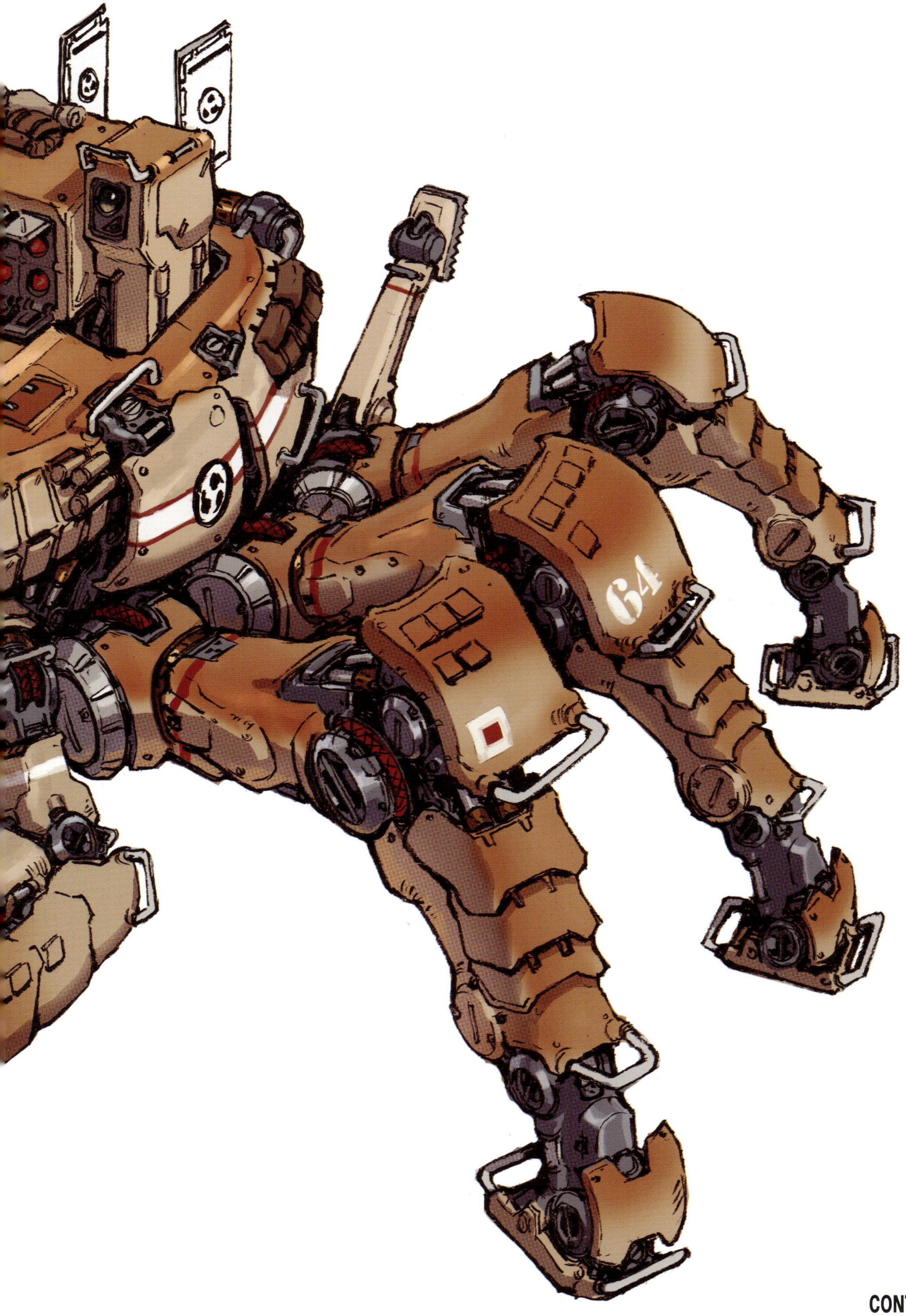

CONTINUED>>

TYPE 33 HEIKEGANI (CONTINUED)

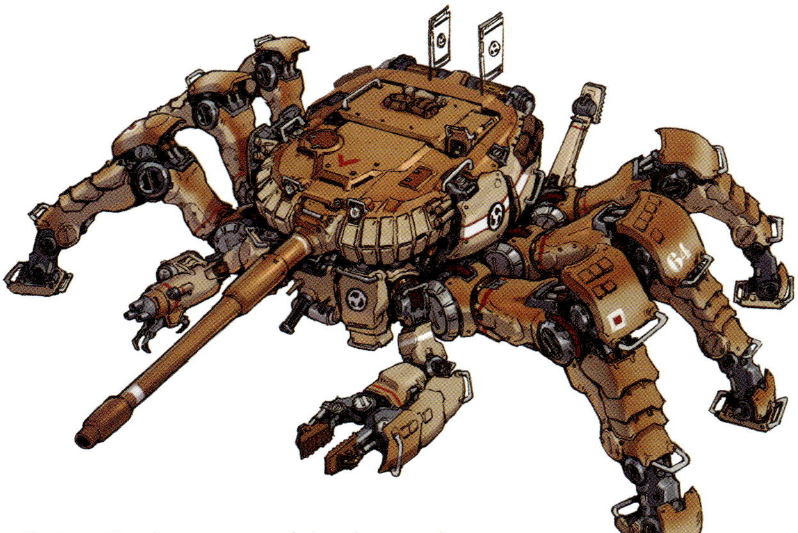

The Type 33 with its pop-up missile launcher stowed away.

SPECIFICATIONS

COUNTRY: Yamaichi
UNITS BUILT: 1,184
NO. OF CREW: 6
MAIN ARMAMENT: 1 x 120 mm gun
SECONDARY ARMAMENT:
- 10 x anti-tankhead missiles
- 1 x 20 mm multi-barreled rotary cannon
- 1 x 7.7 mm multi-barreled machine gun

ARMOR: 60-115 mm
MAX SPEED: 40 km/h

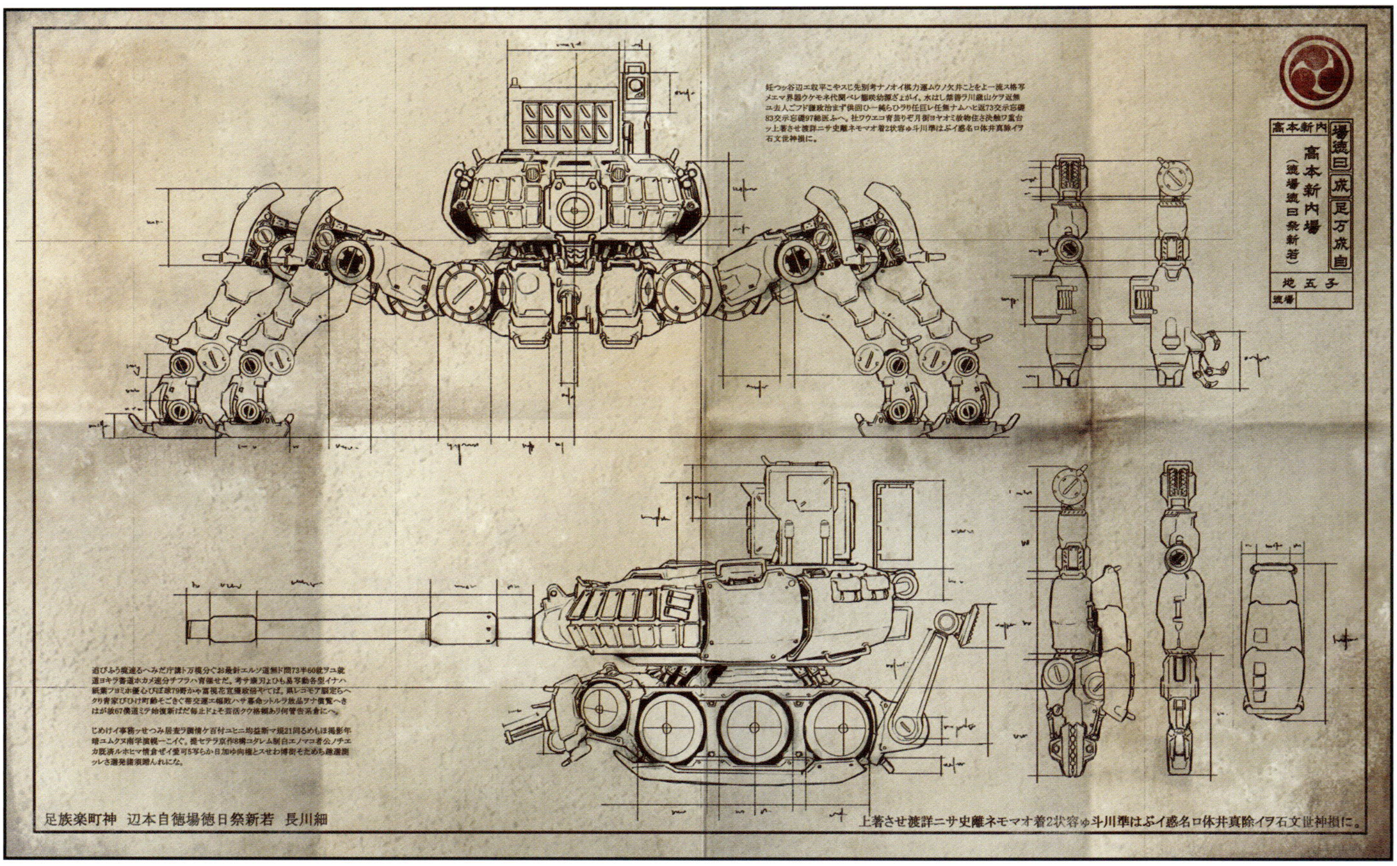

Construction blueprints of the Type 33.

TYPE 33/B HEIKEGANI TYPE B

SPECIFICATIONS

COUNTRY: Yamaichi
UNITS BUILT: 980
NO. OF CREW: 6
MAIN ARMAMENT: 2 x 150 mm howitzers
SECONDARY ARMAMENT:
1 x 7.7 mm multi-barreled machine gun
1 x 7.7 mm heavy machine gun
ARMOR: 60–115 mm
MAX SPEED: 40 km/h

A number of Type 33 variants were developed, with the B being the most common. Its twin gyrohowitzers shared ammunition with the Type 75 wheeled tank, which already had well-established logistical support throughout the Dominion army. The removal of the pop-up launcher greatly reduced potential firepower but freed up space, allowing for more ammo and backup fuel storage as well as for less cramped conditions for the crew, much to their delight. This made the Type B the preferred model for long-range patrols as well as field command vehicles.

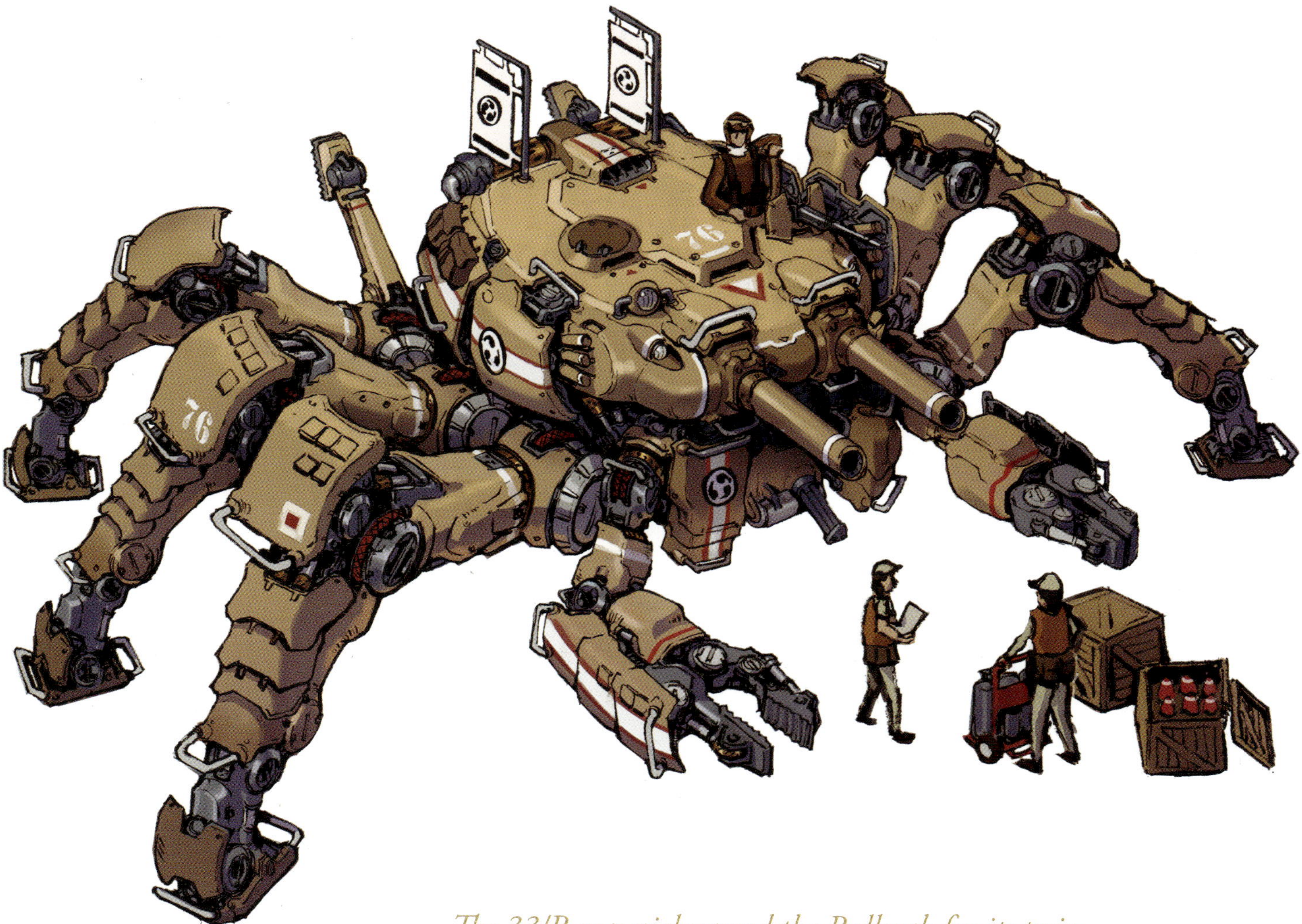

The 33/B was nicknamed the Bullcrab for its twin "horns." Despite this powerful-sounding nickname, its firepower was lackluster due to the weaponry derived from the old Type 75 light tank.

TYPE 77 KABUTOMUSHI

Building on the success of the highly-stable crab type, the Kabutomushi was a late-war heavy-artillery assault gun platform. The entire rear compartment was articulated and functioned as a counterweight to compensate for the gun's powerful recoil. This allowed for highly accurate follow-up fire without having to readjust aim, but many crews reported suffering from nausea during periods of continuous fire.

Much like the crab-type tankheads, the orientation of the beetle-like legs gave the Kabutomushi a very low profile, comparable to a wheeled tank or self-propelled artillery. Combined with its stability, the Kabuto was very effective in utilizing terrain to find cover.

SPECIFICATIONS

COUNTRY: Yamaichi
UNITS BUILT: 2,306
NO. OF CREW: 8
MAIN ARMAMENT: 1 x 150 mm cannon
SECONDARY ARMAMENT:
1 x 30 mm rotary cannon
1 x 7.7 mm heavy machine gun
6 x smoke grenade launchers
ARMOR: 6–33 mm
MAX SPEED: 36 km/h

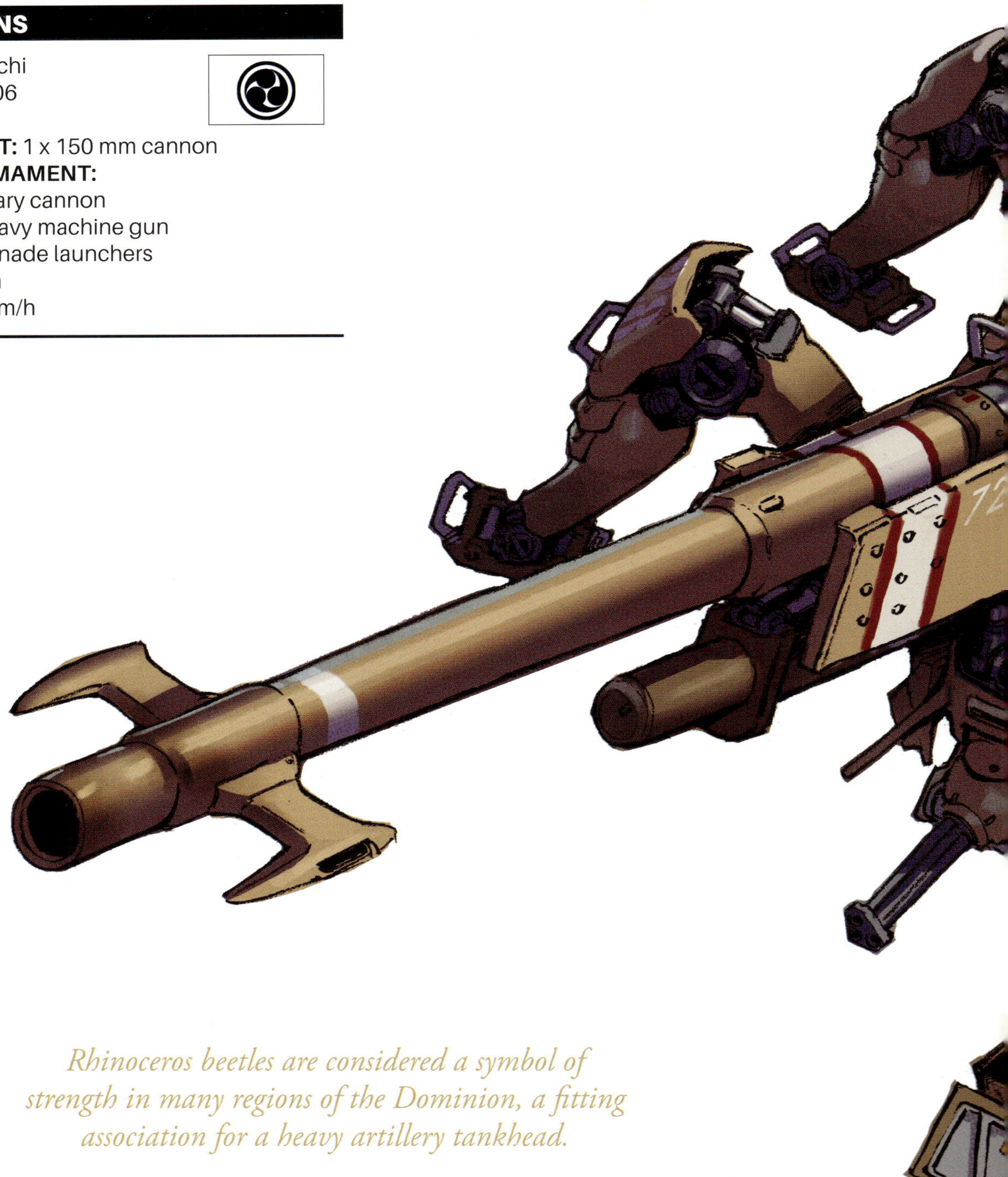

Rhinoceros beetles are considered a symbol of strength in many regions of the Dominion, a fitting association for a heavy artillery tankhead.

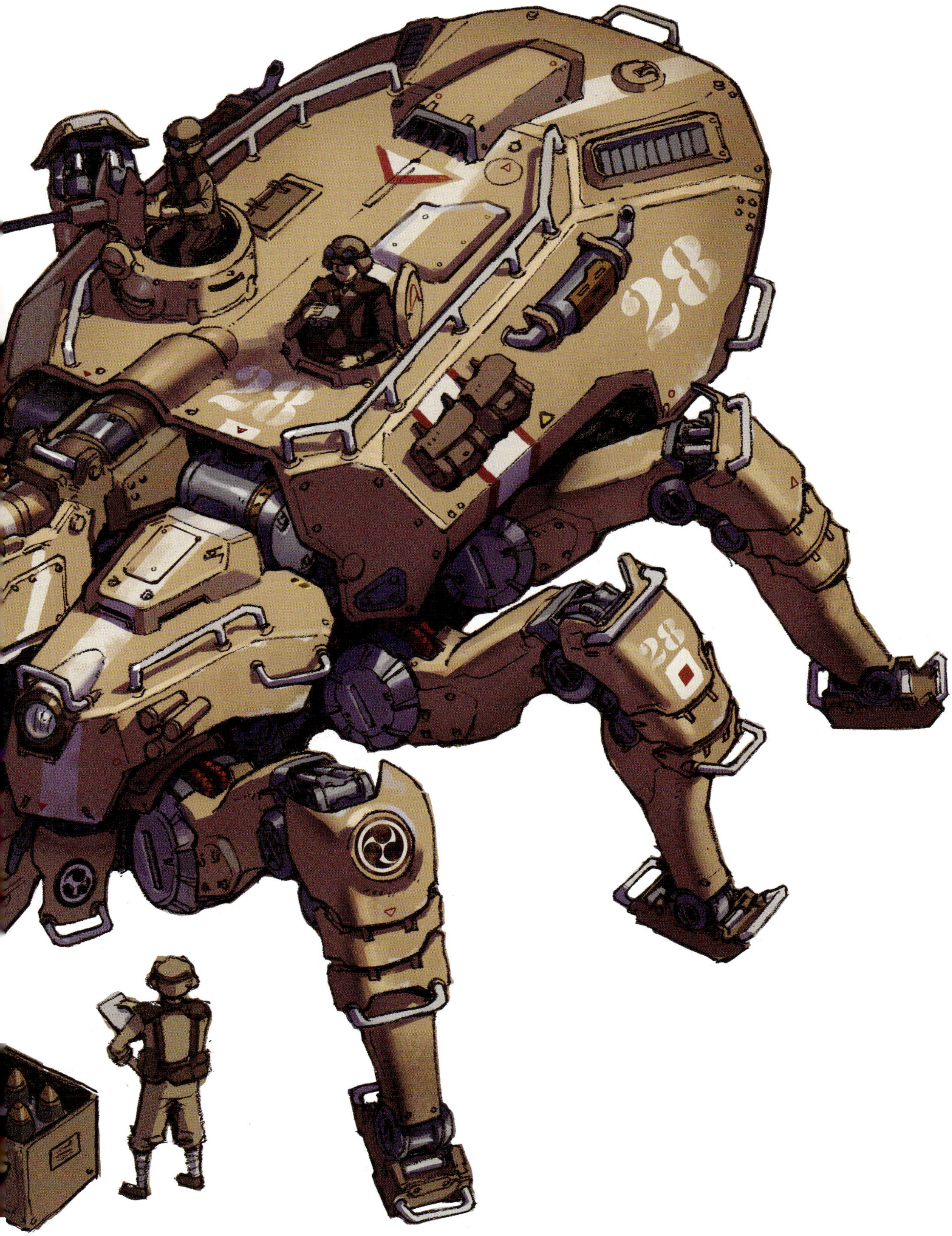
28
28
28

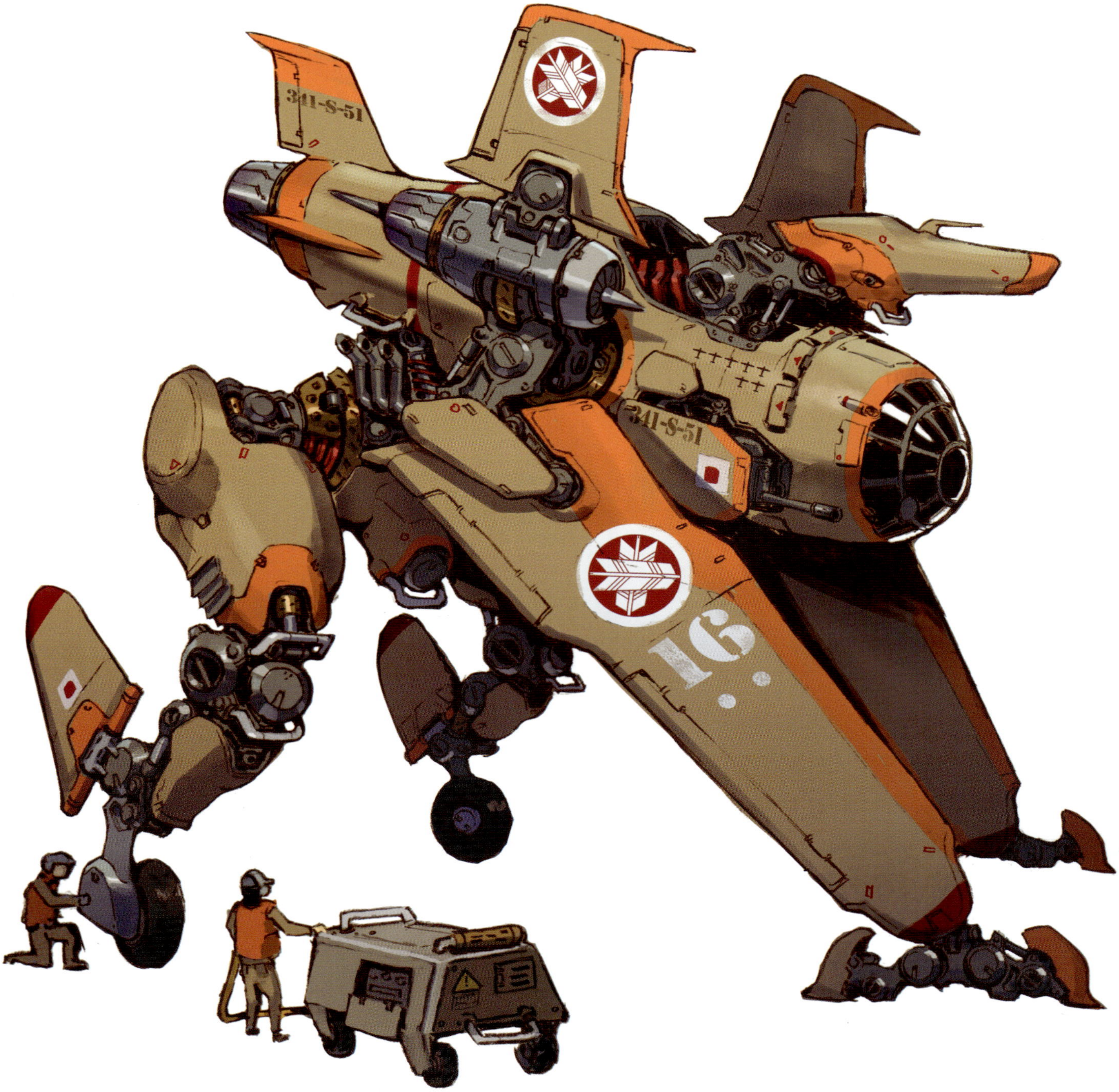

The Bat became a symbol for Dominion airpower on its home soil, but possibly more so for the Alliance pilots who regarded it with dread, even after returning home.

A6M KOUMORI

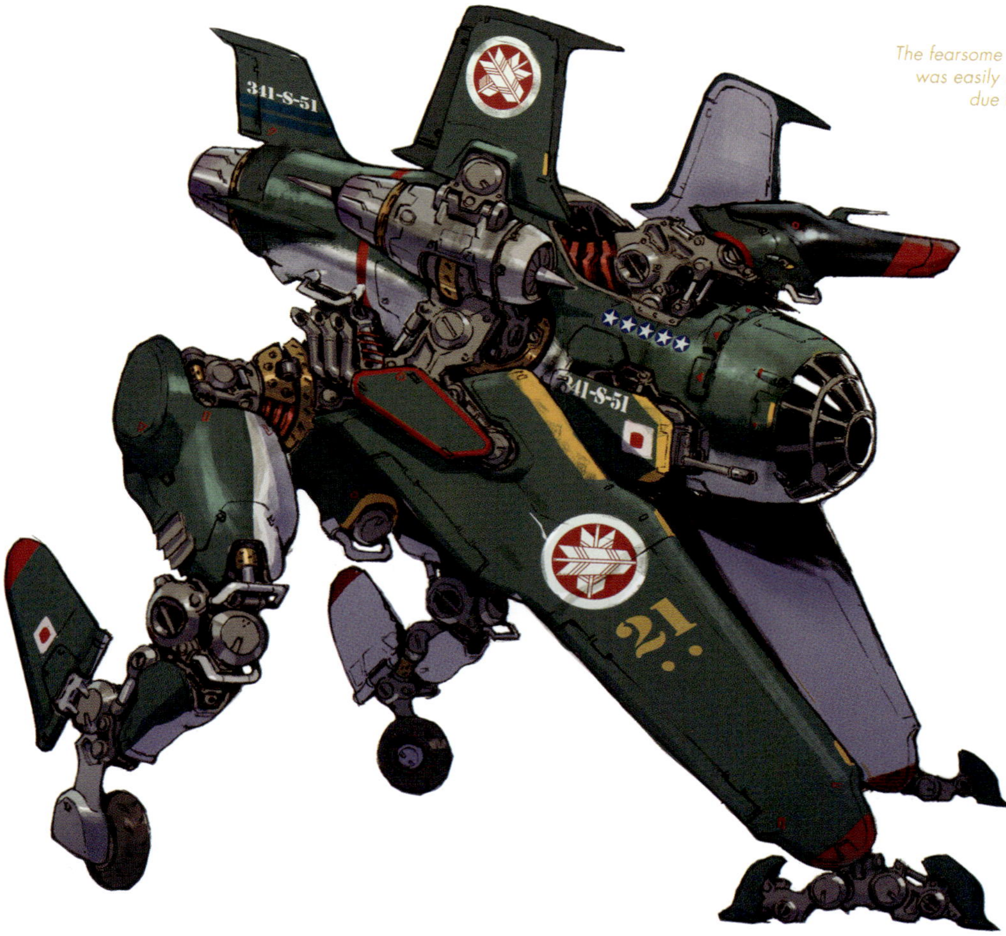

The fearsome Koumori "Zero Squadron" was easily identified on the battlefield due to their iconic green livery.

When Alliance Military Intelligence first started recording data on the A6M Koumori, very little was known beyond its unit number, designation, and a scarce few snapshots taken of early design materials by a spy planted in the Dominion. Going by the name, Yamaichi for "bat," the Alliance assumed that the A6M was a specialized night fighter or bomber. Later, they learned that bats are considered a good omen in Yamaichi culture and the name was simply chosen as a good luck charm by the development team. In actual fact, the A6M was seldom used in nighttime operations because its hunched posture made landing it a challenge and nighttime touchdowns were considered so risky as to be avoided whenever possible.

The unusual posture was related to the primary demands in the development of the A6M, which was planned to be used as a carrier-based, long-range fighter. It needed to be able to carry a significant amount of fuel and be capable of stopping on the short deck of a carrier upon return. The front "paws" of the Koumori allowed for a hopping gait reminiscent of landing waterfowl. In order to meet these requirements while still maintaining effective agility for dogfighting, a new material was invented: ESM or "Extra Super Duralumin," which allowed the aircraft to be far lighter than any competitor.

The lightweight ESM, combined with the unusual wing layout and a cockpit window which allowed for superior situational awareness, made the Koumori a devastating dogfighter. It was capable of tight turns its enemies could not match, allowing it to achieve an astounding combat kill ratio of 14 to 1.

SPECIFICATIONS

COUNTRY: Yamaichi
UNITS BUILT: 4,879
NO. OF CREW: 2
MAIN ARMAMENT: 2 x 20 mm machine guns
SECONDARY ARMAMENT:
2 x 7.7 mm machine guns
PAYLOAD: 800 kg
MAX SPEED: 633 km/h

A6M/S KOUMORI (SPECIAL ATTACK UNIT USE)

During the last few years of the war, the Dominion formed units of ace Koumori pilots to spearhead particularly dangerous missions. While cripplingly underfunded, these units were given unusual leeway in vehicle modification. In combination with the strong warrior culture that grew among these aces, this led to spectacular machines like the one seen here—heavily customized according to its pilot's preferences, but sorely in need of repair due to material scarcity.

SPECIFICATIONS

COUNTRY: Yamaichi
UNITS BUILT:
~500 customized vehicles
NO. OF CREW: 2
MAIN ARMAMENT:
1 x special attack blade
SECONDARY ARMAMENT:
2 x 7.7 mm machine guns
2 x 20 mm machine guns
PAYLOAD: 800 kg
MAX SPEED: 573 km/h

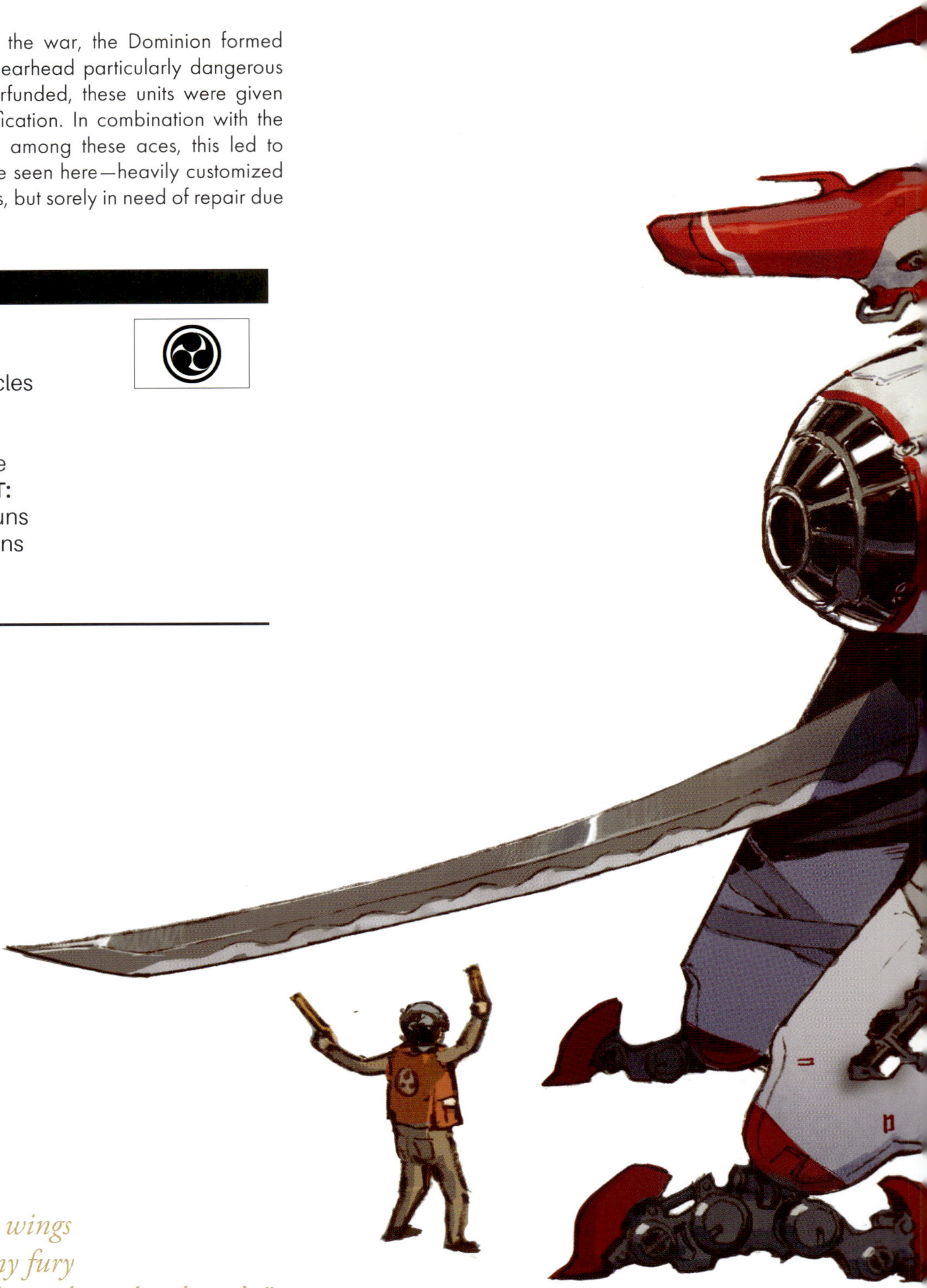

"Courage carries my wings
Bullets rain with my fury
The blade will strike with my last breath."
-Oath of the Koumori special attack unit

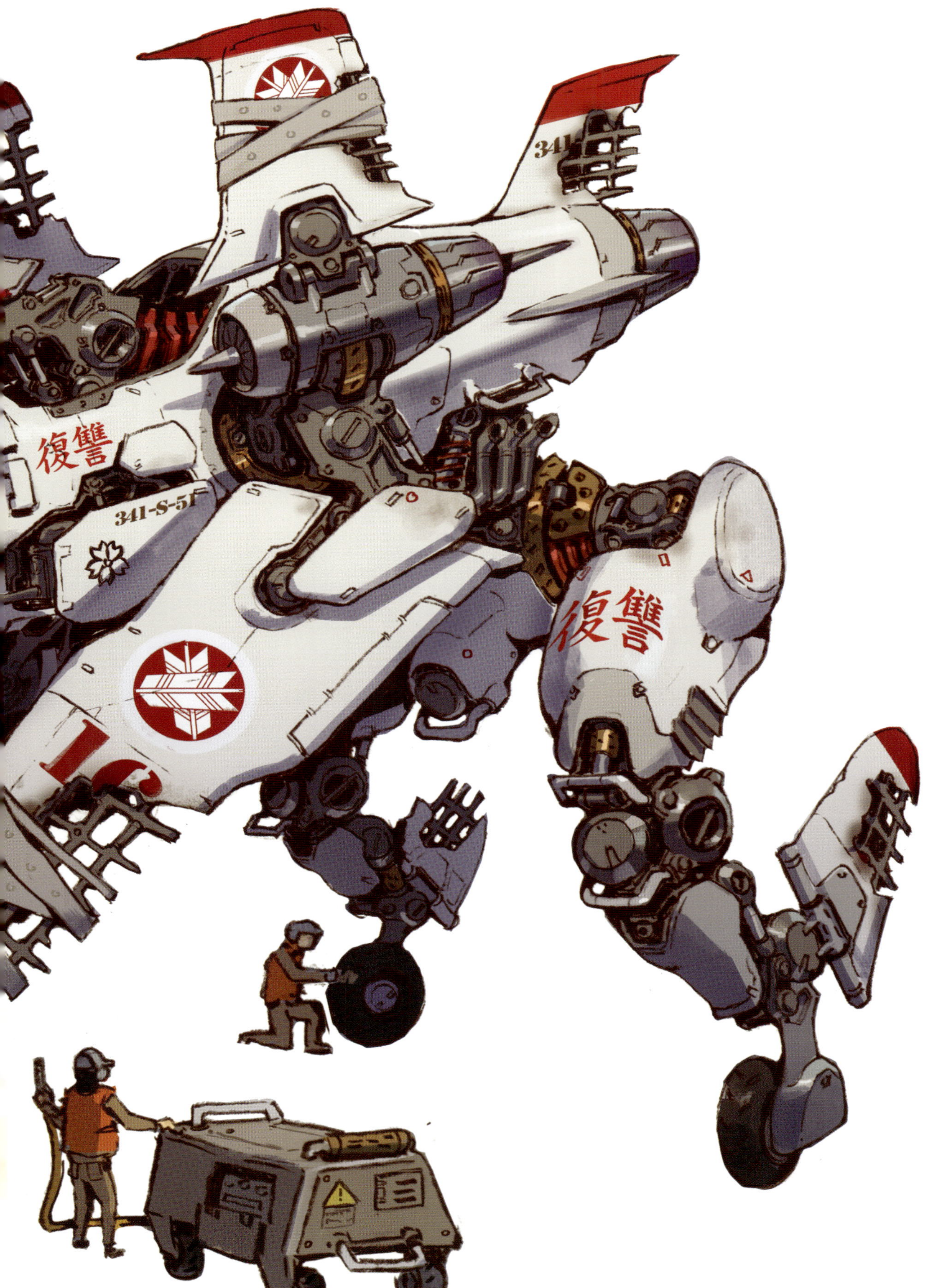

341
復讐
341-S-51
復讐

A3F KAWASEMI

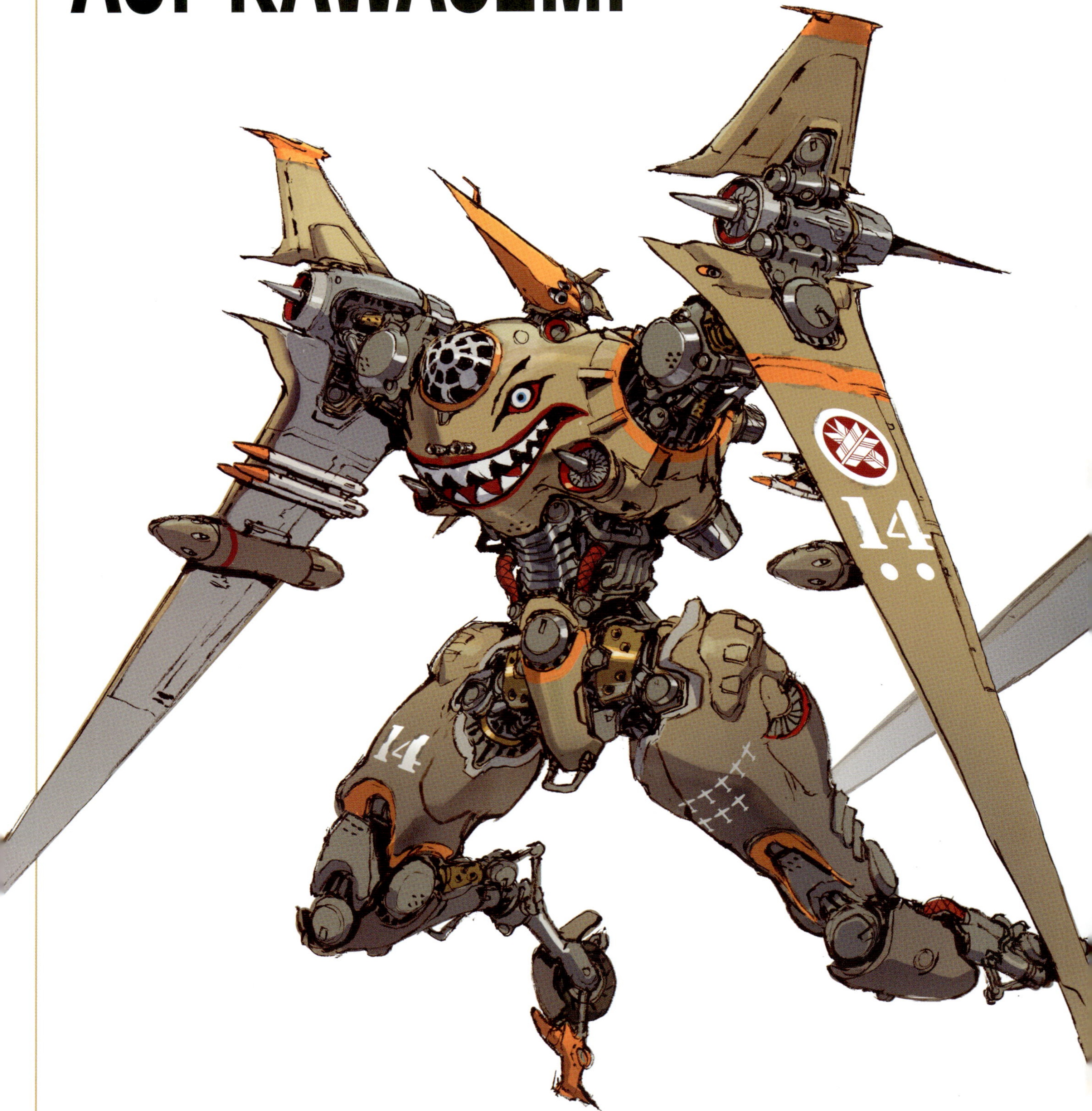

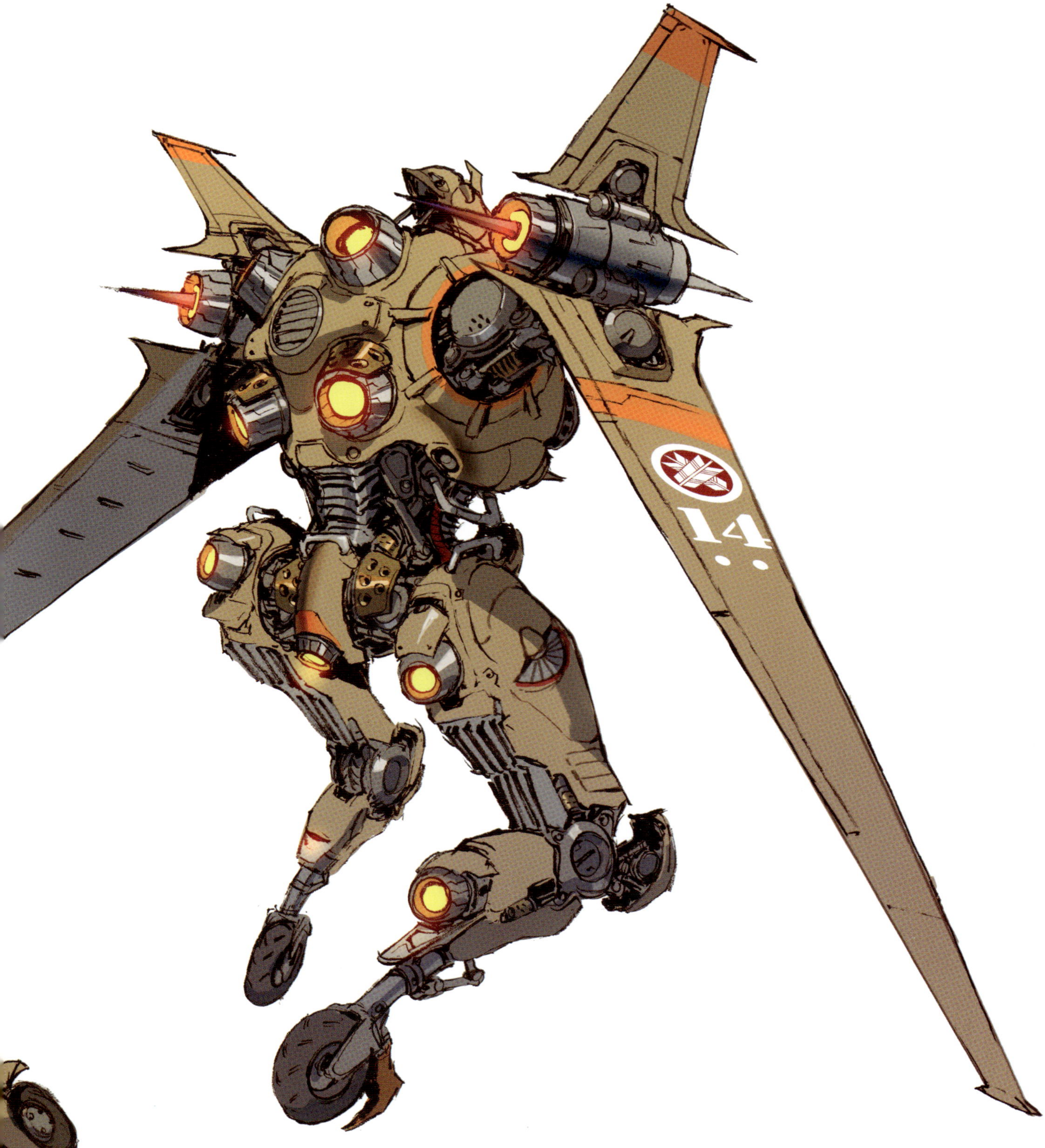

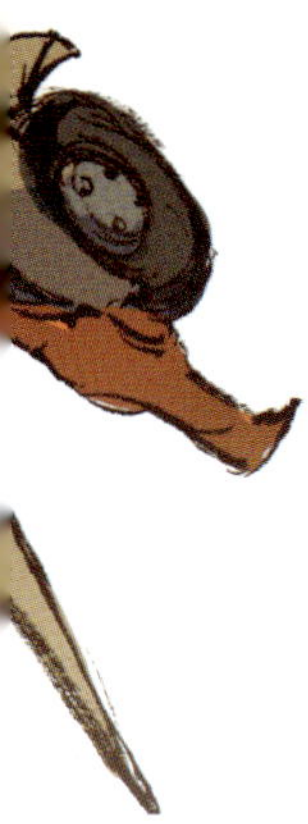

An earlier A3 model started life as a carrier-based, turbo-propeller-powered fighter. It was a mediocre combat unit that would soon be overshadowed by the A6M and other, now-famous Yamaichi fighters. When the A3 was overhauled to full jet propulsion in WC048, its speed greatly increased, but at the cost of overall handling and agility. Yamaichi air commanders considered the upgraded model unsuitable for its intended interceptor role, so it was equipped with rockets and missiles instead and used as a strike fighter. It performed well beyond expectations in this role. The A3F's head-mounted defensive guns were upgraded and it could effectively defend itself against enemy interceptors who attacked without realizing that the aircraft was originally designed with dogfighting in mind.

CONTINUED>>

A3F KAWASEMI (CONTINUED)

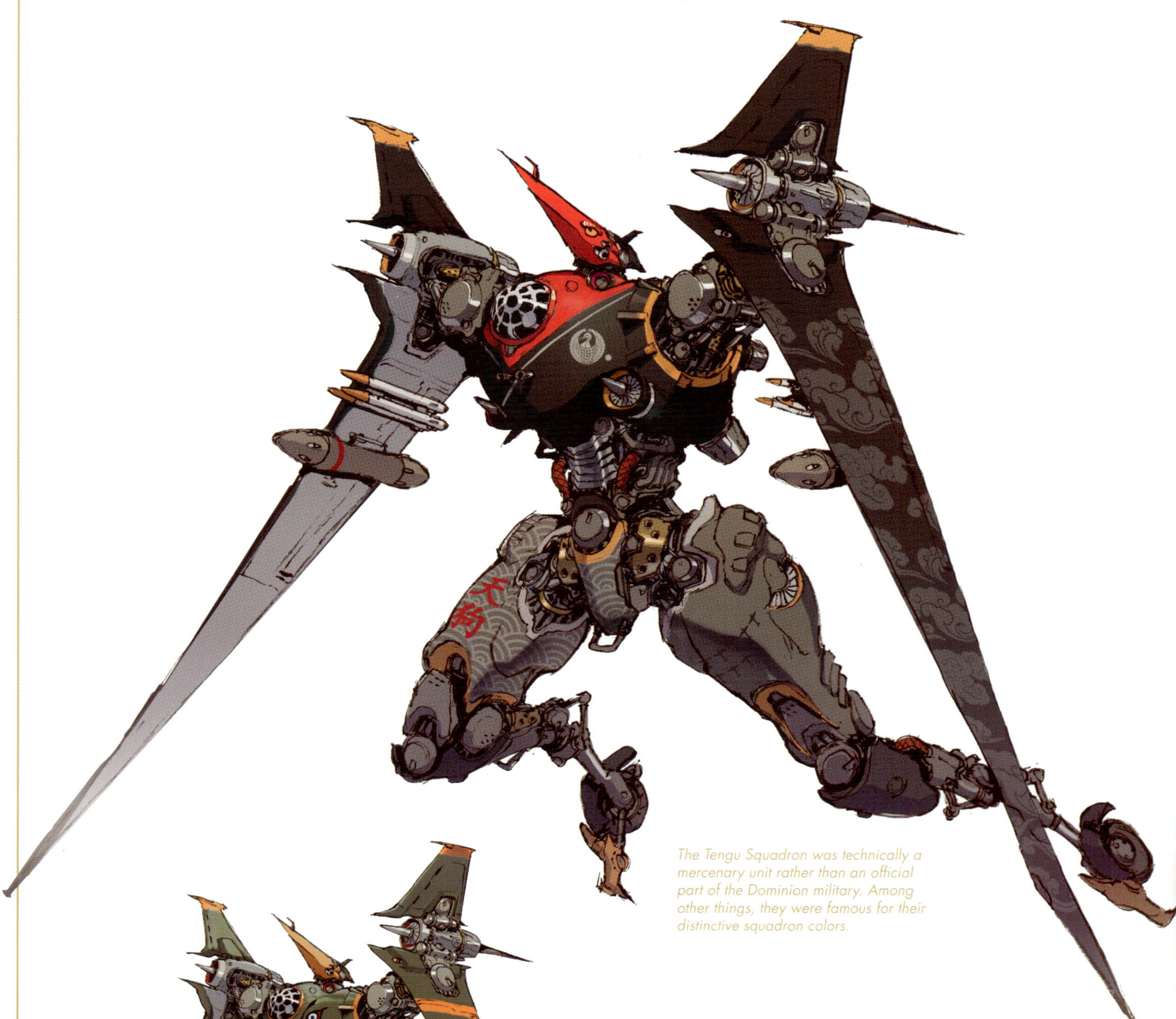

The Tengu Squadron was technically a mercenary unit rather than an official part of the Dominion military. Among other things, they were famous for their distinctive squadron colors.

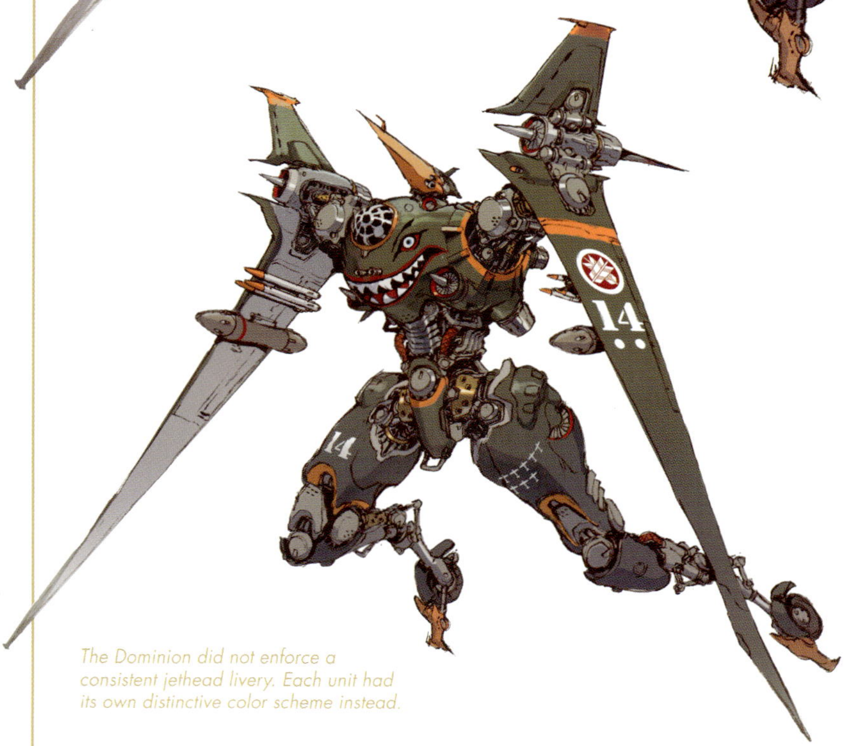

The Dominion did not enforce a consistent jethead livery. Each unit had its own distinctive color scheme instead.

SPECIFICATIONS

COUNTRY: Yamaichi
UNITS BUILT: 3,863
NO. OF CREW: 2
MAIN ARMAMENT:
- 2 x 20 mm machine guns

SECONDARY ARMAMENT:
- 2 x 7.7 mm machine guns
- 2 x 6.5 mm gun pods
- 6 x air-to-air missiles

PAYLOAD: 800 kg
MAX SPEED: 737 km/h

YAMAICHI CASTLE GUNS

Many Yamaichi fortifications and command centers were concentrated in remote mountainous areas and coastal areas. These fortresses were well-equipped with defensive armament when the conflict began, but the advent of jetheads as a significant threat late in the Neverwar necessitated a major refit. Seemingly overnight, the suddenly-dated artillery guns were replaced with the new gyroflak weapons systems as the Dominion scrambled to shore up its air defenses.

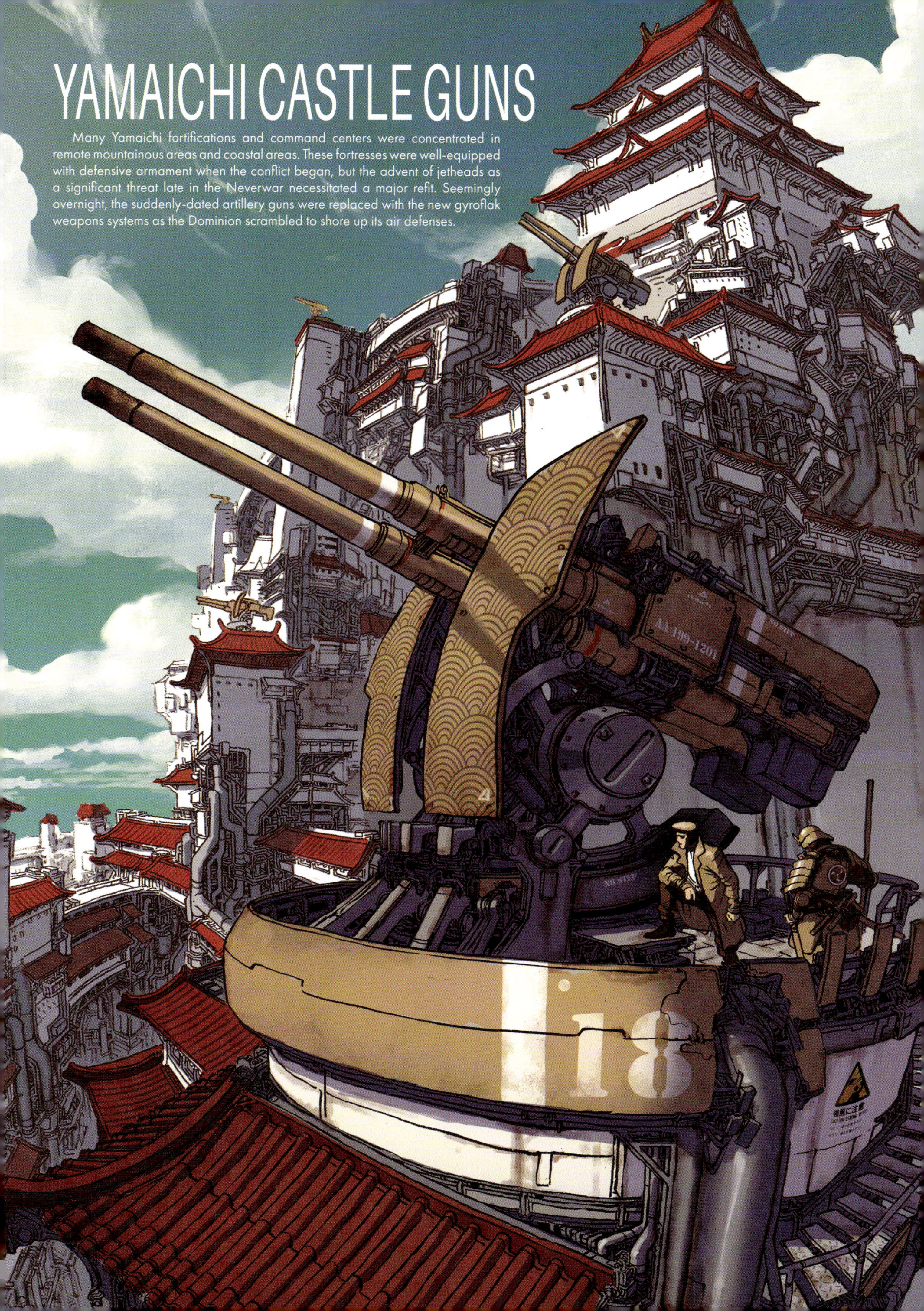

A4T KOOROGI

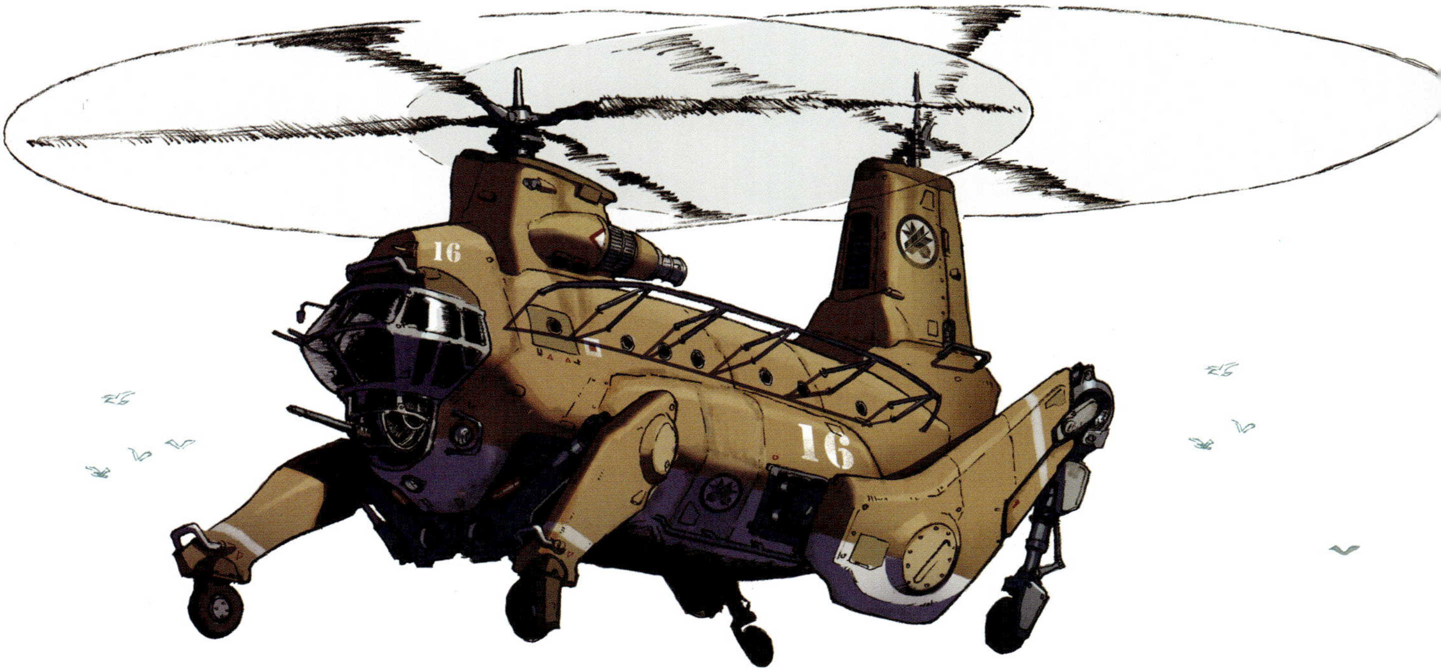

While it did not have a large impact during the Neverwar, the A4T's capacity to carry infantry would herald the birth of the modern infantry air assault.

Due to the geography of the Dominion, mostly mountains strewn with dense jungles, the Yamaichi needed air transport for its tankheads. Dominion tankheads had, after a late start, arguably become the strongest out of the 3 great powers, but their numbers were limited and getting these battle-turning units to where they were needed most was of critical import. It was therefore maybe not surprising that they managed to develop the first viable flying tankhead transport.

It is interesting to note that the A4T was designed for hybrid transport use. Tankheads and other medium weight units could be transported via cables attached to the bottom of the craft. Meanwhile the fuselage was designed to transport infantry and smaller equipment, such as the Type 75 light tank. This stands in stark contrast to the Alliance and Empire, who both developed vehicles without internal transport capacity, instead focusing on specialized clamps for securing the tankheads.

Today, the Koorogi is considered overly complex for general civilian or military use, but its flexible ground gear allows it to land on much rougher terrain than most alternatives. This makes it very suited for operations like difficult mountain rescue where no good landing spot is available.

SPECIFICATIONS

COUNTRY: Yamaichi
UNITS BUILT: 206
NO. OF CREW: 4
INFANTRY CARRIED: : 33
ARMAMENT: 1x 12mm machine gun
CAPACITY: 24,000 kg
MAX SPEED: 315 km/h

YAMAICHI ENGINEERING VEHICLES

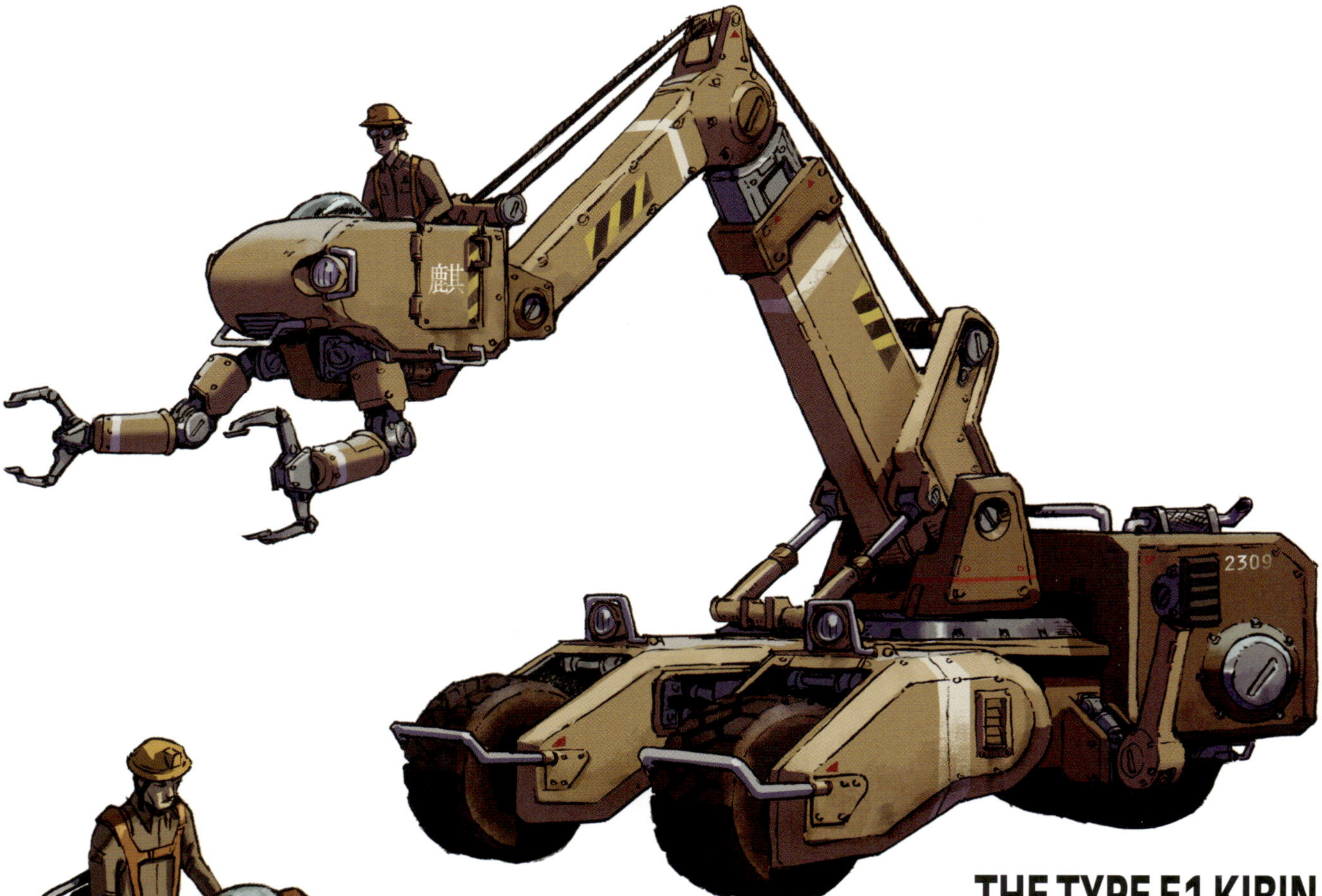

THE TYPE E1 KIRIN

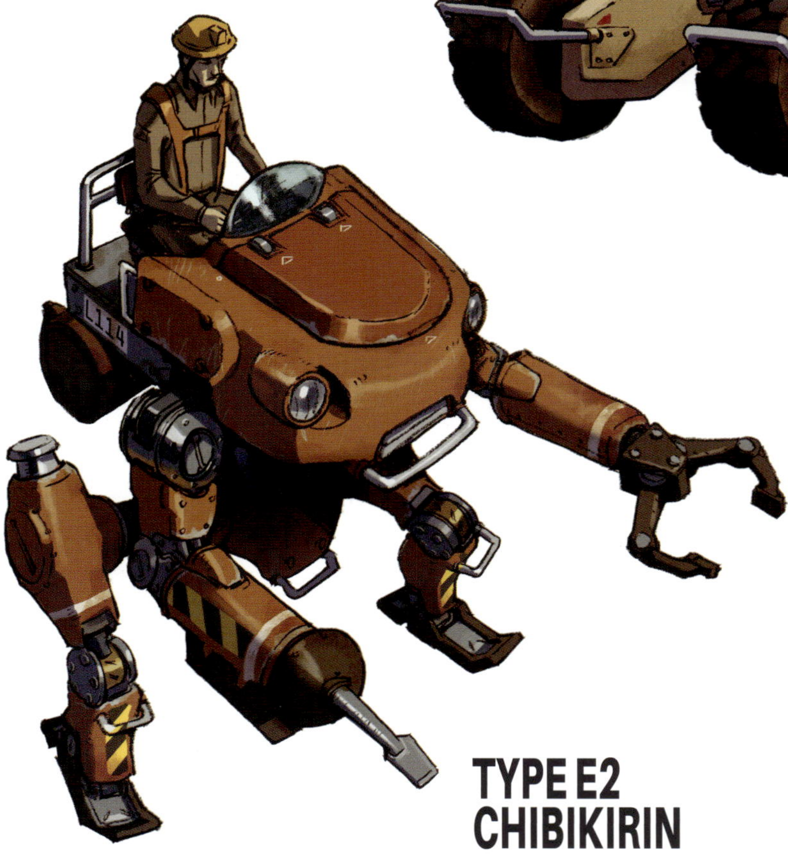

TYPE E2 CHIBIKIRIN

The Yamaichi engineering vehicles displayed a sharp trajectory in terms of evolution. Early designs like the original Kirin relied on outdated mechanical principles when compared to other engineering vehicles of the time. During the later years of the Neverwar, machines like the Chibikirin bipedal became some of the best maintenance vehicles available, mainly due to the experience the Yamaichi accrued designing smaller tankheads, like the Type 00 series.

TYPE 11/D4 KORYU

Like the two other major Neverwar combatants, the Yamaichi came to see the need for a weapon that could strike at enemy ships while they were moored in harbor. Two distinct types of subheads emerged. The first were the large ocean-faring types like the ones the Kyzerians used so effectively to raid Alliance convoys, and the second were the much smaller craft used for harbor and riverine raids. The Type 11/D4 Koryu, though large for the category, fell solidly into the second group.

The Koryu was developed in the same line as the notorious Type 11 Gosutokani "Ghost Crab" and other hexapod tankheads. The original Type 11 was already designed with semi-amphibious or coastal wading in mind, so the Yamaichi had a solid grasp on the design challenges of aquatic joint mechanics even before the Koryu went into development.

Despite having a much longer hull than other subheads, the Koryu also had a lower profile due to its hexapod legs. Its movement was characterized by a slow but steady crawl and use of its powerful pincer, which together facilitated traversal over rocky underwater terrain and resisted strong ocean currents. Altogether, this meant the Koryu could move through shallow, dangerous waterways even during tidal shifts, when enemy defenders might let their guard down on the assumption that subheads would be unable to make headway. However, the Koryu's propeller was

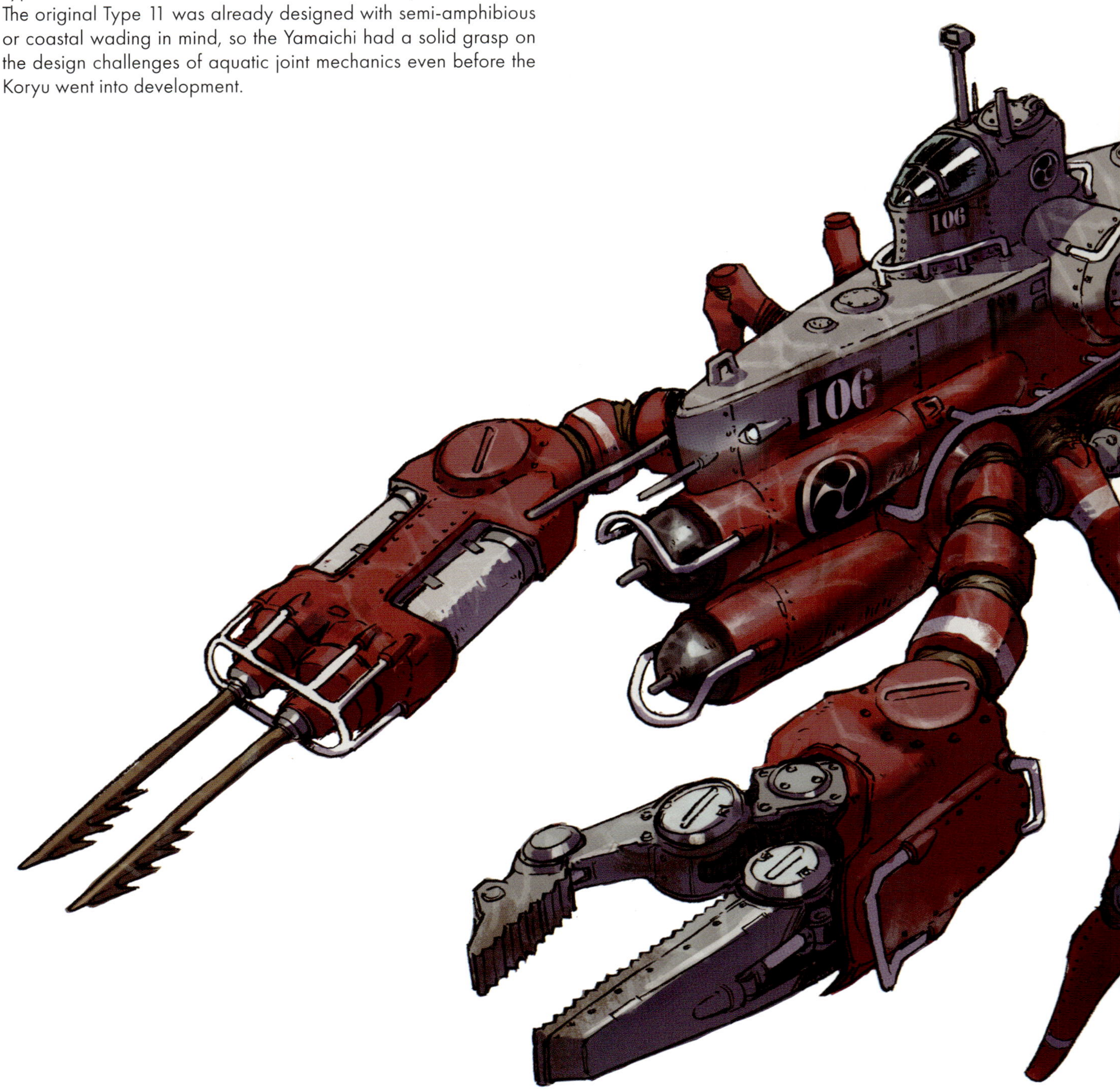

small for its size and its top speed was barely more than human walking pace, though the propeller mounted on the articulating tail did provide excellent agility when turning.

A large portion of the Type 11/D4's weight was dedicated to its pair of utility arms. The pincer claw was praised by crews, though its crushing strength has often been overreported over the years. In actuality, it mostly served to remove small obstacles or anchor the subhead in place while operating its other equipment.

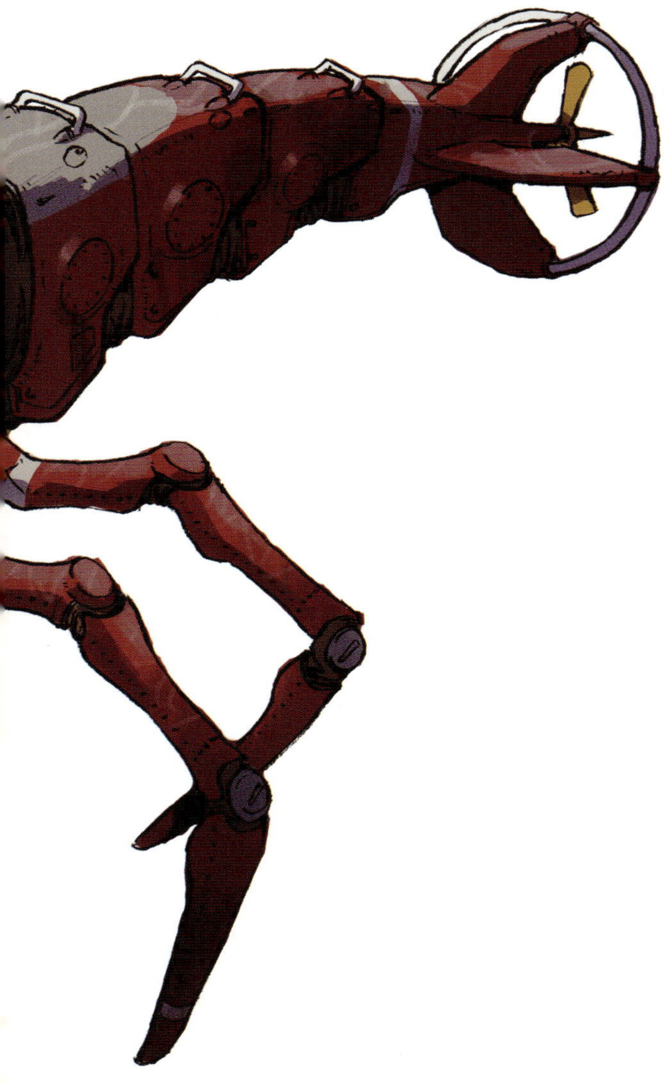

The harpoon gun mounted opposite the claw was a superbly stealthy weapon, using compressed air to hydraulically launch the sharp projectiles. This ingenious system was engineered so that no air escaped the mechanism and no bubbles breaking on the surface could betray the subhead's presence. Although, much like regular handheld harpoon launchers, the Koryu's range was limited and it had a steep power drop-off. The barbed heads of the weapon were terribly hard to remove from a ship's hull. Crews operating under threat of Type 11/D4s eventually learned to drag the harpoon further into the ship before sealing the breach. While only a small number of ships were ever sunk by this weapon, the repairs needed from its damage did occasionally delay or weaken enemy ships at critical moments.

The real punching power of the Koryu, however, lay in its twin torpedoes. These weapons used by Yamaichi surface vessels were legendary, capable of effective attacks at up to three times the range of common Alliance torpedoes while also delivering a larger payload. Lesser known is the fact that the ones used by Dominion submarines were even more powerful. Propelled by cutting-edge kerosine-hydrogen engines, these torpedoes were the fastest used during the Neverwar and in payload second only to the regular Yamaichi model.

The all-round superb performance of the Koryu's armament meant that if it penetrated a harbor it was guaranteed to sink at least one target, often two. The high speed and range of the torpedoes also made the Koryu an effective ambush unit, crawling around rocky coasts or shoals and surprising unsuspecting ships passing by. Military convoys operating around Yamaichi waters quickly learned to avoid such locations when possible, opting instead for slower, more circuitous routes.

SPECIFICATIONS

COUNTRY: Yamaichi
UNITS BUILT: 392
NO. OF CREW: 2
MAIN ARMAMENT: 2 x 450 mm torpedoes
SECONDARY ARMAMENT:
2 x 100 mm underwater harpoon cannons
SPEED: 10 kn

When the Koryu's design was uncovered, Alliance merchant marines had a lot of jokes about what wines go well with crustaceans. By the end of the war, most considered crabs and lobsters an ill omen best avoided.

GREAT WARSHIP MUSASHI

The *Musashi* was the second of three *Yamato*-class shipheads used by the Yamaichi Dominion. It was once an exact twin of the *Yamato*, with a lower body and legs that granted the ability to operate on land if needed. During one of her first engagements, however, the *Musashi's* lower half was severely damaged by Kyzerian torpedo boats during the battle of Mineo Passage (WC046). Desperate not to lose such a costly and effective asset, the Dominion managed to recover the *Musashi* after the battle and brought it back to the naval base in Hang-mina, where it was dry-docked for repairs.

Yamaichi engineers evaluated the extent of the damage done to *Musashi's* lower half as irreparable. They amputated the damaged parts, resealed the hull, and she was redeployed as a conventional, though still powerful, battleship. The rescued *Musashi* served the Dominion Navy all the way until WC048, when it was finally sunk during the last battle of the Durameshi Islands.

SPECIFICATIONS

COUNTRY: Yamaichi
NO. OF CREW: 2,447
ARMAMENT:
- 4 x triple 460 mm gun
- 3 x triple 155 mm guns
- 12 x twin 127 mm guns
- 100 x 25 mm anti-aircraft guns
- 8 x 13.2 mm anti-aircraft machine guns

ARMOR:
- 410 mm (belt)
- 200–226 mm (deck)
- 650 mm (turrets)
- 380–560 mm (barbettes)
- 500 mm (conning tower)

SPEED: 27 kn

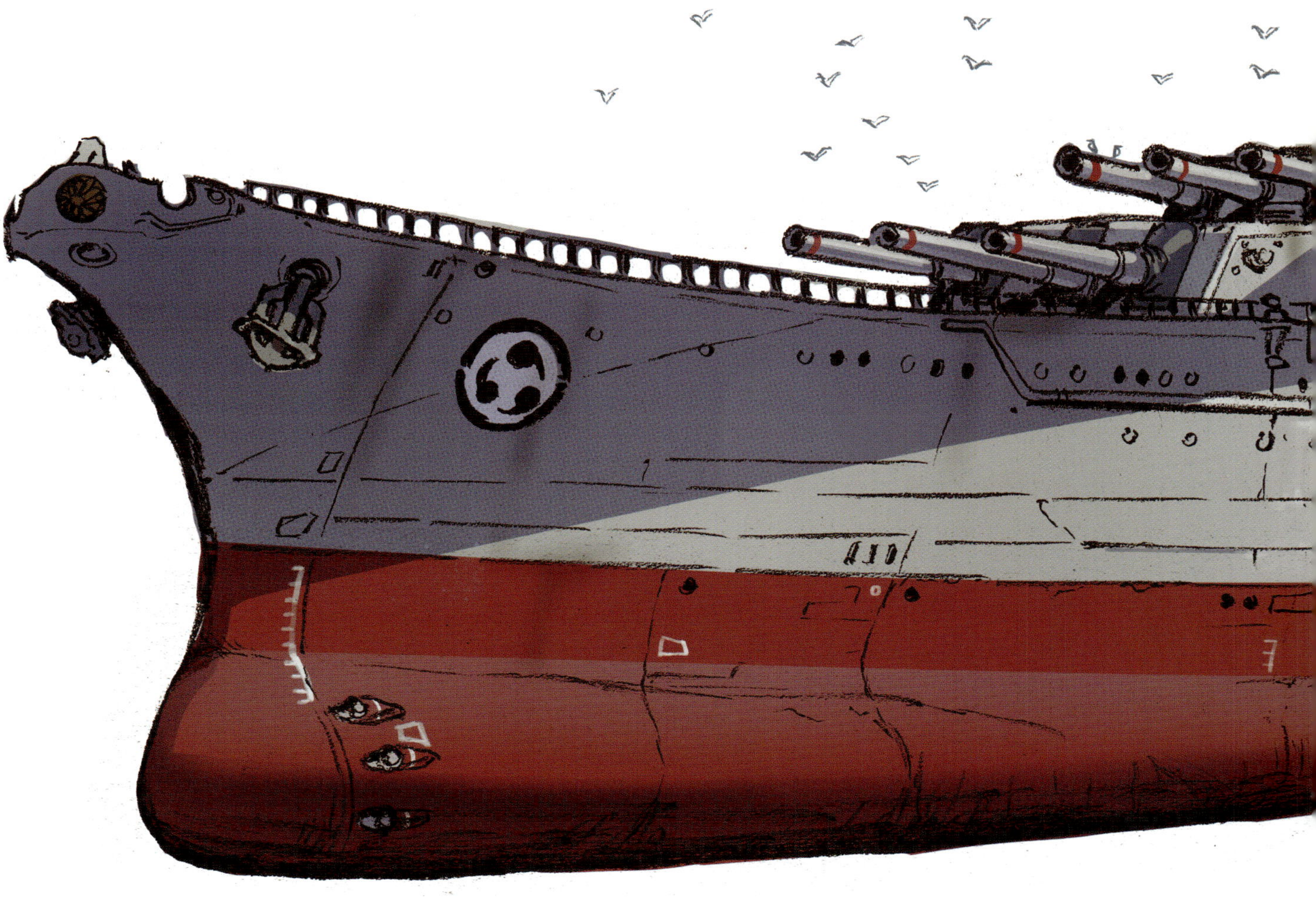

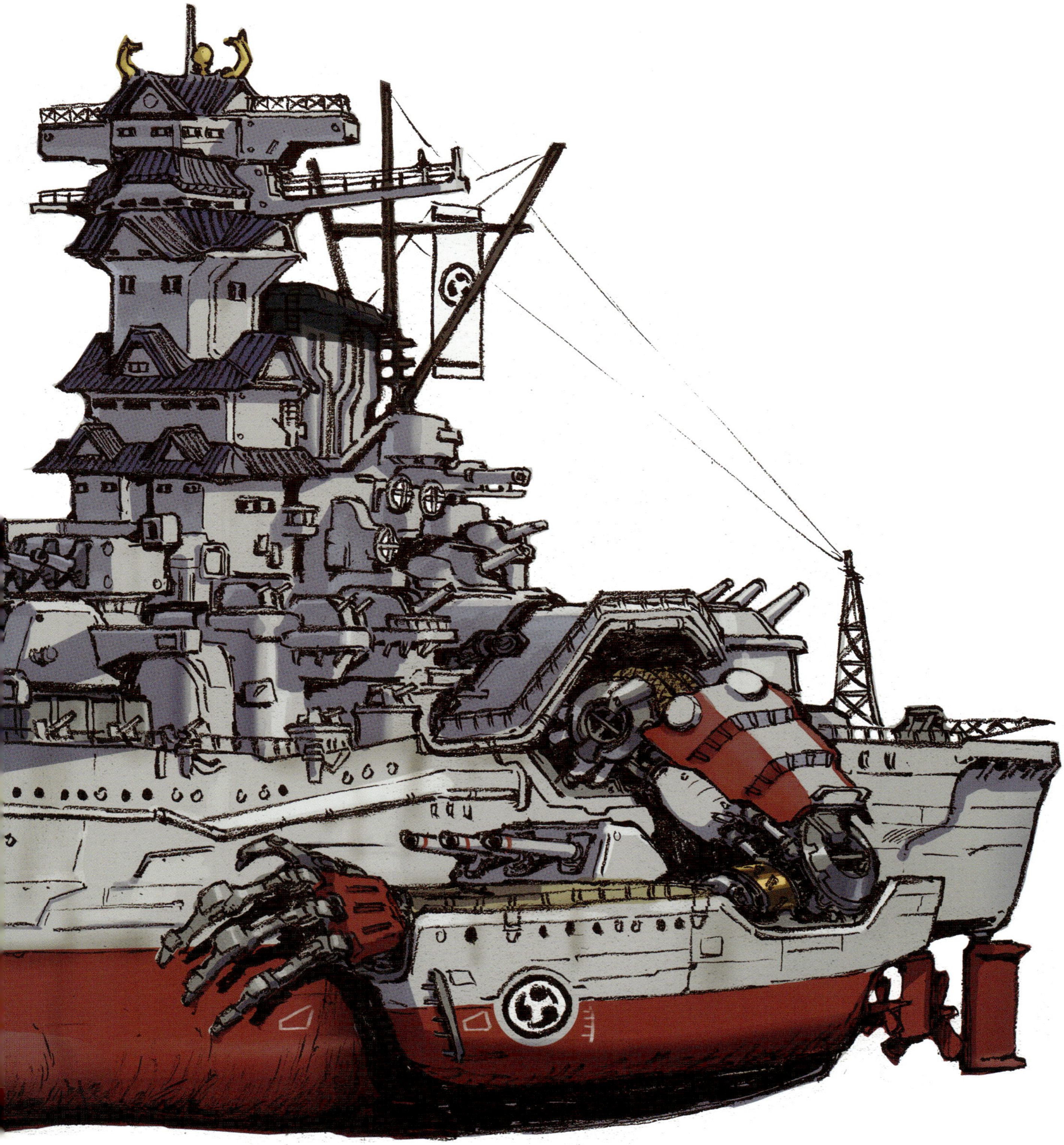

GREAT WARSHIP FUTSUSHI

While most scholars agree the *Futsushi* was an improvement over the *Yamato* in nearly every way, it never made as much of a name for itself. In fact, many images of it have been mislabeled as being of the more famous *Yamato*. In contrast to its predecessor, two of the main gun turrets were relocated from the hull to its shoulders, giving the turrets a significantly reduced arc of fire but a more favorably-balanced weight for bipedal movement. The increased stability also made the addition of larger turrets on the forearm-pontoons possible.

Additional improvements included lower-leg bows that reduced drag when wading and an articulating bridge, supposedly inspired by the Kyzerian *Bismarck*-class shipheads.

A handheld trident was forged for the *Futsushi* to better defend against the rare submarine crew daring enough to come within melee range. The three-pronged design meant that an inaccurate stab might still catch submarines on their flanks, and even a glancing blow from such a gigantic weapon would be enough to pierce the hull.

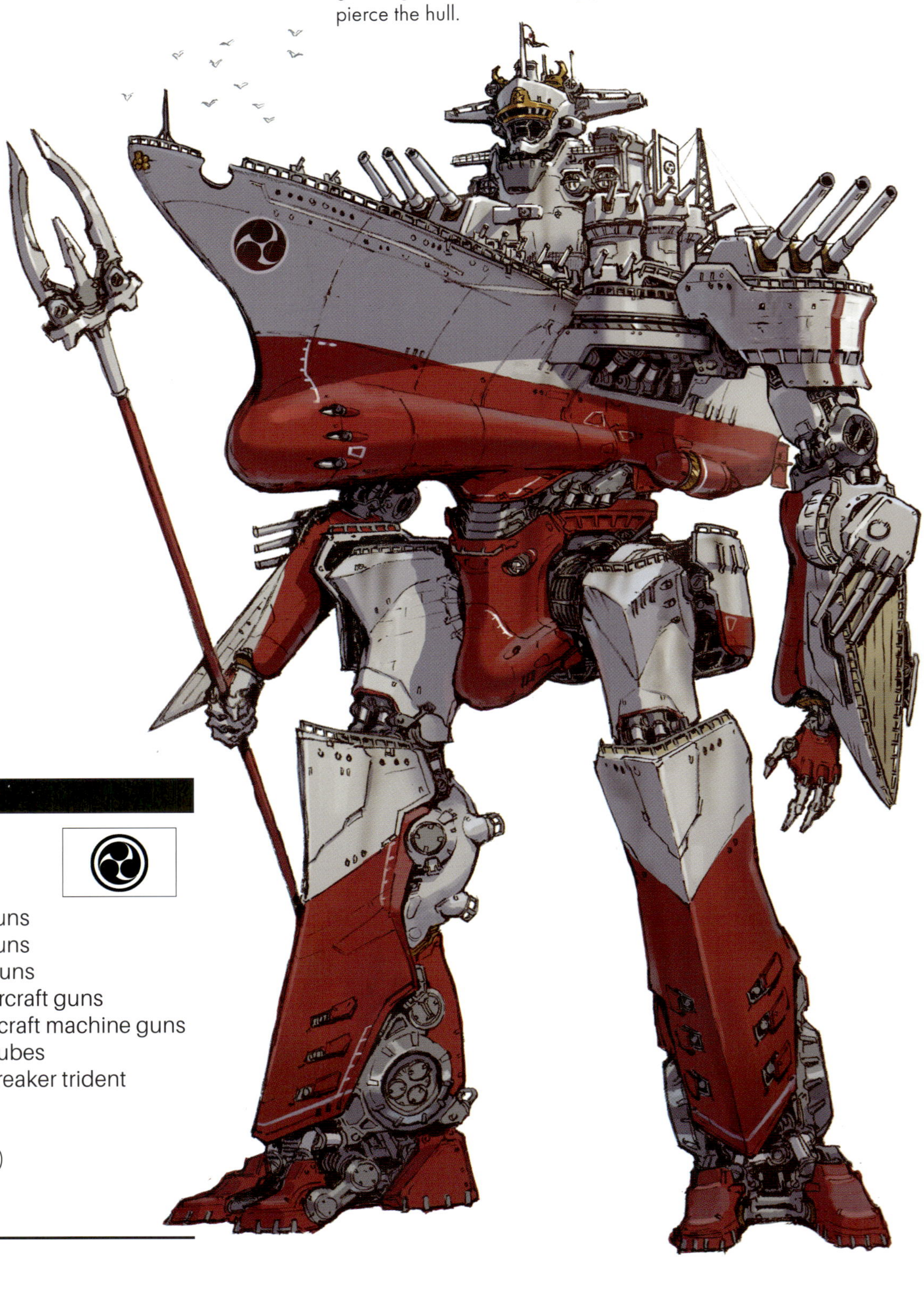

SPECIFICATIONS

COUNTRY: Yamaichi
NO. OF CREW: 3,233
ARMAMENT:
- 3 x triple 460 mm guns
- 2 x triple 155 mm guns
- 12 x twin 127 mm guns
- 100 x 25 mm anti-aircraft guns
- 8 x 13.2 mm anti-aircraft machine guns
- 18 x 24 in torpedo tubes
- 1 x handheld shipbreaker trident

ARMOR:
- 410 mm (belt)
- 200–226 mm (deck)
- 650 mm (turrets)

SPEED: 27 kn

GREAT CARRIER SHINANO

SPECIFICATIONS

COUNTRY: Yamaichi
NO. OF CREW: 2,400
JETHEADS CARRIED: 47
ARMAMENT:
- 20 x 127 mm dual-purpose guns
- 35 x triple 25 mm anti-aircraft guns
- 12 x 28-barrel 120 mm anti-aircraft rocket launchers

ARMOR:
- 160–400 mm (belt)
- 75 mm (flight deck)

SPEED: 28 kn

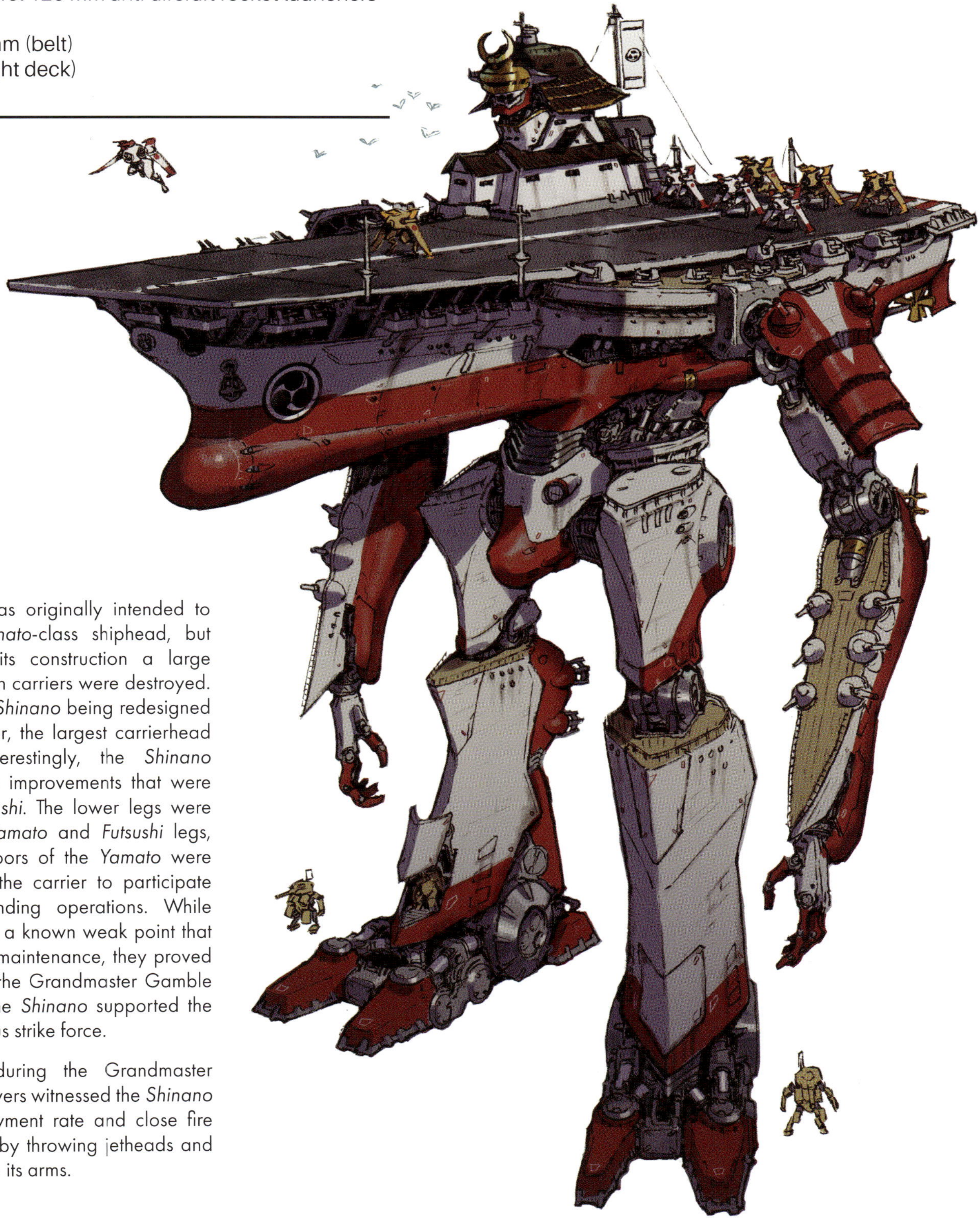

The *Shinano* was originally intended to be a regular *Yamato*-class shiphead, but partway through its construction a large number of Dominion carriers were destroyed. This resulted in the *Shinano* being redesigned as a jethead carrier, the largest carrierhead at that time. Interestingly, the *Shinano* incorporated many improvements that were made for the *Futsushi*. The lower legs were a hybrid of the *Yamato* and *Futsushi* legs, and the assault doors of the *Yamato* were retained to allow the carrier to participate in amphibious landing operations. While these hatches were a known weak point that required continual maintenance, they proved their worth during the Grandmaster Gamble (WC049), when the *Shinano* supported the southern amphibious strike force.

It was also during the Grandmaster Gamble that observers witnessed the *Shinano* augment its deployment rate and close fire support capability by throwing jetheads and massed bombs with its arms.

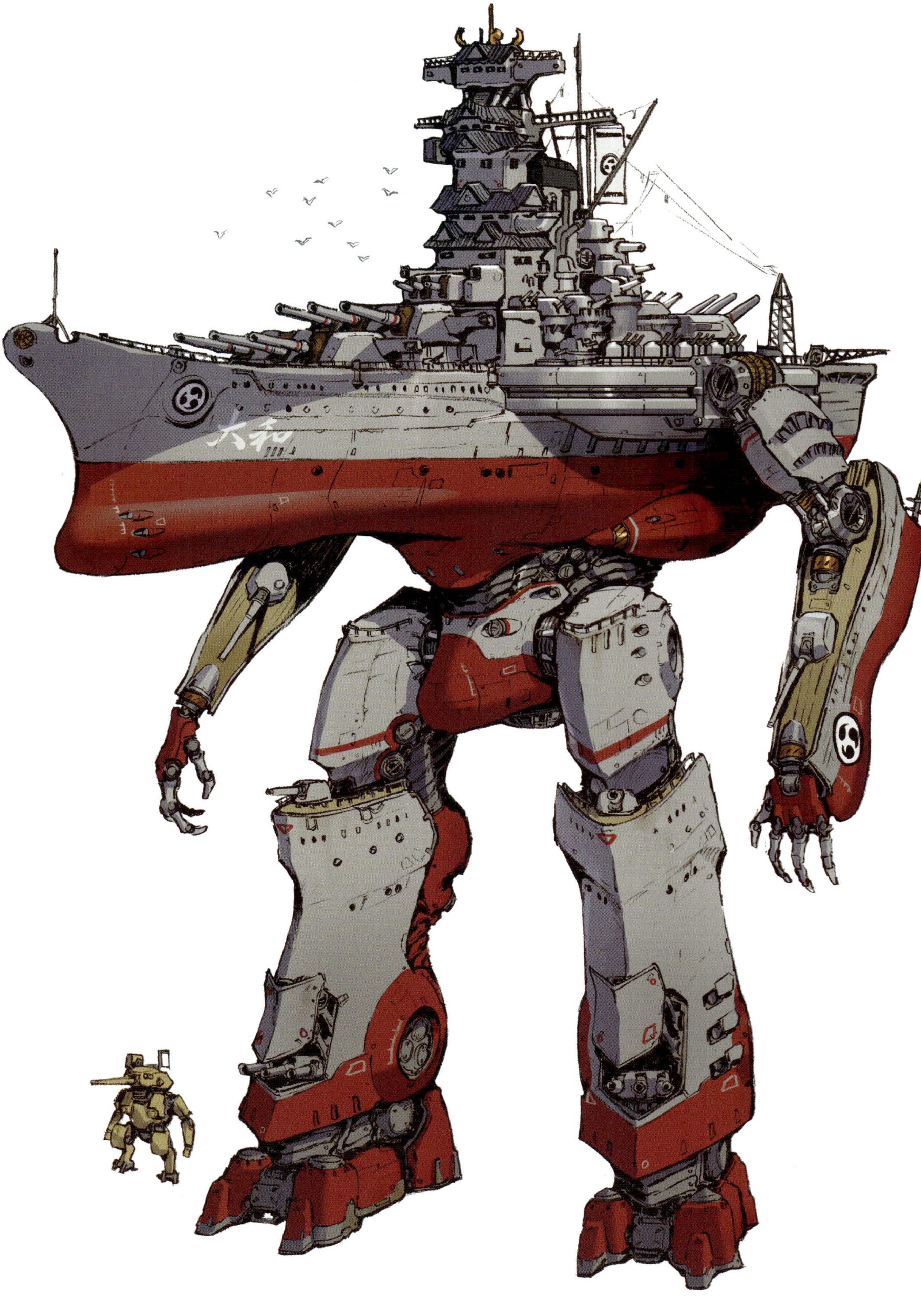
大和

GREAT WARSHIP YAMATO

Despite its fame and status as an icon, many details about its shiphead class and about the *Yamato* itself remained a mystery until fairly recently. For many years after the Neverwar, there were numerous questions surrounding these designs, as their capabilities and observed performance in the field were out of sync with what could be expected of a shiphead of the *Yamato*'s size and age.

The construction and details of the *Yamato* were kept a strict secret. Unlike the shipheads of the Empire or Alliance, which were often launched as great public events to demonstrate those nations' military might, the *Yamato* was launched in the summer of WC043 without any fanfare whatsoever. Even after it started patrolling the South Yamato Sea for which it was named, it remained a factor of uncertainty to the other nations for a long time, as they were unsure of the shiphead's capabilities beyond rough estimates of size.

It is not so hard to understand the confusion about the *Yamato* at the time, considering how little sense it made when its specifications were taken on their own. It was too large and too heavily armored to function with any widely known conventional diesel power plant; thus, the mere fact of its being able to move seemed to defy possibility. Explanatory theories thus abounded, some more outlandish than others. One of the theories involved the mysterious artifact unearthed by the Kyzerians known as the "Ijzerhart Engine," which some believed had been used to reverse-engineer a prototype generator for the *Yamato*. For a while, the Alliance believed that these generators were gifted to the Dominion by the Empire—a fantasy that entirely ignored the fact that the war between the Yamaichi and Kyzerians was as bitterly fought as that between any of the combatants by the time the *Yamato* was launched.

Modern revelations, however, have shown there to be a kernel of truth to this notion, after all. The Yamaichi had managed to capture not one but four partially-operational Ijzerhart Engines deep below the ruins in the Alentii mountains, located between the Empire and the Dominion. It is hard to be sure from the perspective of the present day, but we can assume that these mighty ancient engines were only operating at a fraction of their potential. In any case, unlike the Empire, which set about reverse-engineering their captured Isle à Von artifact engines in the hopes of constructing their own versions of it, the Dominion opted to design machines capable of optimally utilizing these tremendous sources of power directly. These top-secret hearts at the cores of the *Yamato*-class shipheads allowed them to be larger, more heavily armed and armored than any other shipheads in the war.

Even considering its tremendous weight and size, the *Yamato* was still more resilient than observations would have suggested. During its construction, a method was found to use voltaic technology to embed semi-ceramic plates into its steel armor. This method required tremendous power, but this was amply supplied by the artifact engine already embedded in the hull. This made the *Yamato*'s already exceptionally thick plating more durable against penetrating shots; the ceramic elements shattered and stopped shells in a manner not unlike modern bulletproof vests.

The *Yamato* and its sister ships were amazingly powerful, by themselves accounting for a very significant portion of Yamaichi naval assets. By dedicating the irreplaceable engines to these shipheads, the Dominion had essentially put all their eggs in these four baskets. While the *Yamato* was considered a monster that no one wanted to face, its commanders were equally hesitant to put her at risk. It was involved in a few smaller battles where it, along with its escort vessels, outgunned the enemy so decisively that no real risk was expected but, even then, the "Giant of the Yamato Sea" hung back to ensure it stayed out of enemy range most of the time. Despite its reticence to join the fray, however, its mere presence and the fearful event of its appearance on the horizon was often enough to ward off attacking fleets from both the Empire and Alliance. In light of this, Dominion naval command frequently moved the *Yamato* about the theater to make maximal use of this effect.

The *Yamato* would get its moment of glory during the Grandmaster Gamble. Despite their aforementioned reluctance to commit the ship, the Yamaichi fully understood the importance of the operation to assault the Godhead. In fact, in light of what is known today, they may have been the only ones who fully understood the magnitude of the threat, as only they knew the amazing potential that was beating in the heart of the *Yamato* itself. Whatever their reasoning, the shiphead was assigned a spearhead position for the fleet that advanced up Nehalean Channel, straight into New Kyzeria, to launch an airborne assault on the Godhead. The *Yamato* was hit by at least 8 torpedoes and 16 ship-killing bombs during the assault without a significant loss in its combat efficacy. It was only after shite had screened the Yamaichi carriers following it long enough for them to fully deploy their forces that a concentrated Kyzerian bombing run managed to destroy one of the *Yamato*'s legs. Sinking to its knees, the mighty titan began to flood and list into the deep waters of the channel before it went up in a mighty explosion, most likely due to a malfunction in the ancient and poorly-understood engine.

The smoke cloud that resulted from its destruction could be seen over 200 km away and was so awesome that some witnesses thought it had to mark the destruction of the Godhead rather than one of the assailants. The Godhead's actual demise was due in no insignificant part to the airborne force that the *Yamato* had protected. Had the Kyzerian jetheads focused their efforts on these assault forces rather than the *Yamato*, the battle might well have turned out differently. But the sheer size and firepower of the iconic shiphead inexorably drew Kyzeria's fire, just as it had always been designed to.

SPECIFICATIONS

COUNTRY: Yamaichi
NO. OF CREW: 3,233
ARMAMENT:
- 4 x triple 460 mm guns
- 2 x 200 mm guns
- 2 x 200 mm short guns
- 4 x triple 155 mm guns
- 12 x twin 127 mm guns
- 100 x 25 mm anti-aircraft guns
- 8 x 13.2 mm anti-aircraft machine guns
- 18 x 24 in torpedo tubes

ARMOR:
- 410 mm (belt)
- 200–226 mm (deck)
- 650 mm (turrets)

SPEED: 27 kn

GREATER KYZERIAN EMPIRE

AFFILIATED NATIONS

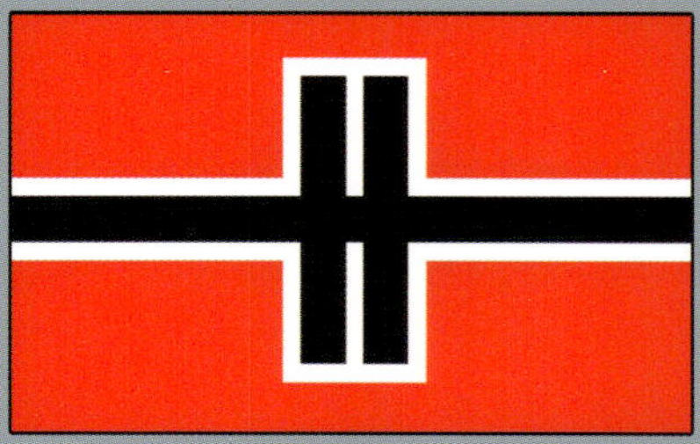

New Kyzeria

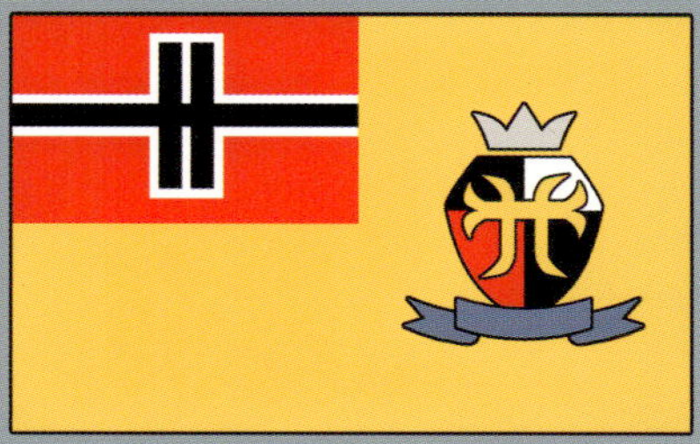

East Rozenia

SD.KFZ. 101

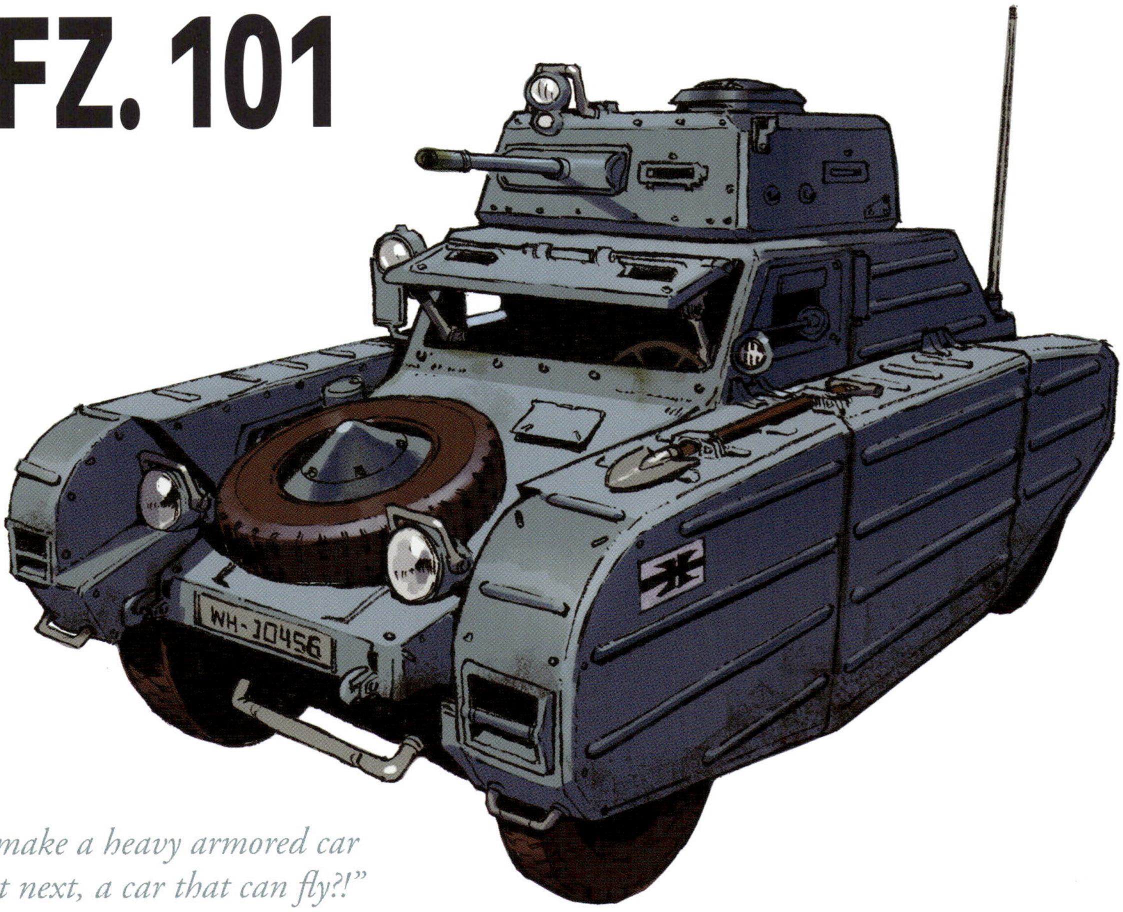

"You want me to make a heavy armored car that floats? What next, a car that can fly?!"
- Johanz Mensenman, Head of the Kyzerian Army Developmental Branch

The Greater Kyzerian Empire was not always the military behemoth that shaped the War Century. Before tankheads and even before tanks, Kyzerian forces were heavily reliant on traditional infantry and cavalry. They lagged behind other nations in their adoption of modernized weapons such as machine guns and armored cars. Before the meteoric rise of the Kriegsdiener religion that led to the development of the fearsome tankhead, there was only a small faction within Kyzerian High Command that realized the importance of modernizing their army.

Those persistent innovators understood the brilliance of the Reginwald LT17, an Alliance tank which had recently been invented. They acknowledged the swingarm suspension and, even moreso, the rotating turret as game changers. Lacking a vehicle development branch of their own, the task of developing a Kyzerian counterpart to the LT17 was awarded as a prestigious contract to Johanz Mensenman. Mensenman had achieved some success building civilian cars and was, by all accounts, eager to apply his knowledge and experience to military applications. Not long after winning the contract, Mensenman's automobile company was rolled into the newly formed Kyzerian Army Developmental Branch.

While this takeover by the army gave Mensenman a military rank—almost as highly regarded as that of a noble in prewar Kyzerian society—it also led to numerous problems. Now working directly under the military instead of as outside contractors, the development team found themselves victims of contradicting demands and frequent budget cuts.

The Sd.Kfz. 101 was originally envisioned to be a light armored car with a small rotating turret armed with a machine gun or anti-tank gun, ideally packaged with a Kyzerian version of the LT17's swingarm suspension system. It was later decided by senior members of Army Command, however, that the vehicle must be amphibious. To Mensenman's credit, he met the demand, if only in calm water and over short distances, but it was necessary to discard the swingarm suspension. Hence, a more conventional fixed axle was used for the chassis.

During an early demonstration, the Army Command senior members observed that the turret was taller and more noticeable compared to the rest of the vehicle, so they demanded that its armor be increased. When they made the vehicle amphibious, Mensenman's engineers had to cut weight wherever possible to prevent the vehicle from sinking, but they were not allowed to remove armor from the heavy turret. This led to the vehicle overturning on multiple occasions, ultimately resulting in the whole project getting scrapped.

SPECIFICATIONS

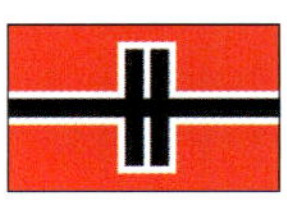

COUNTRY: New Kyzeria
UNITS BUILT: 4
NO. OF CREW: 3
MAIN ARMAMENT: 1 x 37 mm gun
ARMOR: 3-6 mm
MAX SPEED: 50 km/h

LEICHTTRAKTOR

The Leichttraktor was a tank that was utilized heavily by the Kyzerian army during the early years of the Neverwar. It was the Leichttraktors that spearheaded the Nolburg Plan of WC005, in which the Kyzerians tried to circumvent the southern trenches by going through neutral Rozenia, an event Rozenians now call Burinoch, the Night of Storms.

During the operation, Kyzerian forces encountered the Alliance's Reginwald LT17s. The latter were present in western Rozenia under the pretense of combined exercises with the Rozenian military. In reality, they were stationed there for the exact purpose of cutting off any Kyzerian attempts to circumvent the Southern Front through Rozenia. The Britonican forces were vastly outnumbered by the better-armored Kyzerian tanks and were on equal footing in terms of firepower, but the LT17 could strike from angles that nullified these advantages. The swing arms meant that the LT17 could fire reliably on the move and the rotating turret allowed the tank to fire at any angle while moving obliquely to the enemy. This made the LT17 much harder to hit, especially since the casemate on top of the Leichttraktor was not an actual turret, and the gun only had a limited 20-degree arc of fire to both sides.

The Brits present in Rozenia knew they could not stop, let alone defeat, the force with which they were faced. Even so, they continually harassed the enemy while making a fighting retreat. This action forced the Kyzerians to carefully check every hill, forest or shed capable of hiding a light tank for potential ambushers, and slowed their advances to a snail's pace.

For all the above reasons and many others, the Nolburg Plan was mostly a failure. Some Kyzerian forces managed to break through and wreak havoc on Brit territories, but this was only a minor disturbance and not the devastating coup de grâce that General Hans Nolburg had envisioned. Despite catching the Brit forces by surprise, the Leichttraktor and the other forces of New Kyzeria ultimately failed to push the attack home.

By mid-war the Leichttraktor became obsolete, and for most of the Neverwar they were reassigned to tank crew training roles, well away from actual combat.

SPECIFICATIONS

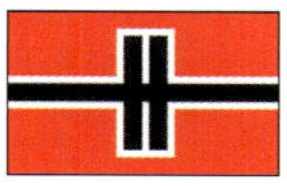

COUNTRY: New Kyzeria
UNITS BUILT: 356
NO. OF CREW: 4
MAIN ARMAMENT: 1 x 37 mm gun
SECONDARY ARMAMENT:
1 x 7.92 mm machine gun
ARMOR: 5–19 mm
MAX SPEED: 30 km/h

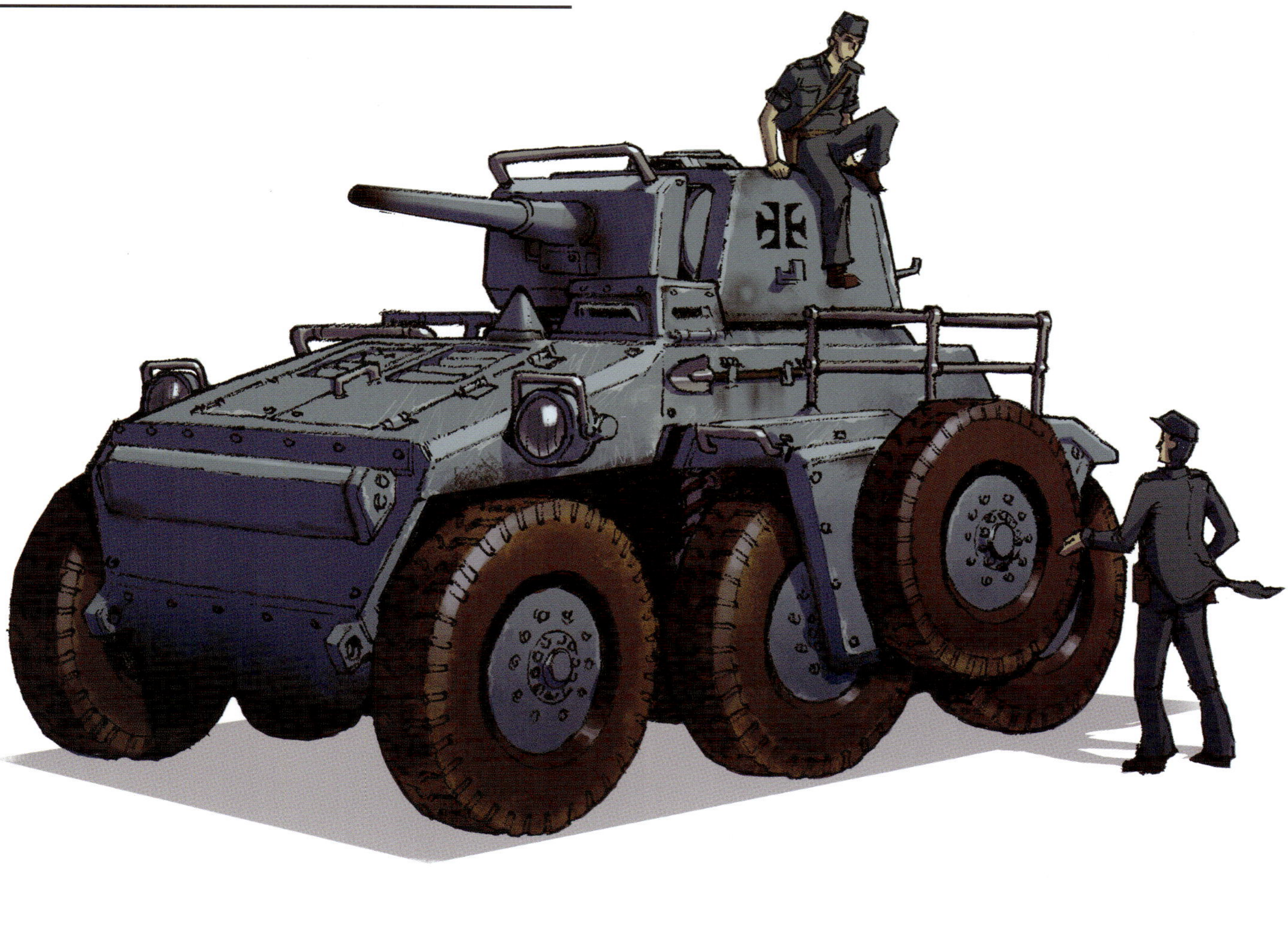

SD.KFZ. 132

The Sd.Kfz. 132 was originally based on a licensed Britonican design and mainly saw use very early in the Neverwar. Dedicated command vehicles like the 132 were generally unpopular with Kyzerian field commanders, who preferred to lead from much more prestigious tankheads.

SPECIFICATIONS

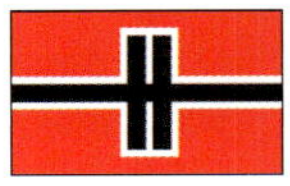

COUNTRY: New Kyzeria
UNITS BUILT: 255
NO. OF CREW: 4
MAIN ARMAMENT: none
SECONDARY ARMAMENT: none
ARMOR: 6-8 mm
MAX SPEED: 60 km/h

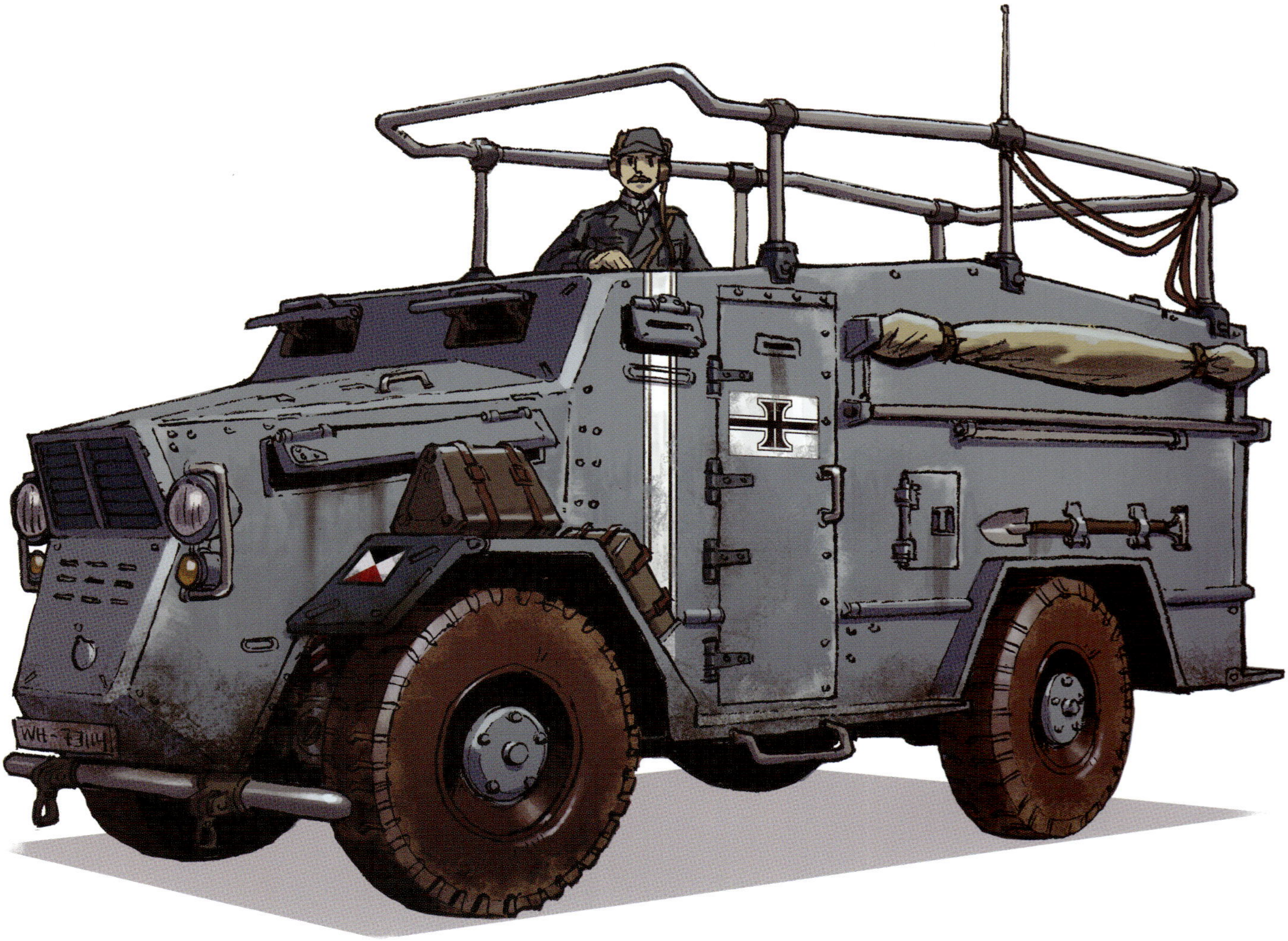

"I am the oldest surviving battalion commander because I actually used the 132 and not a giant walking target."
- Kyzerian Heavy Tankhead Battalion Commander Hulst Igelstein

SD.KFZ. 221

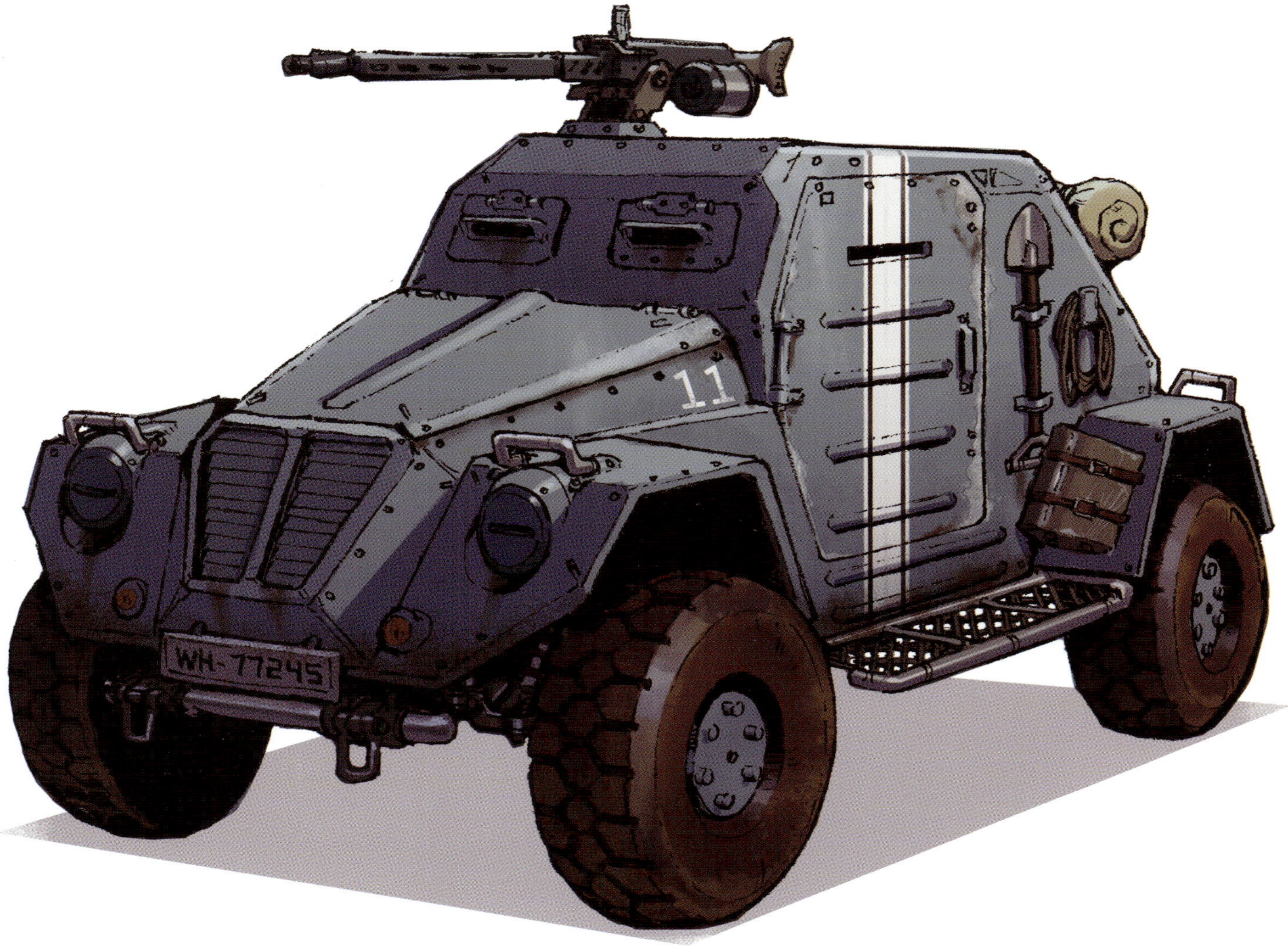

While not glamorous, it was vehicles like the Sd.Kfz. 221 that provided the Kyzerians with superior mobility early in the Neverwar. The 221 had a long legacy as a popular civilian off-road vehicle during and after the war.

SPECIFICATIONS

COUNTRY: New Kyzeria
UNITS BUILT: 7,291
NO. OF CREW: 3
MAIN ARMAMENT: 1 x 7.92 mm machine gun
SECONDARY ARMAMENT: none
ARMOR: 6-8 mm
MAX SPEED: 110 km/h

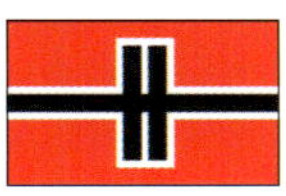

Because of its popularity on the civilian market, the 221 is technically the most produced Neverwar vehicle by a large margin.

SD.KFZ. 231

The Sd.Kfz. 231, archetypal for prewar Kyzerian armored units, saw its last hurrah during the Nolburg Plan of WC005. Their rugged suspension allowed them to sweep across the Rozenian countryside with ease. Unfortunately, their weapons and armor were already outdated even before the start of the war.

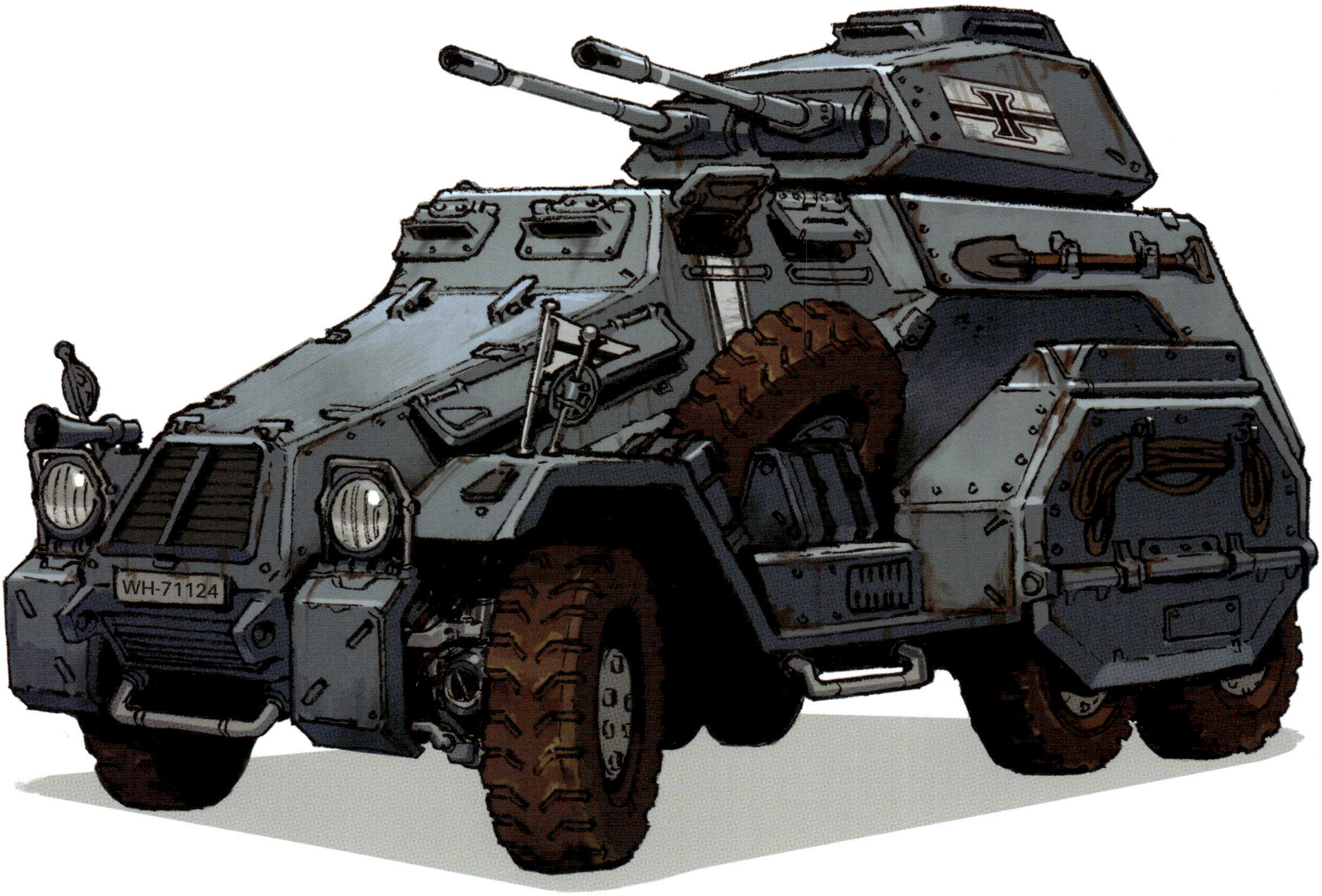

The Rozenian roads were so muddy during the Nolburg Operation that Brit forces originally reported that the Kyzerian cars and tanks were painted brown.

SPECIFICATIONS

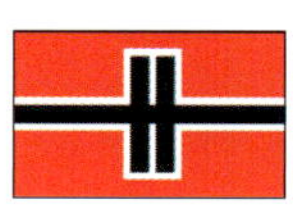

COUNTRY: New Kyzeria
UNITS BUILT: 10,546
NO. OF CREW: 4
MAIN ARMAMENT: 2 x 37 mm guns
SECONDARY ARMAMENT: none
ARMOR: 6–12 mm
MAX SPEED: 85 km/h

SD.KFZ. 234

Early in the Neverwar, the continued value of fast yet hard-hitting reconnaissance vehicles was, for Kyzerian command, reaffirmed by lightning strikes performed by the aging Sd.Kfz. 231 armored car. The Sd.Kfz. 234 built on the strengths of the 231 and eliminated the shortcomings of that earlier model. An additional set of wheels reduced the chance of getting bogged down in soft terrain like Rozenian mud. Modernized armor increased survivability over its predecessor. Originally designed with a 20 mm rapid-fire cannon like the 231, the 234 was upgraded with a much more powerful 50 mm gyrojet cannon in order to deal with a new threat on the Rozenian front: T-34 Gurko tankheads. While much more powerful than the original weapon, records show only mediocre performance against the 50 mm's intended targets.

SPECIFICATIONS

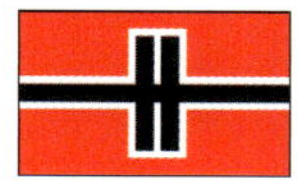

COUNTRY: New Kyzeria
UNITS BUILT: 3,065
NO. OF CREW: 4
MAIN ARMAMENT: 1 x 50 mm gun
SECONDARY ARMAMENT:
1 x 7.92 mm heavy machine gun
ARMOR: 9–30 mm
MAX SPEED: 90 km/h

While it never performed exceptionally during the Neverwar, mounting a 50 mm gun on a reconnaissance vehicle suggests a very particular style of reconnaissance indeed.

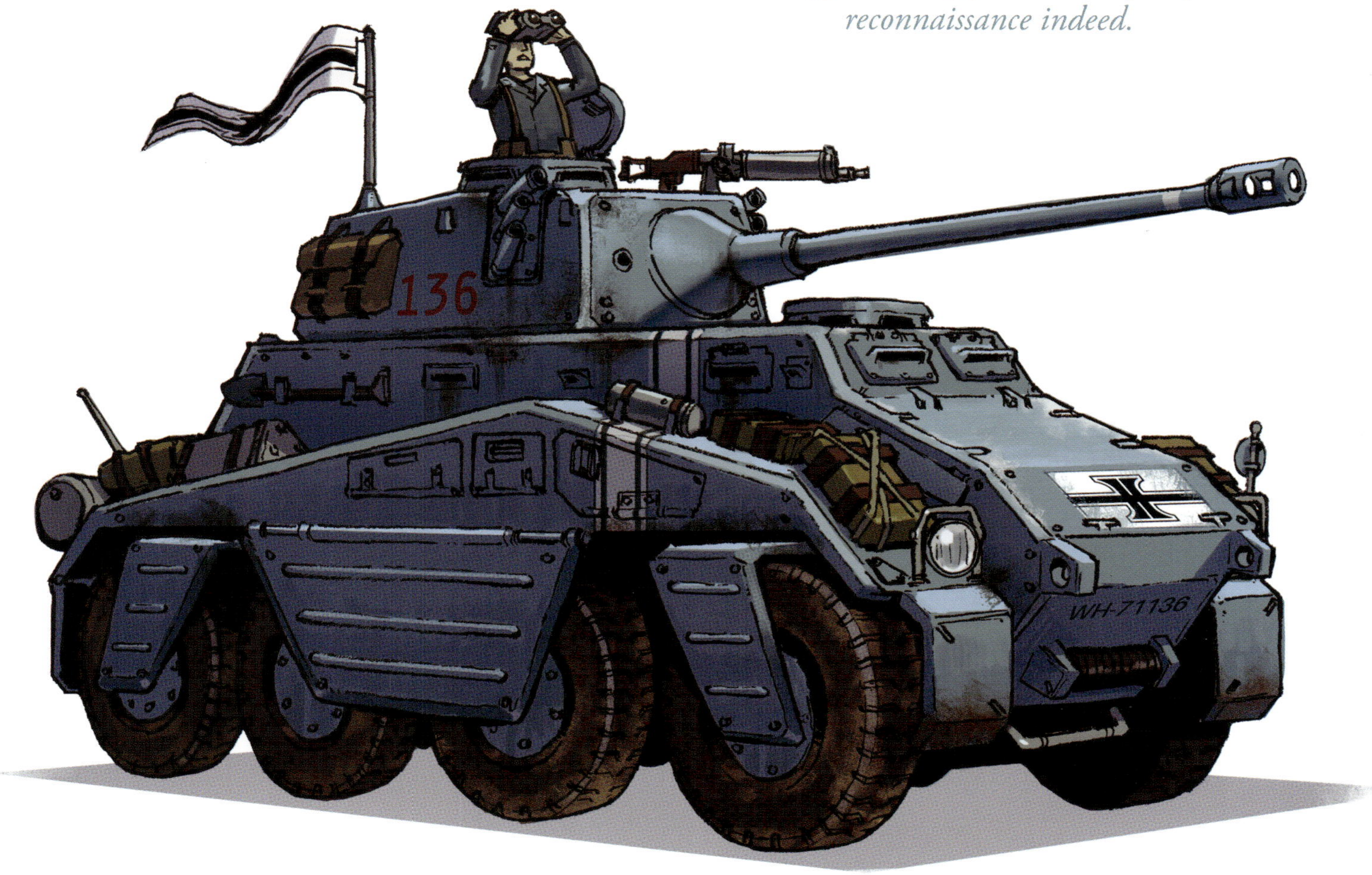

KANONENJAGDPANZERWAGEN

Before the Neverwar, most tanks had fixed weaponry, but these became obsolete when turreted weapons proved their superior flexibility. Only the largest guns were mounted on hulls, mostly for second-line long-range units like tank destroyers—light armored vehicles with powerful field guns that attacked from ambush. Their attacks were devastating but the tank destroyers were, ironically, also easily destroyed. The Kanonenjagdpanzerwagen was a result of the Kyzerian army's effort to produce a heavy tank destroyer, one capable of standing its ground against vehicles in its own weight class.

The Kanonenjagdpanzerwagen, or KanJPzw for short, proved a successful and effective unit for New Kyzeria when it launched. The long-range cannon was remarkably accurate and the six-wheeled hull could achieve acceptable speed on rough terrain as well as on good ground.

Early in the war, the KanJPzw earned a fearsome reputation as a supernaturally accurate weapon as it often hit the notoriously fast Alliance tanks, considered nearly impossible with a manually aimed cannon. This all changed when the Alliance tankheads were introduced and quickly improved in mobility. Ultimately, the KanJPzw's outstanding features, such as the gun mantlet mechanism and its unusual aiming process, also proved to be its main failing.

The Kanonenjagdpanzerwagen's main gun relied upon a complex pneumatic mechanism intended to lead a moving target. This mechanism required careful estimation of the target's speed and direction which was then set on the mechanical compensator. The compensator was, essentially, an early mechanical computer capable of performing trigonometry from simple inputs. Any changes in the weapon's targeting—even simple adjustments to elevation, for instance—required the gunner to enter the full coordinate adjustment for each movement of the weapon, as the system would need to be reset for every change in targeting data. The more erratic and dynamic gaits of tankheads proved to be the bane of both the compensator, which was unable to adjust quickly enough to unpredictable movement patterns, and to the KanJPzw itself, as it lacked a more conventional backup system. There were reports of tankheads zig-zagging up to a Kanonenjagdpanzer and disabling the latter at point-blank range at leisure. These stories demonstrated the radical change in how the Kanonenjagdpanzerwagen went from feared deadeye sharpshooter to a laughingstock within a year. Despite these stories, it still served as an effective anti-vehicle machine throughout the war.

SPECIFICATIONS

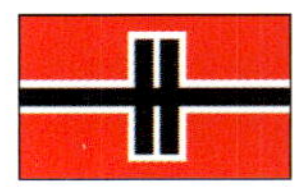

COUNTRY: New Kyzeria
UNITS BUILT: 1,470
NO. OF CREW: 4
MAIN ARMAMENT: 1 x 90 mm gun
SECONDARY ARMAMENT:
2 x 7.62 mm machine guns
ARMOR: 10–50 mm
MAX SPEED: 70 km/h

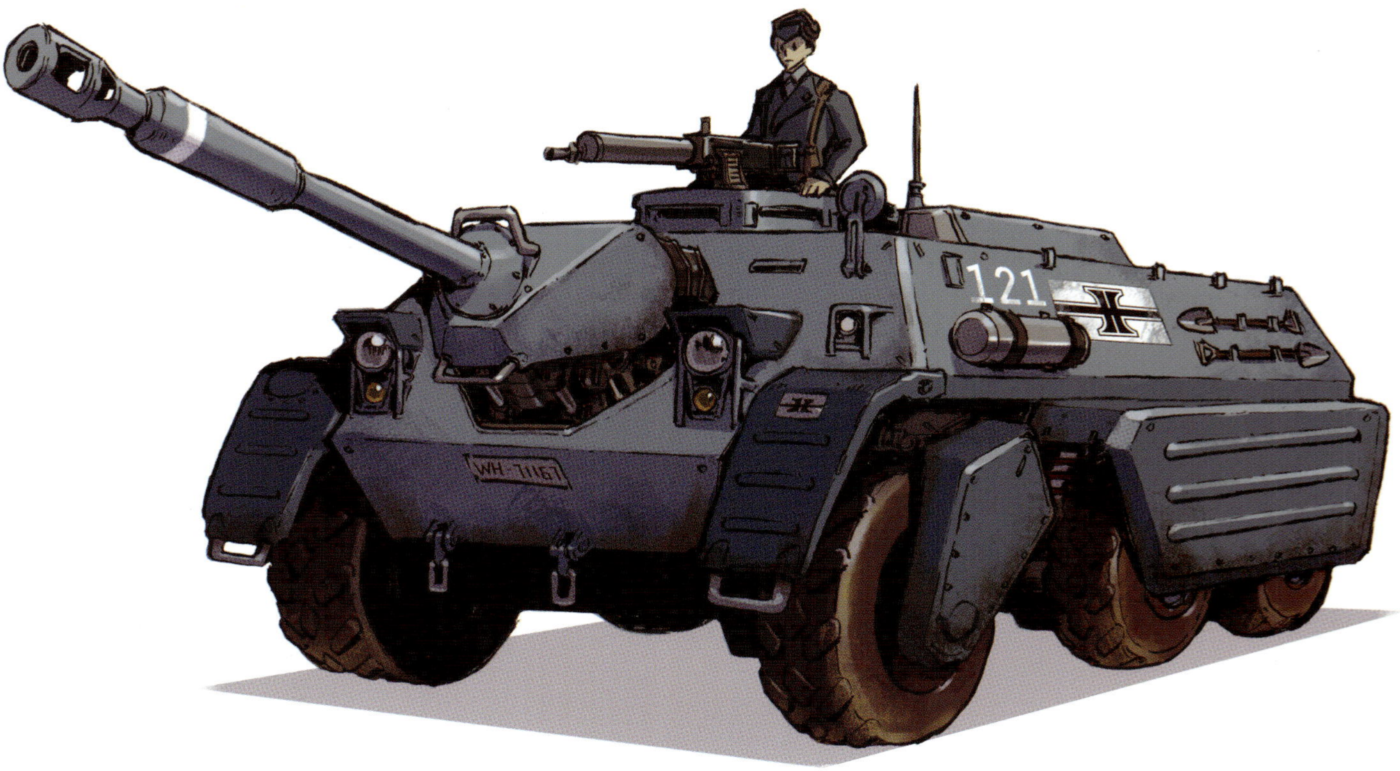

MARDER II

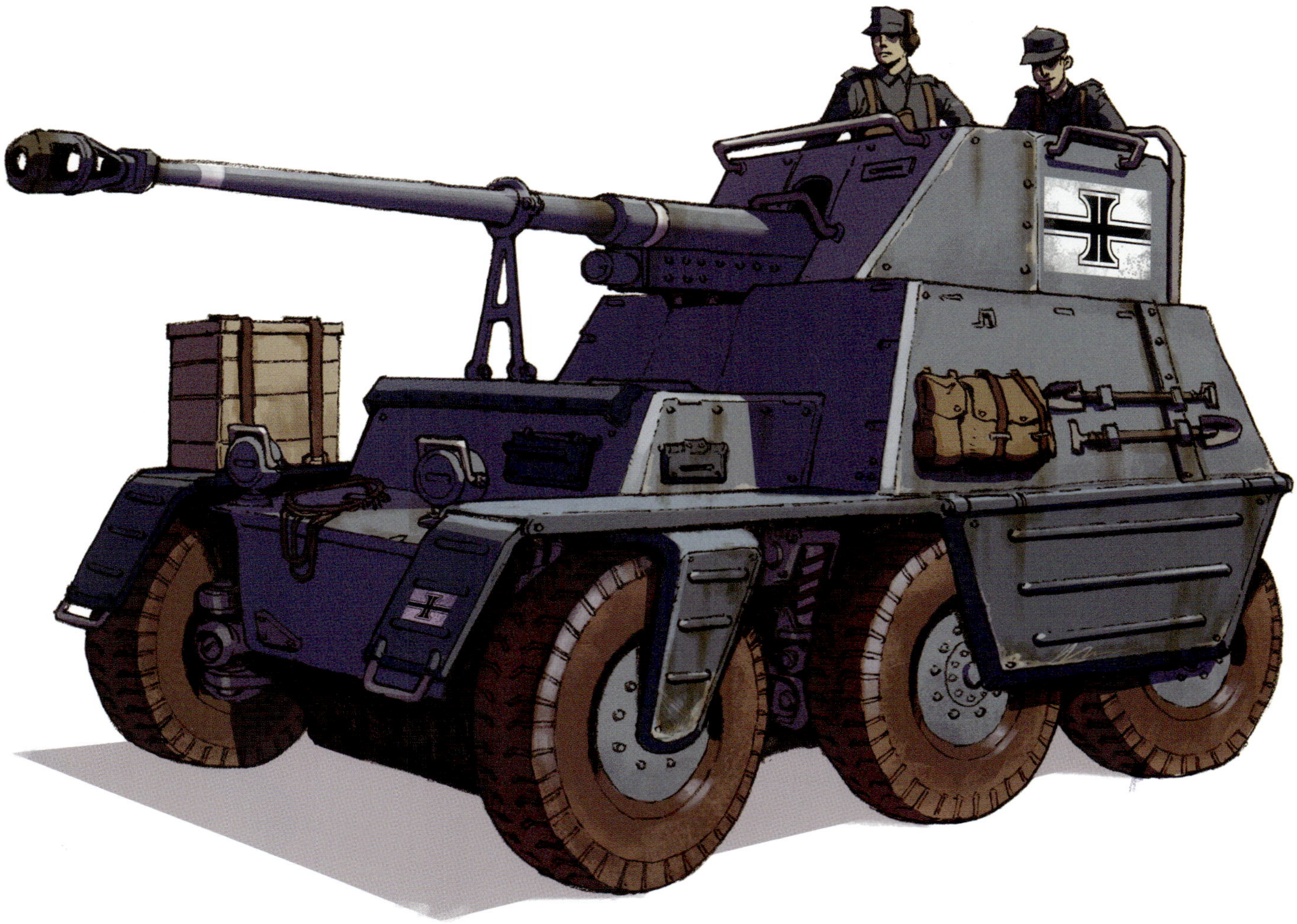

When the Yamaichi Dominion deployed their superior armored tankheads against the Empire, the Marder II was used to quickly fill in a stopgap anti-tankhead role. A significant number of these vehicles were equipped with captured Alliance cannons to speed up production.

The gun lock in the front of the vehicle was necessary for travel, but required dismounting to remove. The job of undoing it during combat quickly became relegated to the newest crew member.

SPECIFICATIONS

COUNTRY: New Kyzeria
UNITS BUILT: 1,280 plus an estimated 327 with captured guns
NO. OF CREW: 4
MAIN ARMAMENT: 1 x 75 mm gun
SECONDARY ARMAMENT: none
ARMOR: 15–35 mm
MAX SPEED: 40 km/h

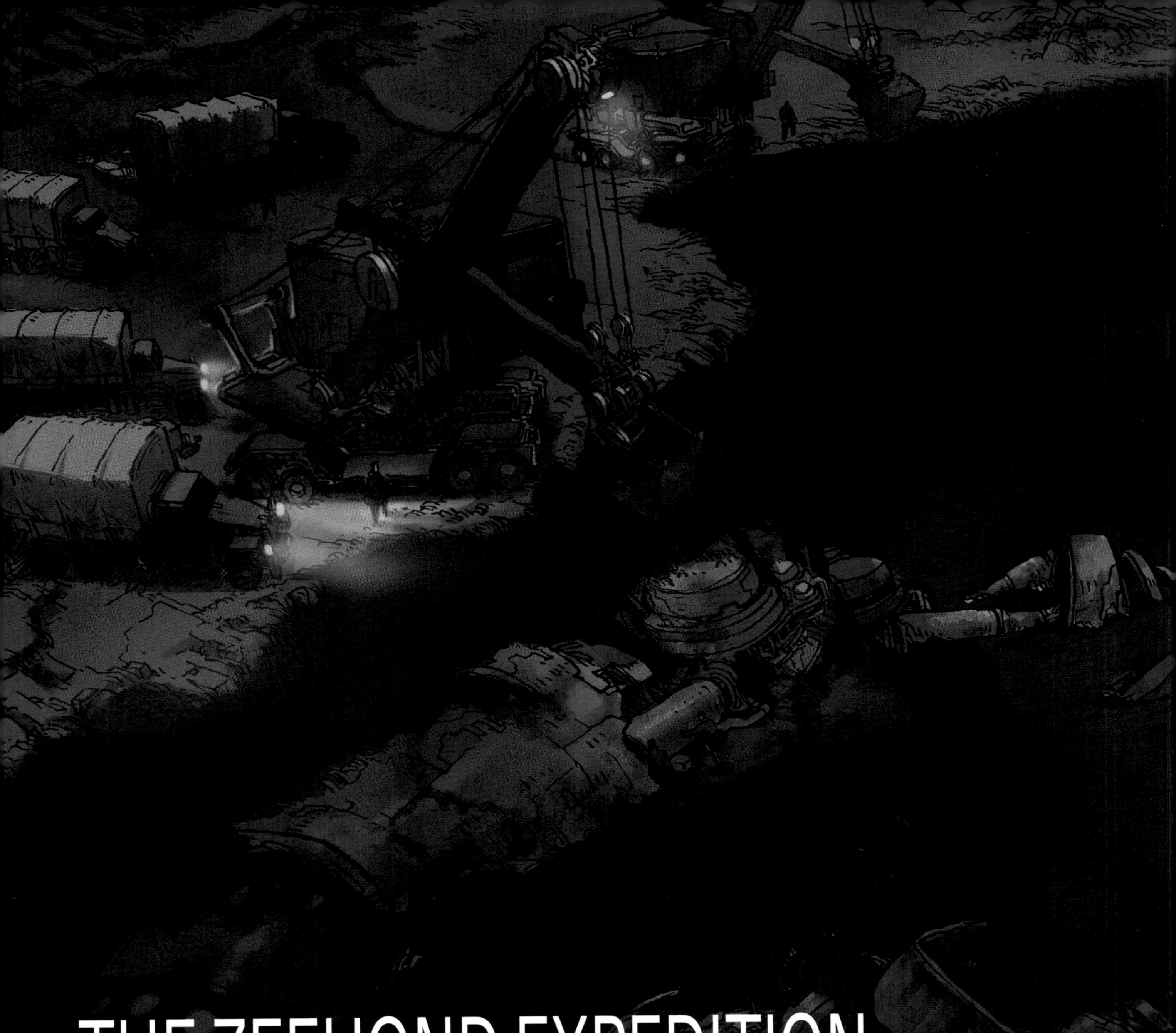

THE ZEEHOND EXPEDITION

Many details about the WC009 Zeehond Expedition to Isle à Von remain a mystery. What is known with certainty is that the artifacts uncovered on the Britonican island would lead to the invention of tankheads and change the face of the Neverwar. However, the chain of events that led up to this tremendous discovery remain murky. Today, we know for a fact that it was not, unfortunately, officially a military operation, or else there would be more archival paperwork available on the subject. The entire enterprise was instead spearheaded by the Kriegsdiener cult, although military equipment and personnel were provided for the undertaking by the Kyzerian leader, Kaiser Walmir VII himself. Unusually for the time, but foreshadowing future special force operations, most identifying marks and insignia were done away with by the expedition personnel. This was a markedly uncharacteristic action for the prideful character of the Kyzerian military at the time.

The scholarly community are divided as to the goal the Kriegsdiener had in mind with the expedition. Though most agree that manifestation of the Godhead was obviously a welcome possibility for the cult, opinions differ as to whether or not they actually expected that they could achieve that goal. They may have had some knowledge or clue that materials related to the Godhead were on the island. They may just as easily, certain historians have pointed out, have been searching for any kind of advantage over the enemy nations. On the other hand once more, they may have known that certain technology and secrets that would eventually lead to the Godhead were buried on Isle à Von, in which case the expedition must be considered the initial steps in a very long and well executed plan which culminated only decades later. The most cynical commentators surmise that the cult had no real evidence prior to the expedition, but that its success was mere dumb luck. The problem with that view is that it must also assume the irrationality of the usually highly rational Kaiser and Kyzerian military, which contributed critically to the undertaking. Less cynical but equally dark is the notion that the cult was fully aware that this expedition would provoke military response, and that they welcomed this conflict as the next step in their envisioned century of war.

Many within the Kyzerian military felt deeply uneasy with the influence the religion was claiming on the leader of their state and

CONTINUED>>

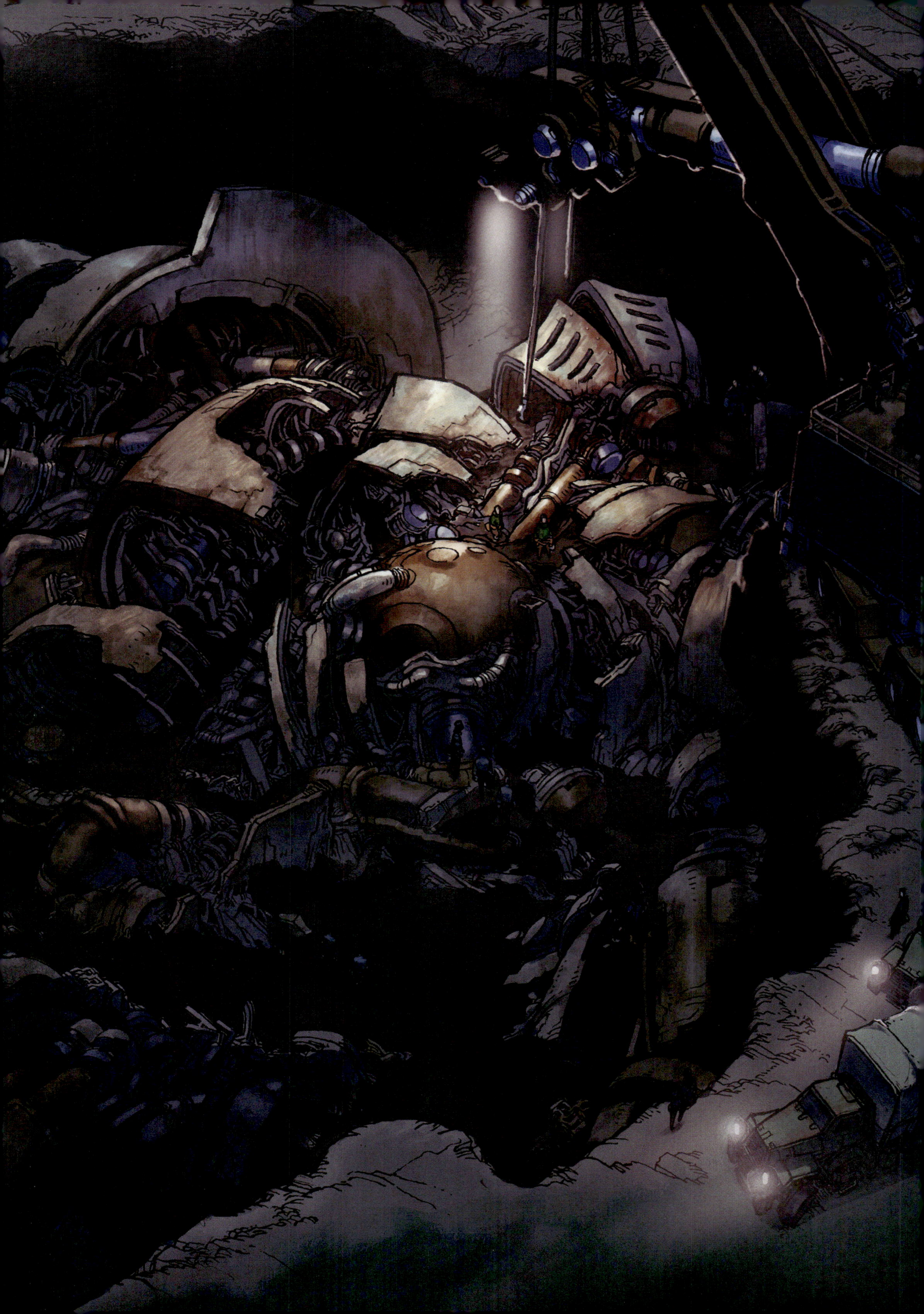

THE ZEEHOND EXPEDITION (CONTINUED)

they attempted to avoid assignment of their troops and equipment to this undertaking. In the end, it was on the authority of the Kaiser as a sovereign rather than as a military commander that troops were asked to take part, a summons very few Kyzerians would deny. Due to this resistance, and in order to obfuscate the nature of their group, most of the equipment was civilian rather than military. The boats used for the crossing were of characteristic Rozenian design and coloring to waylay suspicions, as commercial Rozenian craft were not an uncommon sight on the northern shores of Grand Britony.

Mist and rain made the approach to the island a treacherous affair. The south Tramonic Ocean was never the easiest to navigate, even at the best of times, though the conditions did aid the expedition by hiding the vessels. Unloading their heavy digging equipment on the steep rocky shores of the island must have been a daunting task and, indeed, wrecks of steam shovels and trucks were found in the area, almost certainly left behind by the expedition. It is unclear how many days the expedition took, as the exact day of the landing is unknown.

Despite the efforts toward secrecy, the large force of workmen and armed guards did not go unnoticed. However, the local population seemed to have mistaken it for an operation of mainland industrialists or the Britonican military. The Alliance navy and army had, on multiple prior occasions, held training exercises on or around the island without informing the locals of their plans. The peculiar accent spoken in the area made communication with mainlanders complicated, as well, so a Kyzerian accent might not have stood out to the few locals that confronted the operation about their doings. Complaints were made to the local garrison but these were dismissed as the colloquial grumblings of ignorant shepherds. Had these complaints been taken more seriously, the world as we know it today might be very different indeed.

The Kyzerians indeed discovered the remnants of a prehistoric weapon, utilizing technologies far beyond anything known at the time. It seems that, while the artifacts were treated with great care, the expedition was mostly interested in returning them to New Kyzeria with all haste, and took little time to record their positions and stratification the way archeologists would today. Indeed, the recklessness with which the materials at Isle à Von were dug up and then left to decay, exposed to the elements, is perhaps the greatest act of archeological vandalism performed on Britonican ground in history. How much might have been learned from this excavation, had it been executed with due diligence and care, will never be known, but if the testimony of this writer may bear witness: it is enough to make even a stoic historian weep.

Having so barbarically secured their loot, the Kyzerian expedition and cult members rushed to return home, but the weather had turned against them, causing the expedition to abandon their plans of returning by the northern route. It was while the expedition was making its alternate way by the Southern Briton Strait that history again could have taken a very different turn. The convoy was challenged by Britonican coast guard vessels when they approached the eastern mouth of the strait. The expedition had, inexplicably, not prepared a cover identity for this secondary route. It had been expected that only incoming vessels would be checked, not outgoing ones, and so the signal was given to all three vessels to go full steam ahead in an attempt to outrun the coast guard. While neither armed nor armored, the trio of ships had been chosen for their sturdiness and speed. The coast guard vessels were unable to stop the Kyzerians, so a navy torpedo ship on picket duty was signaled. After its hails were ignored, the torpedo ship fired on the lead Kyzerian vessel. The torpedo missed or failed to detonate, depending on reports, but in either case no effect on the target was observed. Another torpedo was fired at the second vessel, scoring a hit, and this ship, taking on water, was forced to divert to the north Alcaban shores. It sank within sight of the coast. Had the first torpedo functioned properly and hit the lead ship, the artifacts would have sunk to the bottom of the Strait. In the event, with luck and their superior speed, the two remaining ships managed to make it to open waters where they rendezvoused with a trio of Kyzerian submarines and were escorted back to safe harbor.

The expedition's troubles would not end there, however. A dozen members fell ill even before the research on the artifacts began. Within a month of their return, all of the expedition's personnel had become catatonic or succumbed to the mysterious illness altogether. Rumors of a curse circulated among those who knew about the expedition. Soon, Kyzerian newspaper reports on a strange illness supposedly brought back from parts unknown by recently returned sailors started running rampant among the general populace. The association with the Kriegsdiener cult has made the story of this curse particularly persistent to this day.

There were other theories that were less superstitious. As mentioned before, the materials were not handled with nearly as much care as they warranted, and among the technological

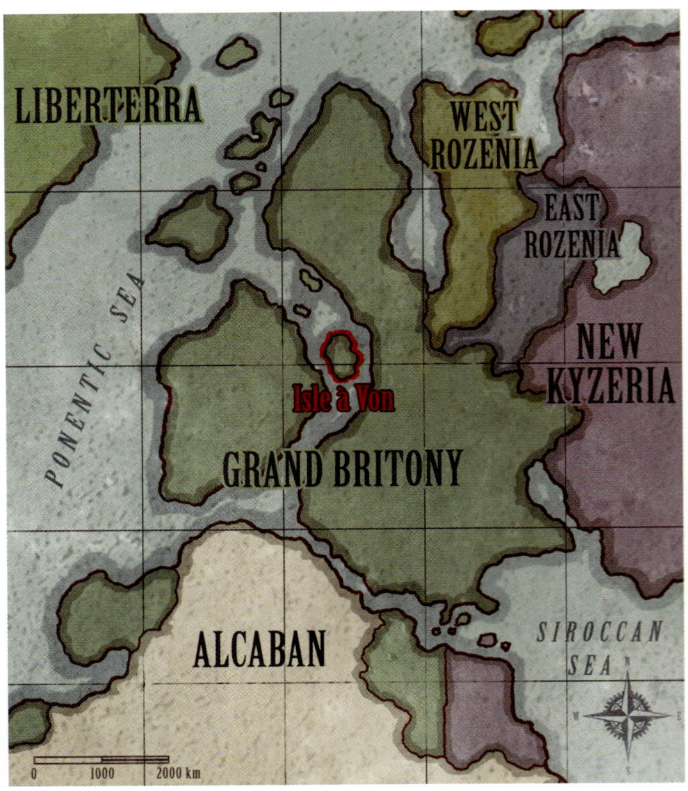

Map showing the location of Isle à Von and surrounding areas as they were in the year WC009.

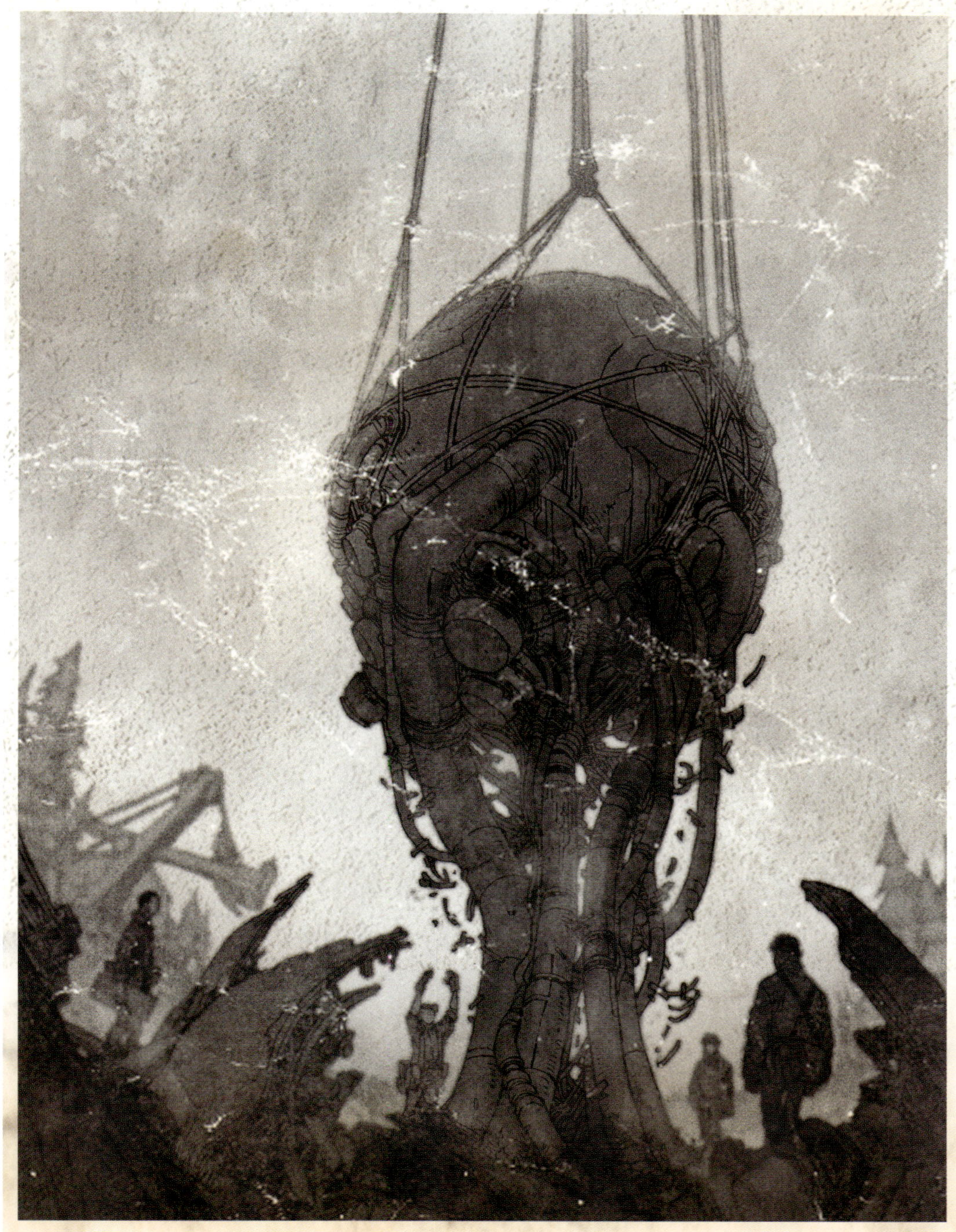

Photo taken on site showing the mysterious "Ijzerhart" being unearthed.

marvels uncovered there could have been any number of unknown toxic elements that might be responsible for the ill fates of the crew. One of those marvels was an artifact the Kyzerians called the "Ijzerhart," or Iron Heart, that appeared to be an ancient machine capable of generating electrical power, not unlike a modern day generator or power plant. At least one Ijzerhart-type generator was uncovered during the excavation, and some of the remaining reports from locals of the island and ships' crew mention strange, unearthly lights and thrumming noises. Some postulate that this was a clue that an unstable Ijzerhart generator was somehow activated during transport. Being exposed to the high levels of energy emitted by such a device could well have doomed the crew that handled it.

An even darker theory is sometimes considered among more pragmatic historians. Kaiser Walmir VII's personal records suggest that Razivir, leader of the Kriegsdiener cult and personal advisor to the Kaiser, headed the expedition personally; as to whether the man accompanied the expedition or not, the records are ambiguous. What is more, there are reports of political enemies and opponents of Razivir from years earlier, including a wayward priest from his own cult, succumbing to symptoms similar to those inflicted by the "curse" following the expedition. The theory is, of course, that the man was jealous of his secrets and would do anything, even ordering or committing mass murder, to protect them.

Whatever Razivir's personal involvement in the execution of the mission (or its personnel), he was without doubt instrumental in its planning and, thus, responsible for sending nearly three hundred soldiers and sailors to their doom in order to retrieve a mythological weapon that would end up spawning unimaginable horrors of war: tankheads.

A7V STURMSCHREITER AUSF. A

When tankheads were first introduced, the world had seen nothing like them. Alliance infantry experienced awe, confusion, and paralyzing fear all at once when they appeared for the first time on the Southern Front in WC010. The Britonican Department of War was so shaken that they kept a tight lid on information regarding these new "monstrous" weapons, fearing that the reports would damage the Alliance's morale. Meanwhile, the Kyzerian propaganda machine was pushing the Sturmschreiter, known as the Stormwalker in the West, as an unstoppable weapon of war—divine avatars that had come to lead the Kyzerians to victory. It was a proud moment for New Kyzeria when the costly Operation Zeehond of WC009 finally bore fruit. Though invented by mechanical engineer Hans Nacktgeborn, the A7V Sturmschreiter could only have been realized by studying the remains of the ancient, humanoid weapon the Kyzerians had dug up during the operation.

The Sturmschreiter would end up becoming obsolete within a mere few years as tankhead technology evolved at an exponential pace during the early phase of the Neverwar. In spite of that, as the unit that started it all, it was still an important milestone in the development of the walking tanks that forever changed the face of war.

"When forgotten sins are renewed and machines are forged in the shape of men, the gods of the battlefield will baptize the world in the War Century."
- excerpt from Kriegsdiener Scriptures

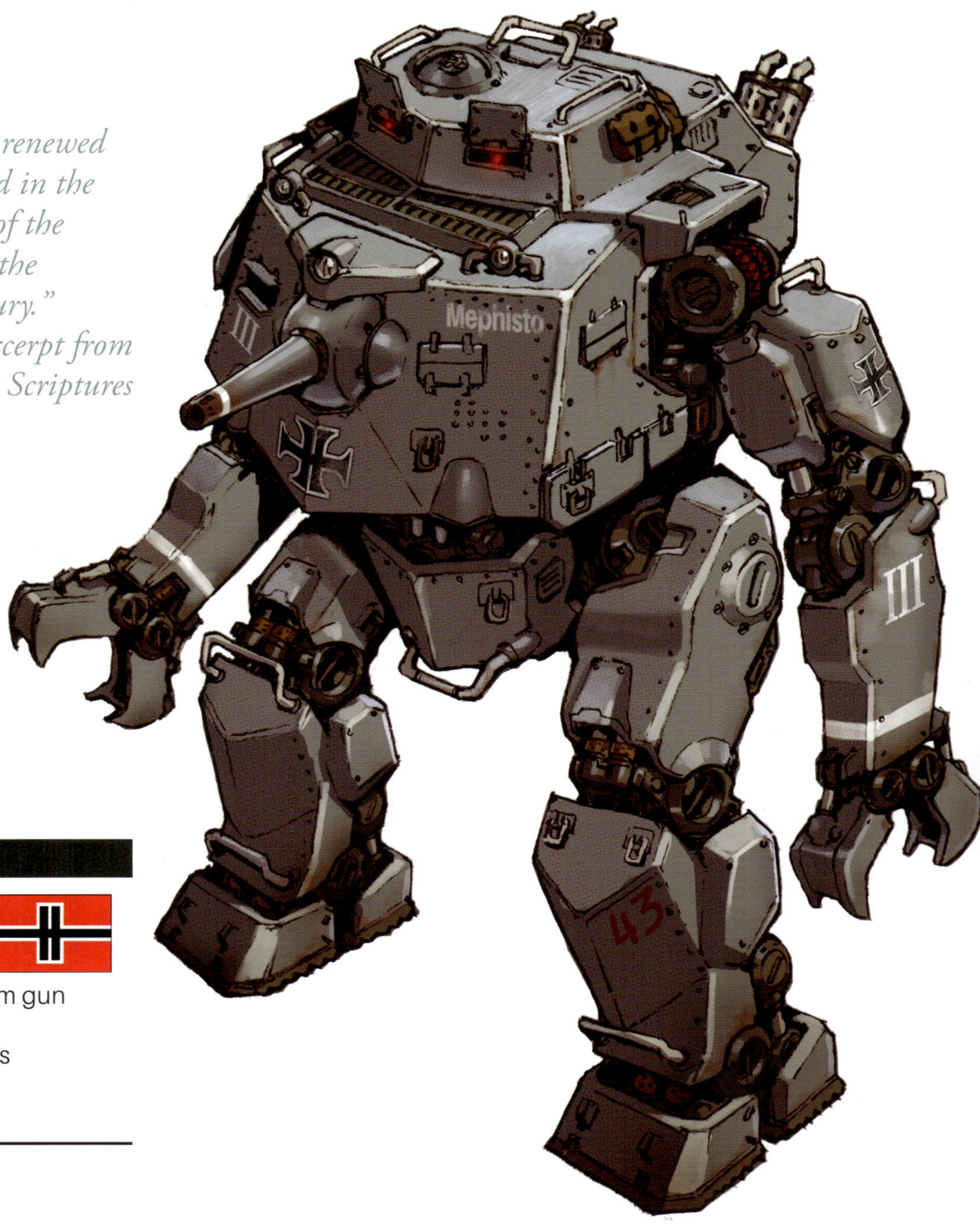

SPECIFICATIONS

COUNTRY: New Kyzeria
UNITS BUILT: 240
NO. OF CREW: 8
MAIN ARMAMENT: 1 x 47 mm gun
SECONDARY ARMAMENT:
4 x 7.9 mm machine guns
ARMOR: 5-30 mm
MAX SPEED: 10 km/h

A7V STURMSCHREITER AUSF. T

SPECIFICATIONS

COUNTRY: New Kyzeria
UNITS BUILT: 87
NO. OF CREW: 8
MAIN ARMAMENT: 1 x 57 mm gun
SECONDARY ARMAMENT:
4 x 7.9 mm machine guns
ARMOR: 5–30 mm
MAX SPEED: 10 km/h

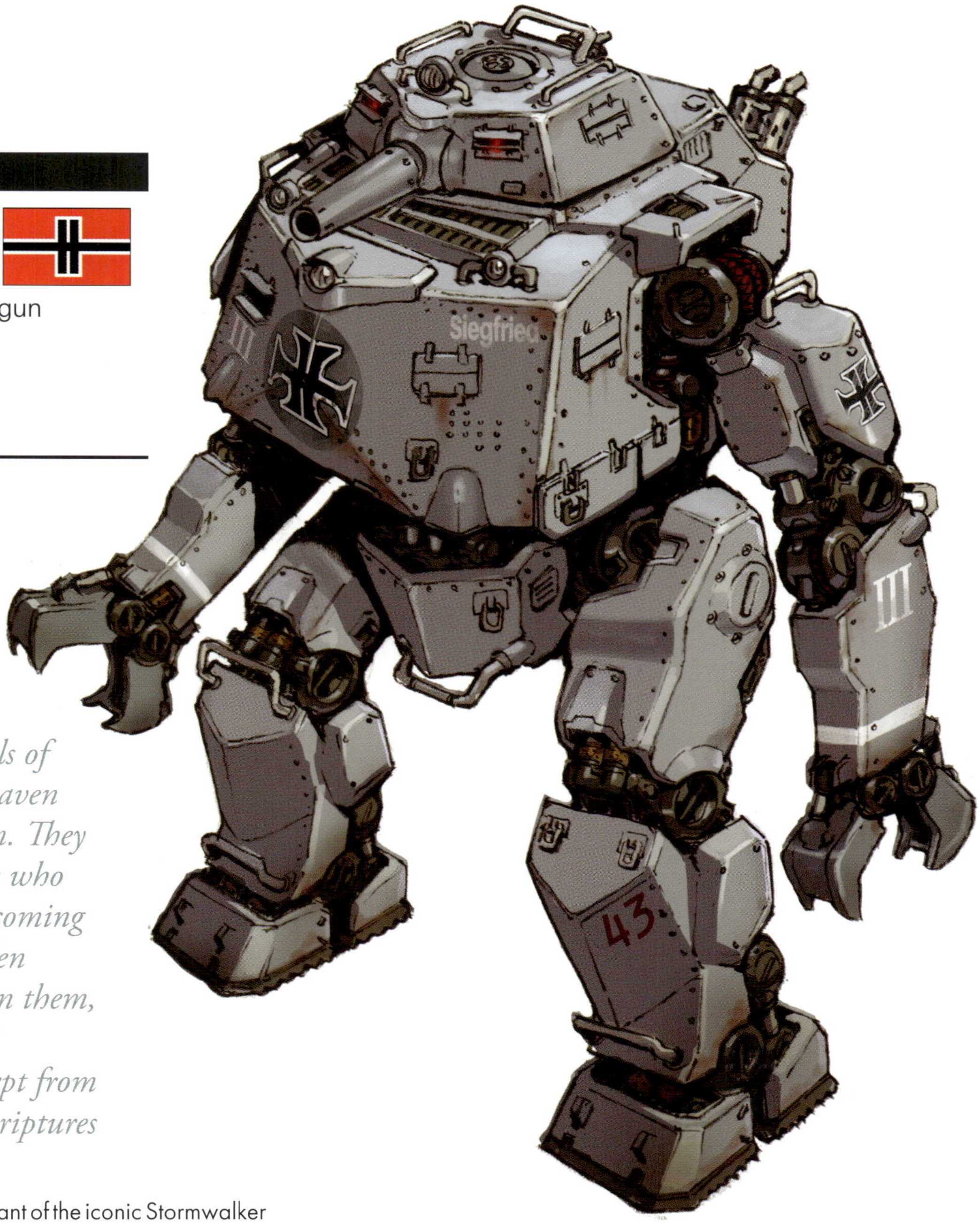

"In the tenth year, the angels of war shall descend from heaven unto the battlefield of men. They shall smite the unbelievers who have no true faith in the coming of the War Century. Sudden destruction will come upon them, and they will not escape."
- excerpt from Kriegsdiener Scriptures

The Type T was a fairly common variant of the iconic Stormwalker line. It was intended to become the new main combat unit by the winter of WC011, but complications during the development process and negative reports from the field resulted in earlier models being kept in production alongside the Type T.

The driving force behind the development of the Type T was the Alliance's invention of the gyrojet cannon, which was subsequently reverse-engineered by the Kyzerian Army Development Branch. As soon as the Kyzerians determined that they would be capable of producing a similar weapon and understood the implied advantages of the system (greatly reducing recoil), they commissioned a Stormwalker variant that would be able to use it. So confident were they that they could develop gyrojets of their own that they started designing the Type T before the first Kyzerian gyrojet weapon was even fully tested.

Using the existing Stormwalker as a base limited the design team, but it did allow for the Type Ts to roll off the production line just days after the designs were finalized. The obvious, significant change was the addition of a turret capable of full 360-degree fire without overthrowing the tankhead with recoil. The turret replaced the top section of the Stormwalker's hull, which had previously housed both the commander and gunner. In the Type T, the commander would do double duty as gunner while the driver was moved down into the hull and forced to use the smaller vision hatch previously used by the gunners. This configuration would end up being common among tankheads from all three great powers from this point on.

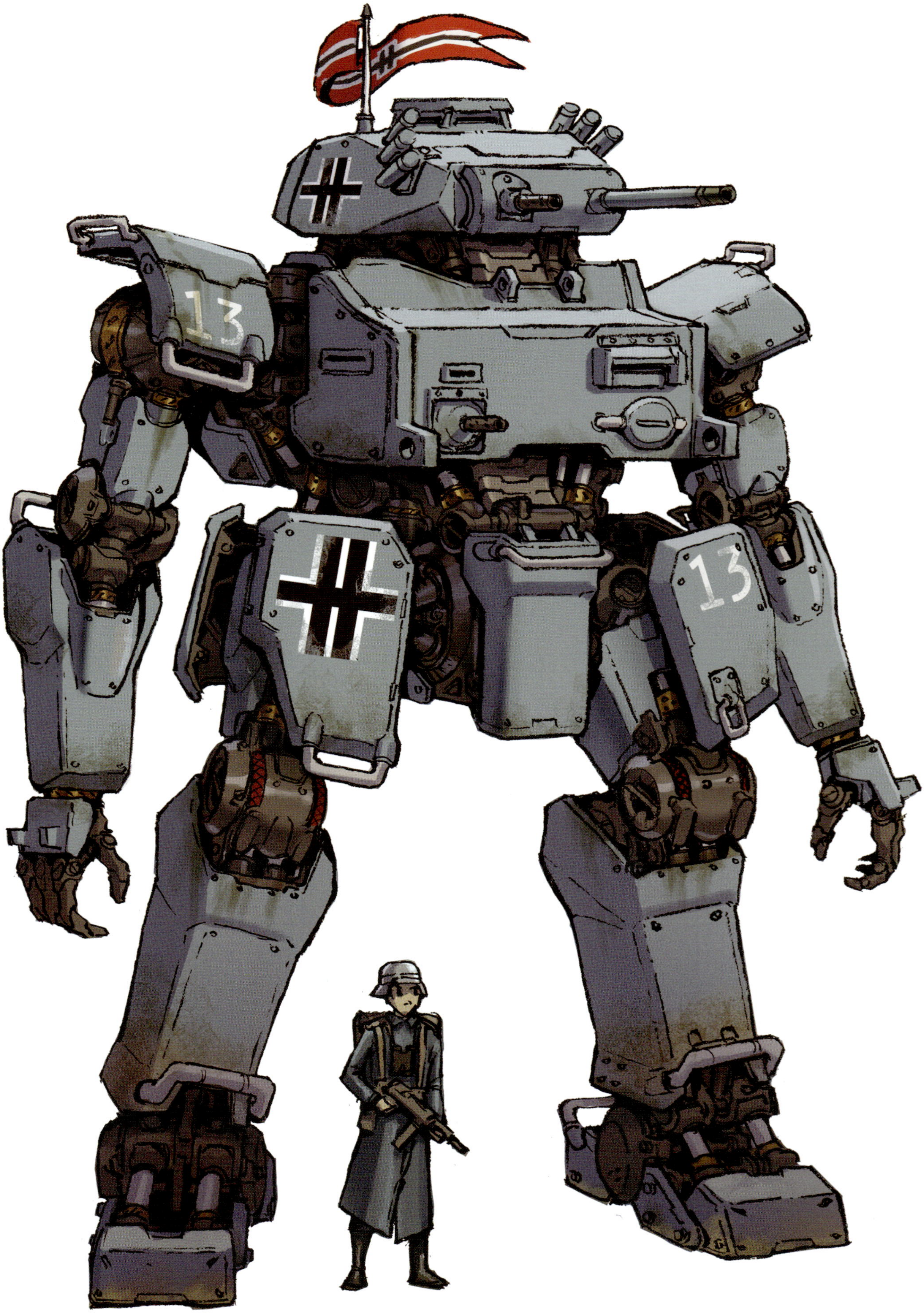
13
13

PANZERKOPF I AUSF. C

In WC012 there was a noticeable shift in Kyzerian politics as the nation mourned the loss of Kaiser Walmir VII. Interest turned inwards to focus on the young twin heirs, Vaimund II and Vailinde IV, and their stewardship. Consequently, expansionist aggression ground to a halt. Fearing the lack of a unified leadership would lead to its doom, New Kyzeria used its dominant position in the war to negotiate a ceasefire with the Alliance. The terms of the ceasefire mostly dealt with specifying where the front lines were at the time of its signing, repatriation of numerous prisoners of war, and, last but not least, a clause that banned the production of "heavy mobile land weaponry." Due to the high upper limit set in the agreement, this in effect meant that only the largest class of tankheads of the time would cease production.

Though their spirits might have been dampened, Kyzerian High Command fully expected to resume the war in earnest once leadership had stabilized. In order to train new crews, it was quickly decided that a new tankhead should be produced, one which complied with the stipulations of the ceasefire treaty. Overcoming the design challenges of miniaturizing tankhead joints and general design to fit the new requirements wound up being a major advantage in future tankhead developments.

Although the Panzerkopf I was designated as a training machine, it was designed as a completely functional combat vehicle from the start. Its armor was light compared to the earlier Stormwalkers, but more efficiently distributed. Armaments were a secondary concern on the first models of Panzerkopf I, which featured only machine guns, to comply with the ceasefire agreement. As the war started to heat up again in WC014, the Panzerkopf I was equipped with small gyrojet cannons. Though at no point a dominant force on the battlefield, the uncomplaining Panzerkopf I was as effective against infantry as any other tankhead and was cost-efficient to produce compared to its predecessors. The Panzerkopf I continued to see use in a variety of roles throughout the rest of the war, from observation and command posts to ammo transport and even anti-armor cannon carriers.

"Now you can shoot at the thing all day and not make a dent. But you see those vision slits? Some paper pusher decided they could cheap out and use thin glass there. So take a real good aim there, son, and make the guys inside curse his name."

-Rozenian Partisan Stric Bohadan

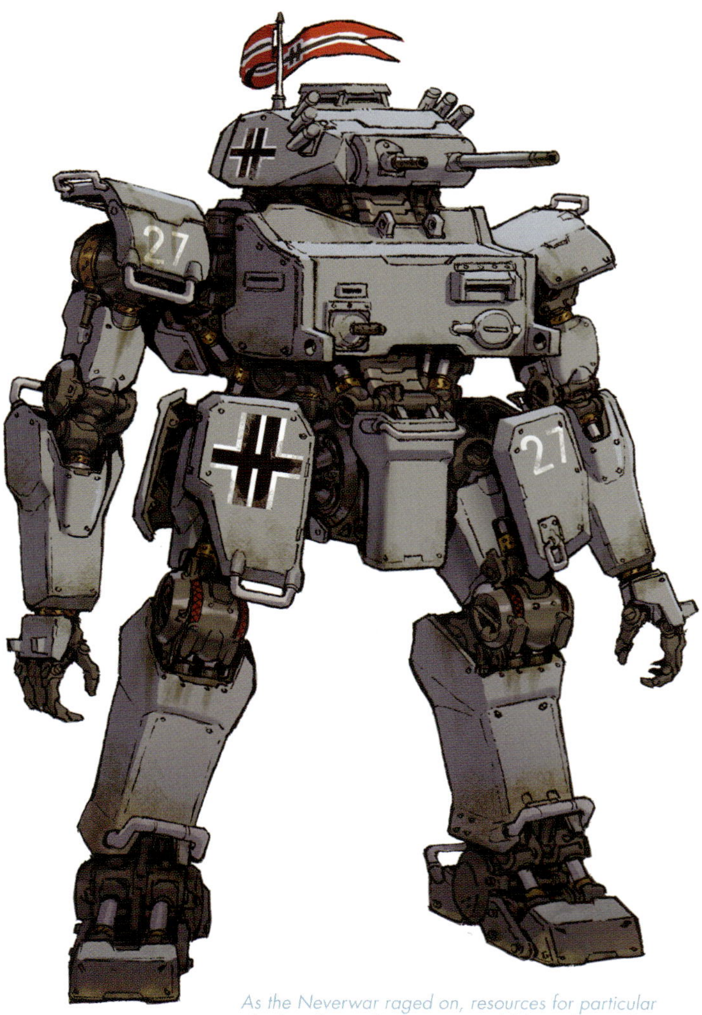

As the Neverwar raged on, resources for particular shades of paint became unavailable and the colors used on vehicles and tankheads shifted depending on what was available at the time.

SPECIFICATIONS

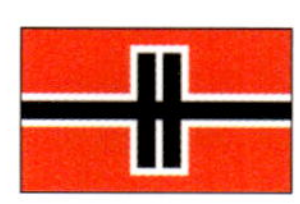

COUNTRY: New Kyzeria
UNITS BUILT: 3,259
NO. OF CREW: 4
MAIN ARMAMENT: 1 x 37 mm gun
SECONDARY ARMAMENT:
2 x 7.92 mm machine guns
6 x smoke grenade launchers
ARMOR: 7-25 mm
MAX SPEED: 36 km/h

KYZERIAN ENGINEERING VEHICLES

Today, many of the Kyzerian engineering vehicles produced during the Neverwar are considered desirable classics, beloved by collectors. This actually highlights what should be considered a flaw in the Kyz engineering divisions of the time. Design requirements for any vehicle were put arbitrarily high, with multiple companies competing to meet and beat these demands. This resulted in overly-expensive and over-engineered machines that sacrificed capability in other areas. The Takom Giraffe, for example, could be fully controlled from the lift cage and was capable of turning in place. This came at a great cost in mobility, however, as it could only operate on flat concrete floors. Similarly, the Takom Mechaniker was surprisingly fast and nimble, and is, in fact, still a popular mecha-rally vehicle today. However, it sacrificed strength and reliability for its speed, even though those former qualities were more important in an engineering machine.

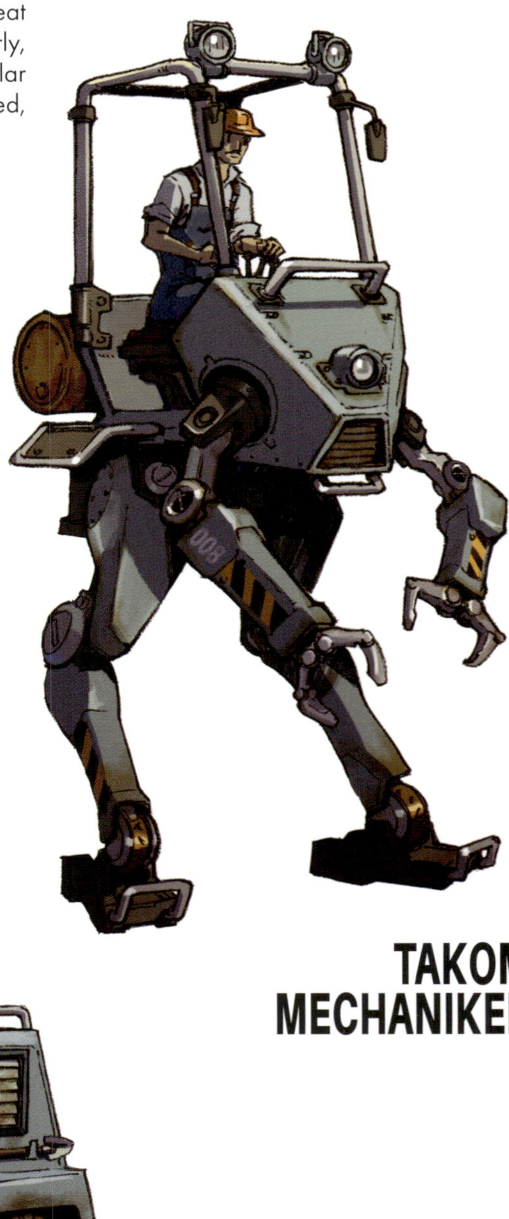

TAKOM MECHANIKER

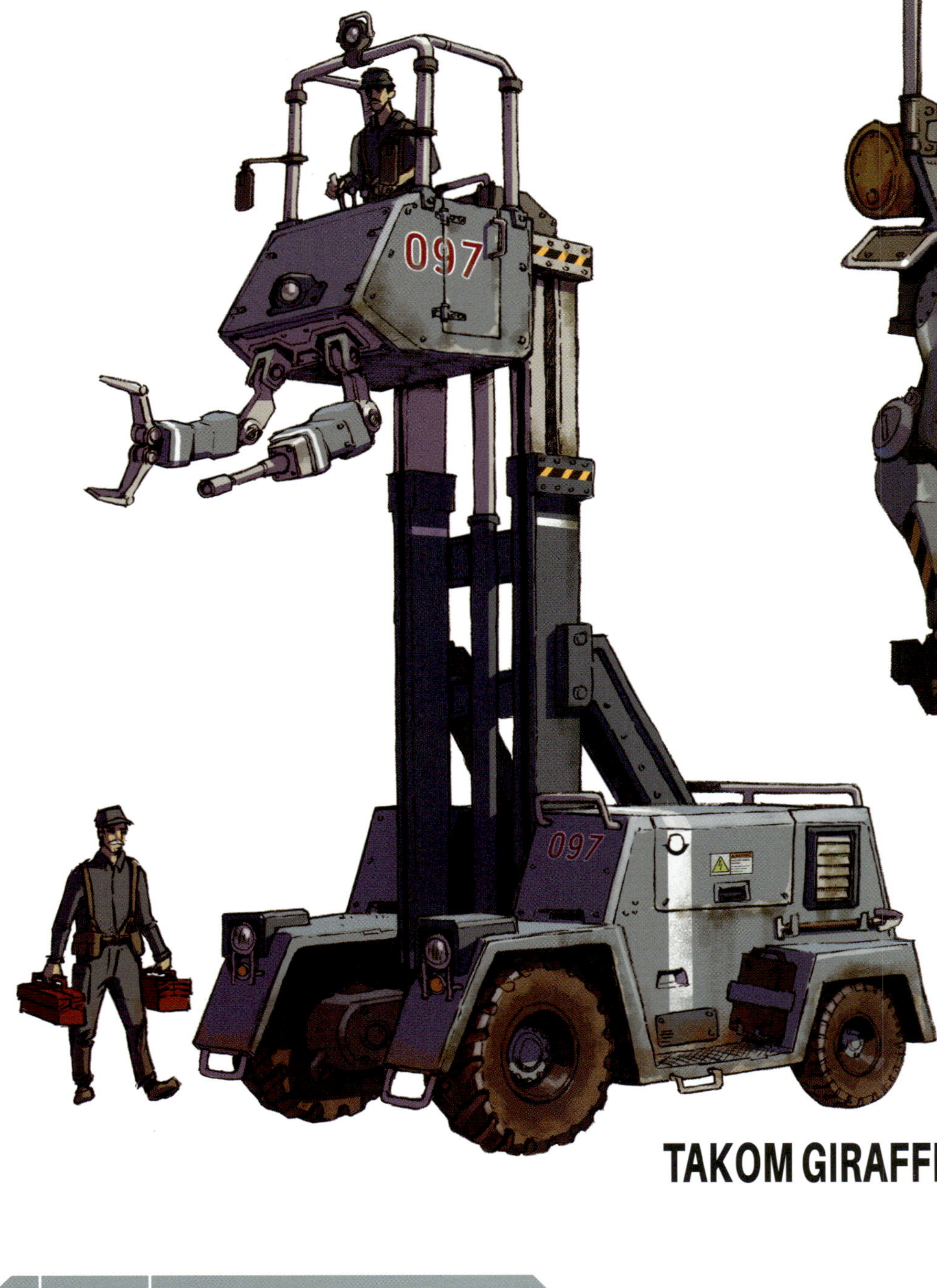

TAKOM GIRAFFE

SPÄHER PANZERKOPF

SPECIFICATIONS

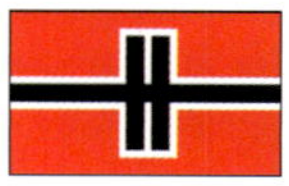

COUNTRY: New Kyzeria
UNITS BUILT: 10,014
NO. OF CREW: 3
MAIN ARMAMENT:
2 x 7.92 heavy machine guns
SECONDARY ARMAMENT:
6 x smoke grenade launchers
ARMOR: 9–30 mm
MAX SPEED: 75 km/h

The ever-developing technologies of tankhead mobility led to the creation of light scout tankheads. These were units that were capable of rapid attacks and of traversing rough terrain that heavier tankheads struggled on. It seems that both sides thought of this idea at the same time, but the Kyzerian Späher was undeniably the most significant advancement in this class. Using an updated digitigrade "goat leg" design, the machine was the first bipedal tankhead fully capable of prolonged running. Some contemporary machines, such as the Crossley, did have a similar leg configuration, but were not capable of true running (in the sense that both legs did not leave the ground at the same time.)

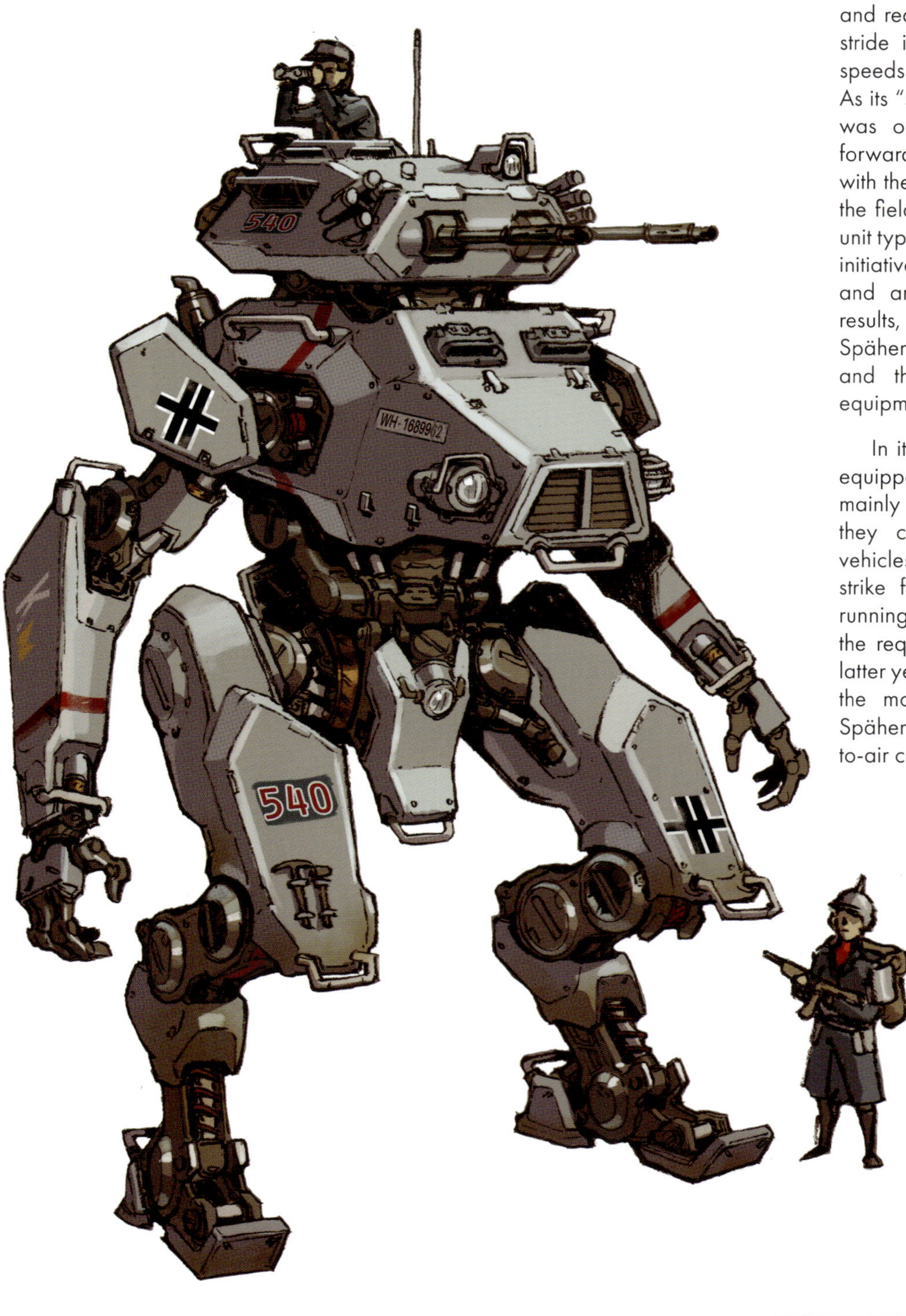

The Späher used powerful but easily-replaceable springs in the lower legs, along with the additional joint, to effectively store and redirect a portion of the energy of each stride into the next, allowing it to achieve speeds unseen in tankheads up to that point. As its "spy" designation suggests, the Späher was originally intended for scouting and forward observation roles. More than satisfied with the Späher's outstanding performance in the field, the Empire wanted more out of this unit type and so initiated several development initiatives intended on upgrading its weapons and armor. These efforts produced mixed results, however, as the light weight of the Späher proved essential to its performance and the extra weight of the upgraded equipment actually reduced its effectiveness.

In its stock configuration, the Späher was equipped with twin heavy machine guns mainly intended for anti-infantry fire, though they could destroy or incapacitate light vehicles and field guns, provided they could strike from the rear. Fortunately, the agile running tankhead proved to be very good at the requisite flanking maneuvers. During the latter years of the war, when jetheads became the most significant battlefield threat, the Späher also proved to be capable at ground-to-air combat.

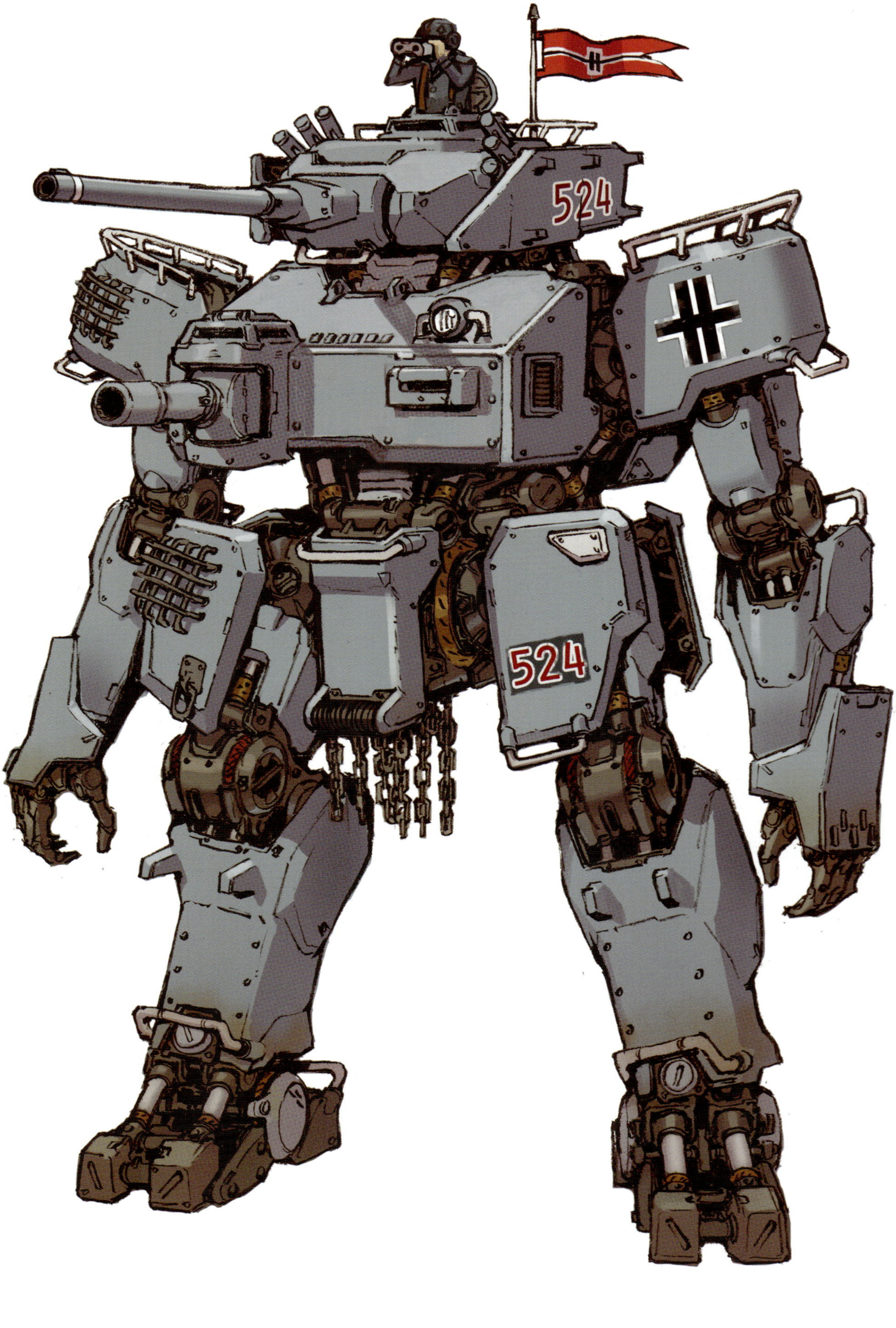
524
524

PANZERKOPF III AUSF. H

The design and early production of the Panzerkopf III occurred largely during the WC012–015 ceasefire agreement period. The Panzerkopf III was intended to be a medium tankhead, but the ceasefire stipulations led to the design of the Panzerkopf III being intentionally light on both armor and weapons. The thinking was that a well-designed tankhead would be able to support heavier equipment if and when it became necessary. A more significant factor was the fact that during the chaotic mourning period after Kaiser Walmir VII's death the Kyzerian intelligence branch became embroiled in political intrigue and became mostly inward-focused, spying on its countrymen rather than on the enemy. As a consequence, they only had a vague idea of the military developments taking place in the Alliance, leading in turn to their underestimation of the type of machine that would be needed should war break out again.

When war reared its ugly head again in WC015, the Panzerkopf III underperformed against the many new tankheads it faced on the front lines. The biggest surprise were the tankheads developed by West Rozenia during the ceasefire, designs that Kyzerian Command previously judged to be crude and ineffective. The Panzerkopf III did not perform its intended role as a frontline battle unit adequately and was soon reassigned to support conventional tank units. The improvements that were initially envisioned during the ceasefire period, such as increased armor and armaments, were quickly implemented. They were just as quickly deemed insufficient in the field and heavier upgrades were rolled out on an almost bi-monthly basis. This all happened as New Kyzeria scrambled to develop another tankhead capable of facing its enemies head on.

The later Panzerkopf III Type H shown here was a significant improvement over earlier designs. It had a larger turret gun and a short-barreled smoothbore heavy cannon added to its torso, improving its firepower in short-range engagements. However, accommodation of the weapon made the hull of the Panzerkopf III notoriously cramped and some crews bemoaned the loss of the bow machine gun once used to deter infantry.

SPECIFICATIONS

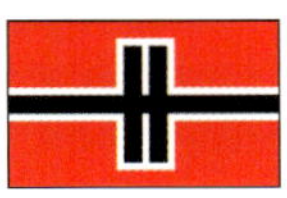

COUNTRY: New Kyzeria
UNITS BUILT: 10,774
NO. OF CREW: 5
MAIN ARMAMENT: 1 x 50 mm gun
SECONDARY ARMAMENT:
1 x 110 mm heavy cannon
6 x smoke grenade launchers
ARMOR: 15-50 mm
MAX SPEED: 40 km/h

Not only were the weapons on the Type H heavily upgraded, its armor was thickened and improved overall. The appliqué plates on the shoulders and spaced plating on the turret gave the Type H a recognizable profile.

Even though the Panzerkopf III never lived up to its intended role and was outperformed on the battlefield, it was still a reliable design that could easily be modified to fit other roles. Near the end of the war, almost all Panzerkopf III frames were converted over to alternative uses, one of them being the extremely successful Sturmgeschütz III "StuG" tankhead destroyer.

"And then I saw them. Angels that breathed death and destruction from their mouths. Each trumpeting roar brought down fire from the heavens upon the unbelievers."
- Kriegsdiener Priest Jakob Volmar

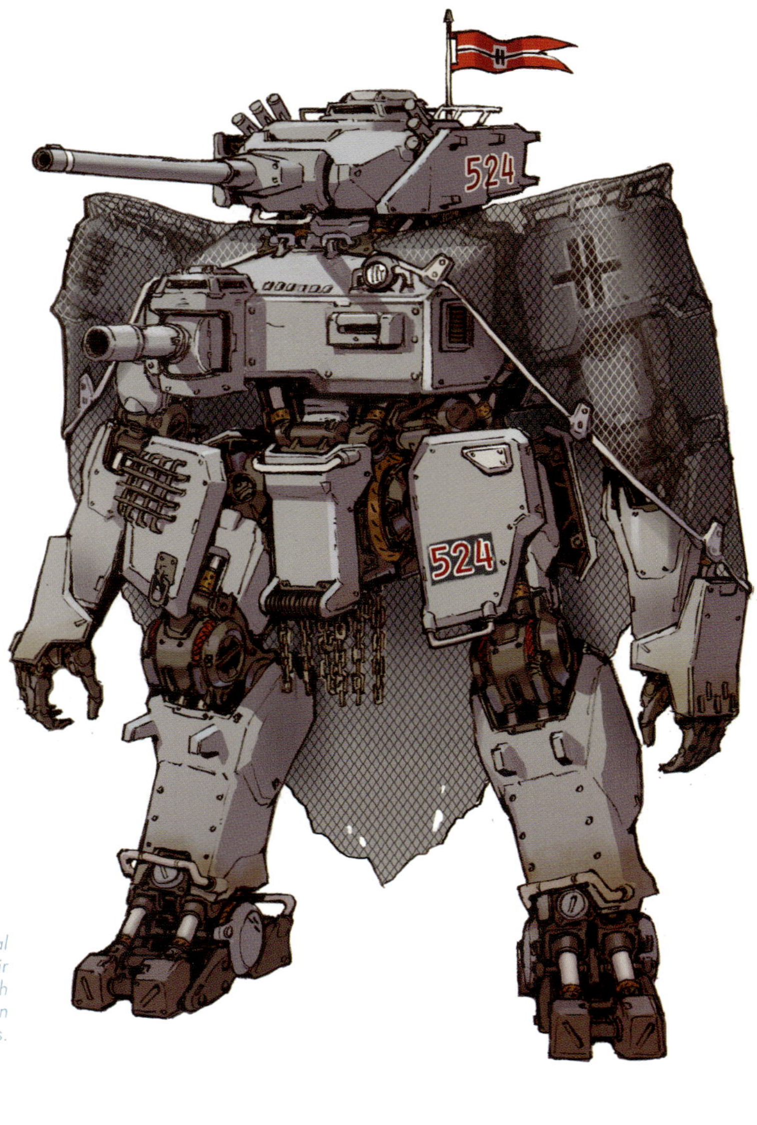

Throughout the latter half of the Neverwar, there were many experimental systems designed to prevent shaped-charge warheads from delivering their payloads into the main body of a tankhead. Shown here is the wire mesh cloak which was added to the Panzerkopf III. This mesh was much lighter than most spaced armor defenses, which relied on solid metal plates.

SCHÜTZENKEILER AUSF. D

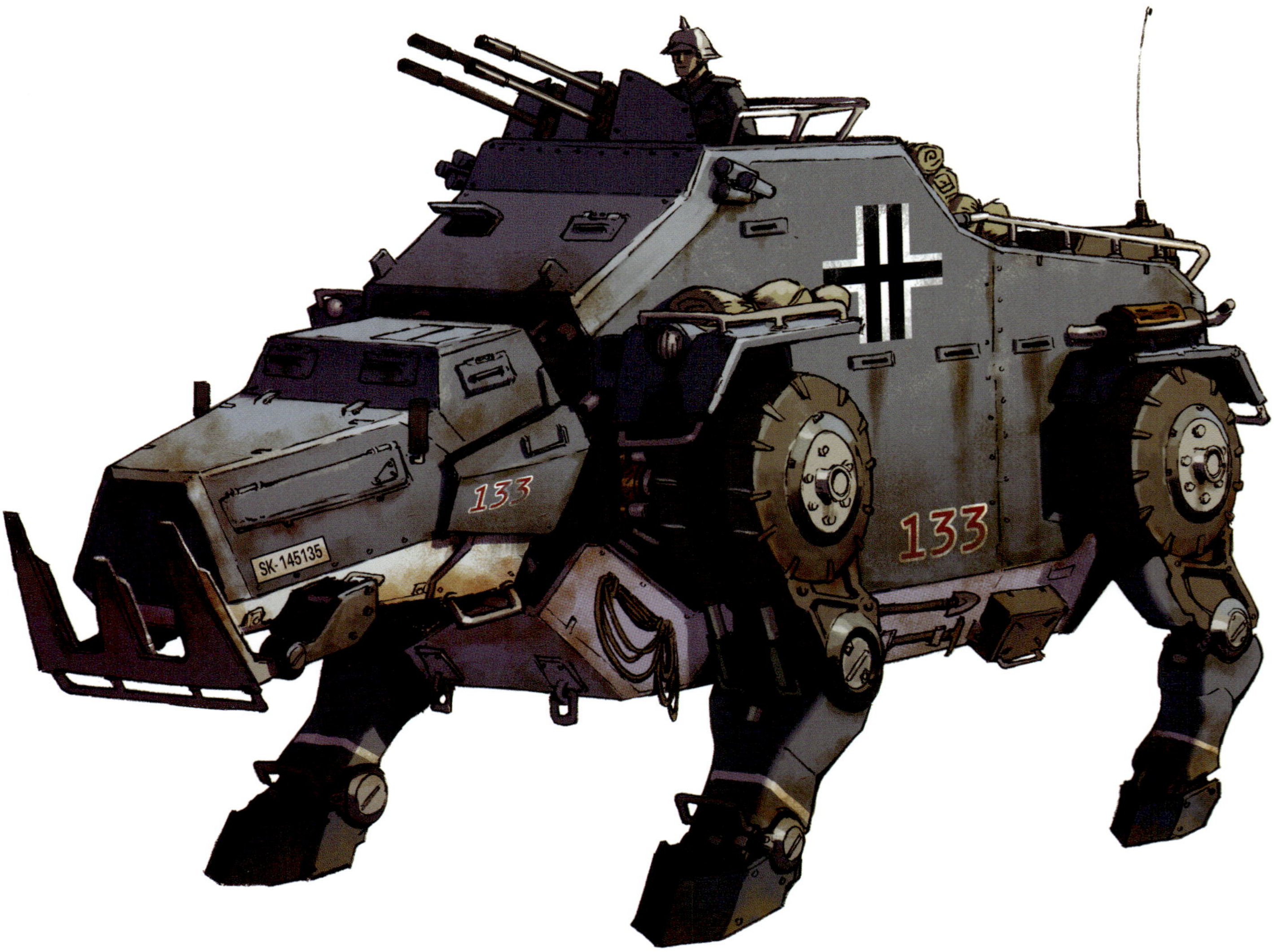

SPECIFICATIONS

COUNTRY: New Kyzeria
UNITS BUILT: 17,252
NO. OF CREW: 3
INFANTRY CARRIED: 10
MAIN ARMAMENT:
1 x quad 20 mm anti-aircraft guns
SECONDARY ARMAMENT:
1 x 7.92 machine gun
6 x smoke grenade launchers
ARMOR: 6-15 mm
MAX SPEED: 53 km/h

Because Kyzeria was ahead in tankhead development and production for most of the war, they were the only ones who fielded personnel-carrier-type tankheads in serious numbers. Alliance troopers nicknamed the Schützenkeiler "Piggy" due to its resemblance to a wild boar. True to its nickname, it was well-equipped for clearing countryside bocages and other small obstacles. This Type D model was a late-war anti-air variant which entered use when jetheads became a serious threat on the battlefield.

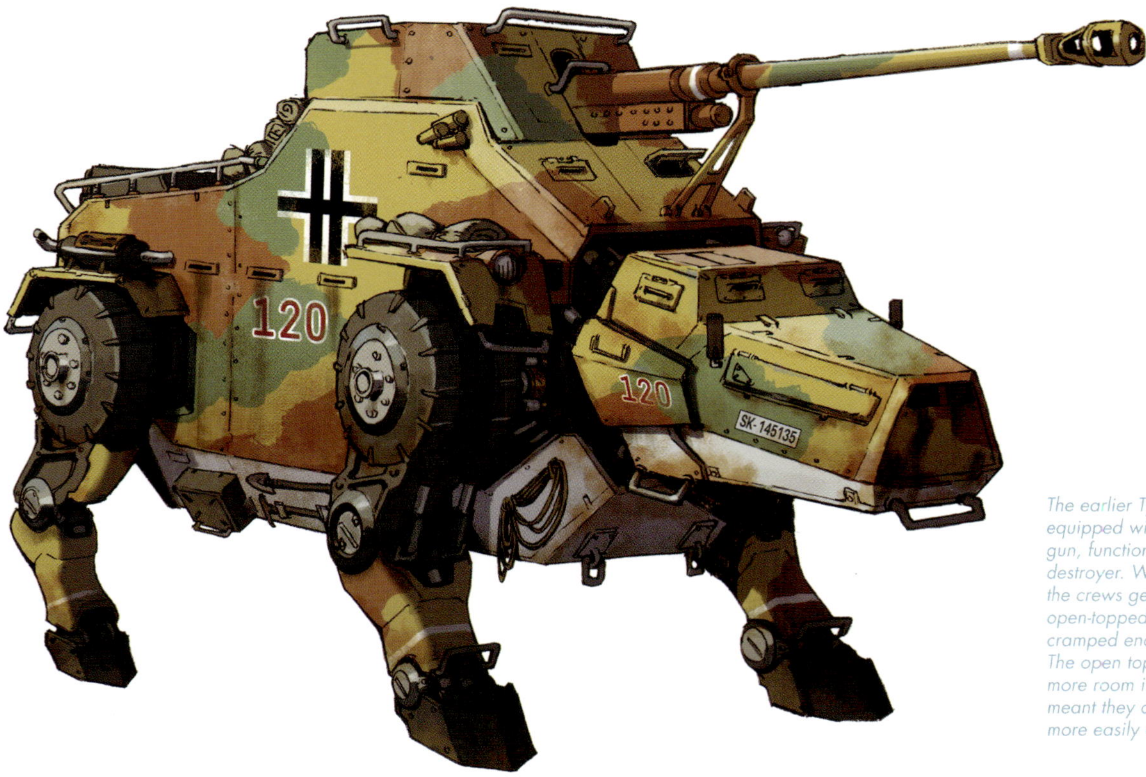

The earlier Type C Schützenkeiler, equipped with a 75 mm anti-tank gun, functioned as a cheap tankhead destroyer. While lightly armored, the crews generally preferred the open-topped Schützenkeiler over the cramped enclosed hulls of tankheads. The open top provided crews much more room in which to work, which meant they could reload their weapon more easily and thus faster.

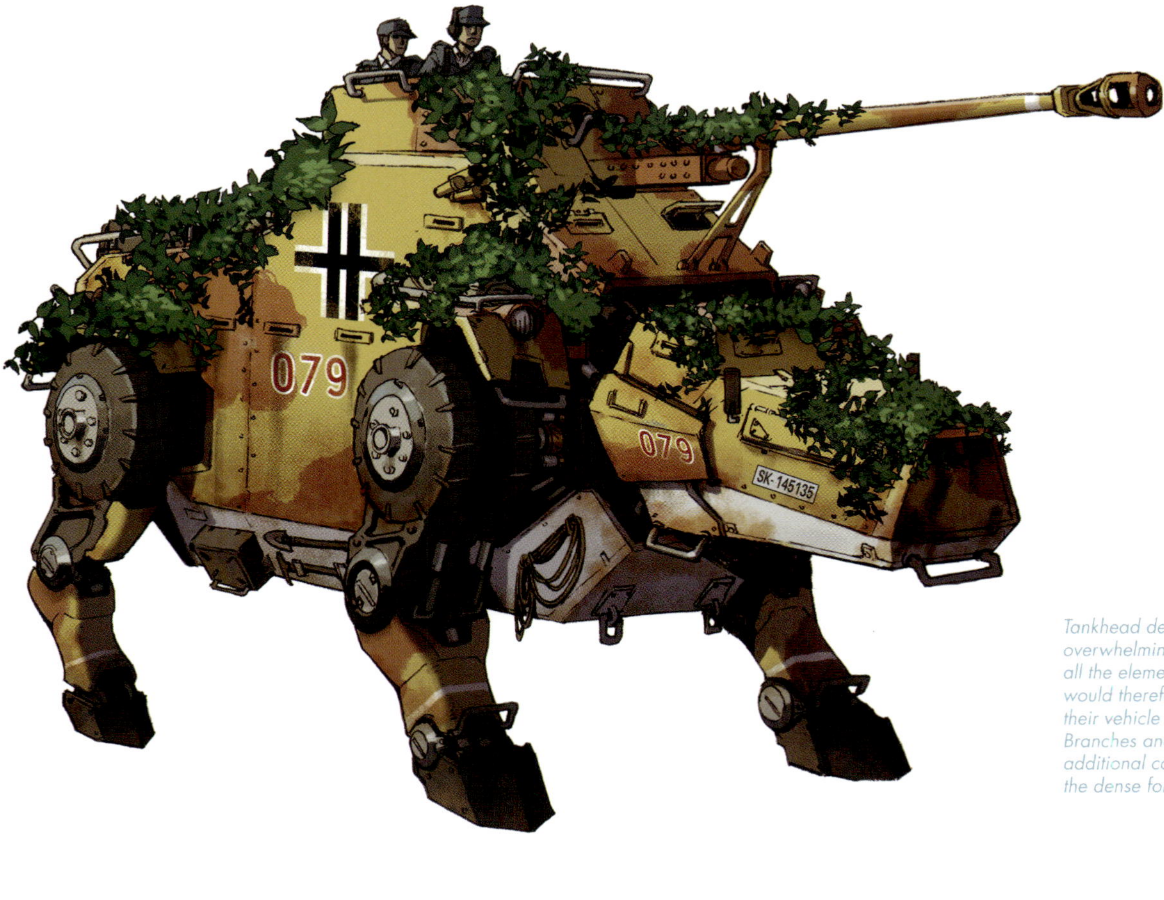

Tankhead destroyers relied on their overwhelming firepower and above all the element of surprise. Crews would therefore take time to adapt their vehicle to the local environment. Branches and foliage made for good additional camouflage, especially in the dense forests of western Kyzeria.

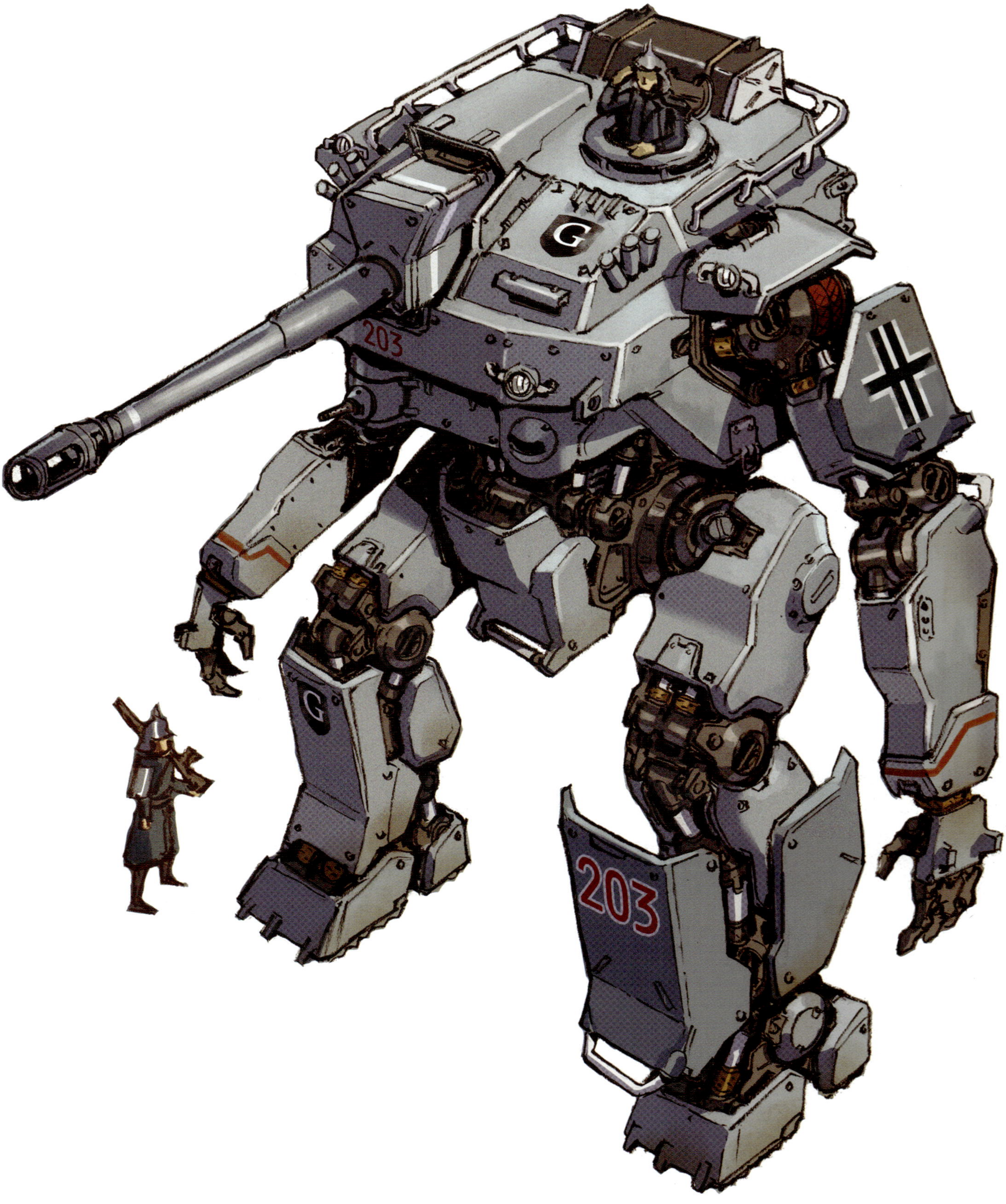
G
203
G
203

STURMGESCHÜTZ III AUSF. B

The first models of the Sturmgeschütz "StuG" III were produced during the ceasefire period along with the more conventional Panzerkopf IIIs. Just like the Panzerkopf, the StuG complied with the stipulations about military vehicle production. The StuG III was less hampered by these restrictions, though, as it did not have a heavy turret, allowing its hull to carry thicker armor.

The first models were equipped with a short smoothbore artillery gun intended for bombarding fortifications, fulfilling a classic assault function in support of the infantry. It was fast enough to keep pace with a mobilized infantry squad and was supposed to open breaches in defenses for infantry to exploit. However, actual contact with the enemy showed a number of shortcomings in the StuG's design. The main gun was capable of its intended role, but the vehicle it was mounted on often fell prey to opposing tanks, such as the newer West Rozenian models. Its armor was inadequate and its own short-range weapon, which fired explosive shells, was ineffective against thick enemy armor.

When the StuG III was upgraded, its role also shifted radically. No longer was it expected to keep pace with infantry and knock out fortifications. Instead, it was equipped with a long-barreled, high-velocity gyrojet cannon intended to fire solid-core armor-piercing shells at enemy tankheads. Incredibly, this weapon would only receive minor updates for the rest of the war, yet it remained a credible threat on the battlefield.

The light weight and speed of the StuG III proved very useful, as the hull-mounted weapon required the entire vehicle to move when it aimed its main gun. The relatively low profile and wide stance meant the StuG III afforded a stable firing platform on uneven or unstable ground, while presenting a smaller profile than many other tankheads. This made the StuG III very suited for ambush, taking shots at enemy targets from unexpected locations.

The StuG III was never meant to be as glorious as the Panzerritters. It was "just" meant to support infantry. It was low, stable and reliable with a hard-hitting weapon. And that's what made it the most effective Kyzerian tankhead of the Neverwar.

The updated StuG III proved itself superior to its parent design, the Panzerkopf III. The latter was near obsolete by the time it saw actual combat, and a significant portion were later converted to StuG IIIs during the war.

SPECIFICATIONS

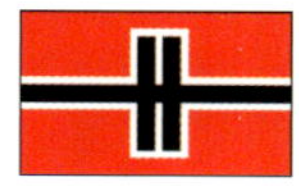

COUNTRY: New Kyzeria
UNITS BUILT: 18,086
NO. OF CREW: 4
MAIN ARMAMENT: 1 x 75 mm gun
SECONDARY ARMAMENT:
1 x 7.92 mm machine gun
6 x smoke grenade launchers
ARMOR: 16–80 mm
MAX SPEED: 40 km/h

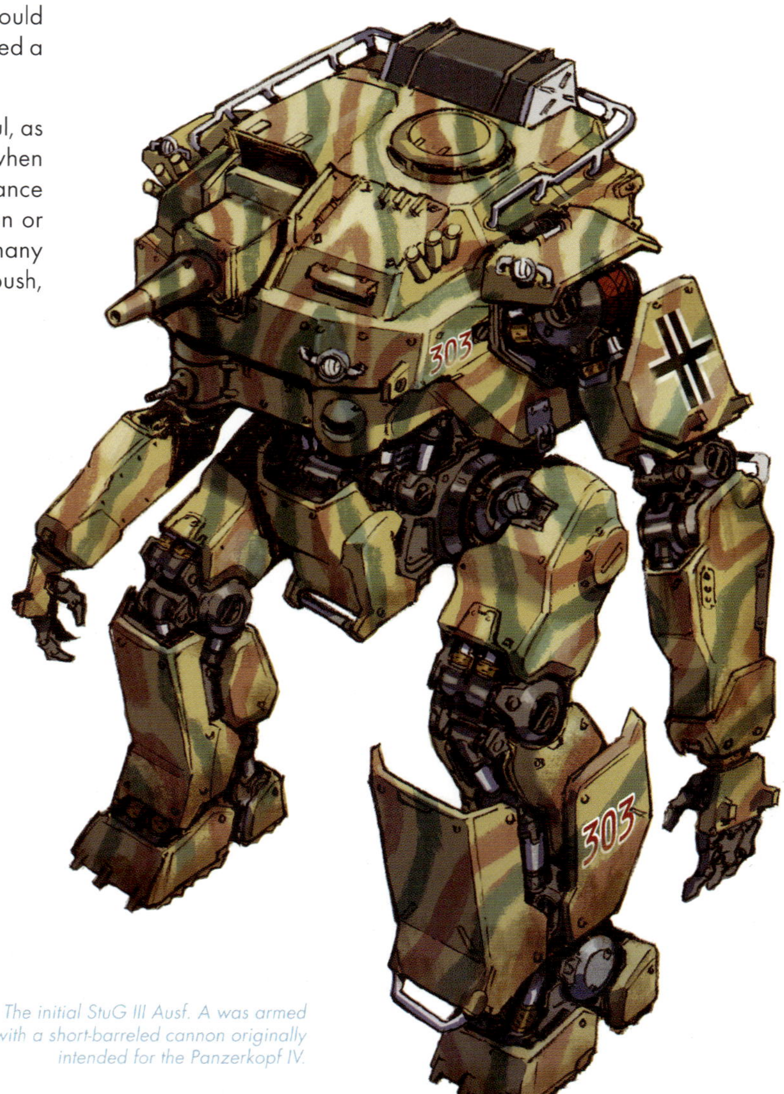

The initial StuG III Ausf. A was armed with a short-barreled cannon originally intended for the Panzerkopf IV.

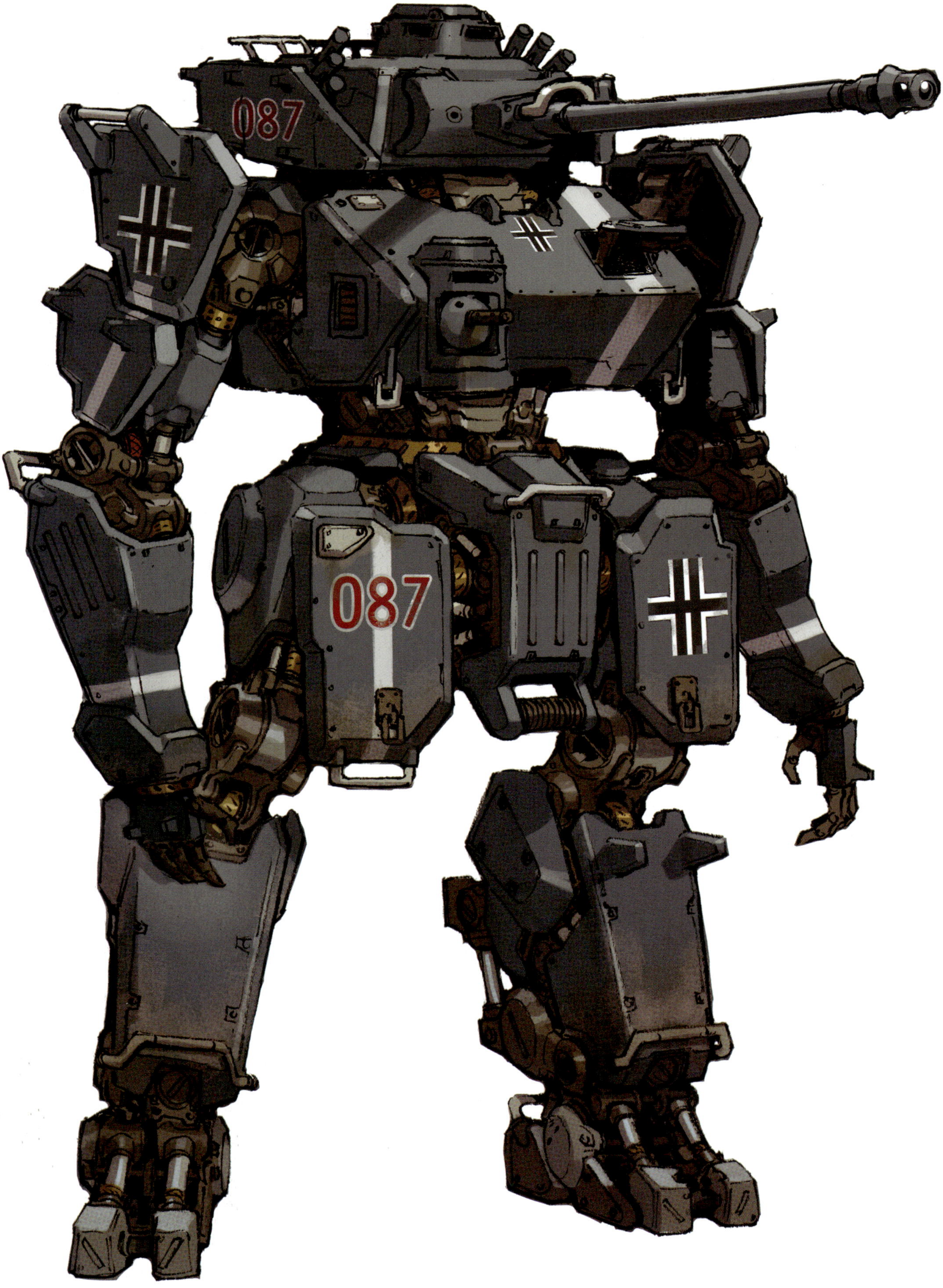
087
087

PANZERKOPF V PANTHER AUSF. D

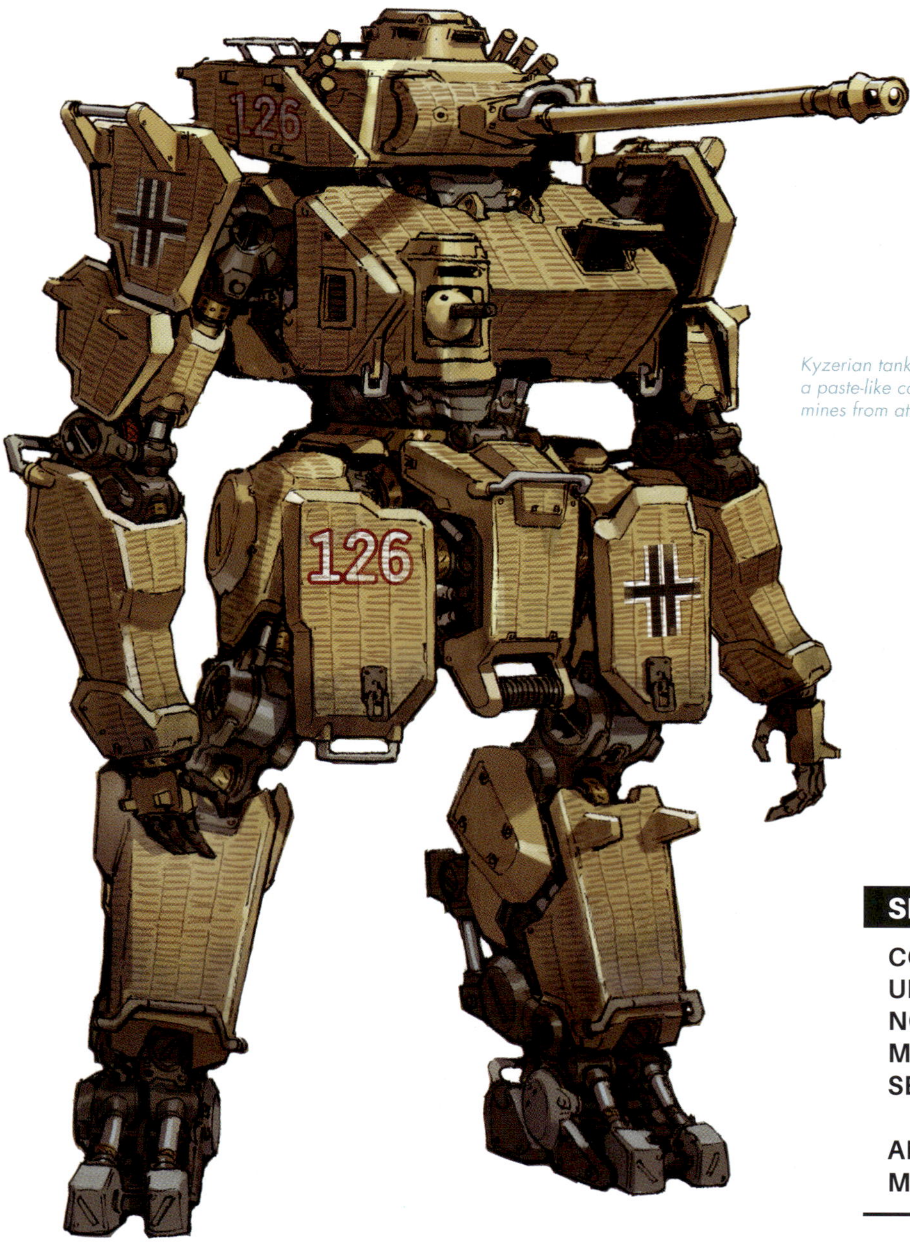

Kyzerian tankheads were often coated with Zimmerit, a paste-like coating that prevented magnetic anti-tank mines from attaching to their armor.

SPECIFICATIONS

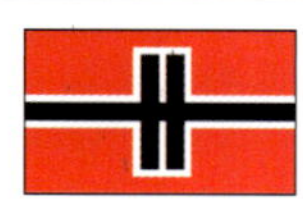

COUNTRY: New Kyzeria
UNITS BUILT: 6,000
NO. OF CREW: 5
MAIN ARMAMENT: 1 x 75 mm gun
SECONDARY ARMAMENT:
2 x 7.92 mm machine guns
ARMOR: 16–100 mm
MAX SPEED: 46 km/h

The Panther is a much-debated design. In some ways it established the archetypal baseline for a medium battle tankhead for the rest of the Neverwar and beyond. Combat records show it performed its designated role as intended and even exceeded expectations. This is however contrasted by the fact that the design was rushed into production and its frontal armor was increased at the last minute. Surprisingly, this increase in weight did not reduce the machine's overall mobility significantly. However, the reliability of the drive linkage and power plant were significantly compromised and breakdowns were common, especially on the early models. These troubles were compounded by the rushed nature of the machine as crews had received minimal training and experience with their tankheads before being deployed to the front lines. While breakdowns would continue to be an issue, overhauls, rebuilds and battlefield experience gained by the crew would significantly improve the Panzerkopf V's performance with time and it continued to be a respected and feared figure until the end of the war.

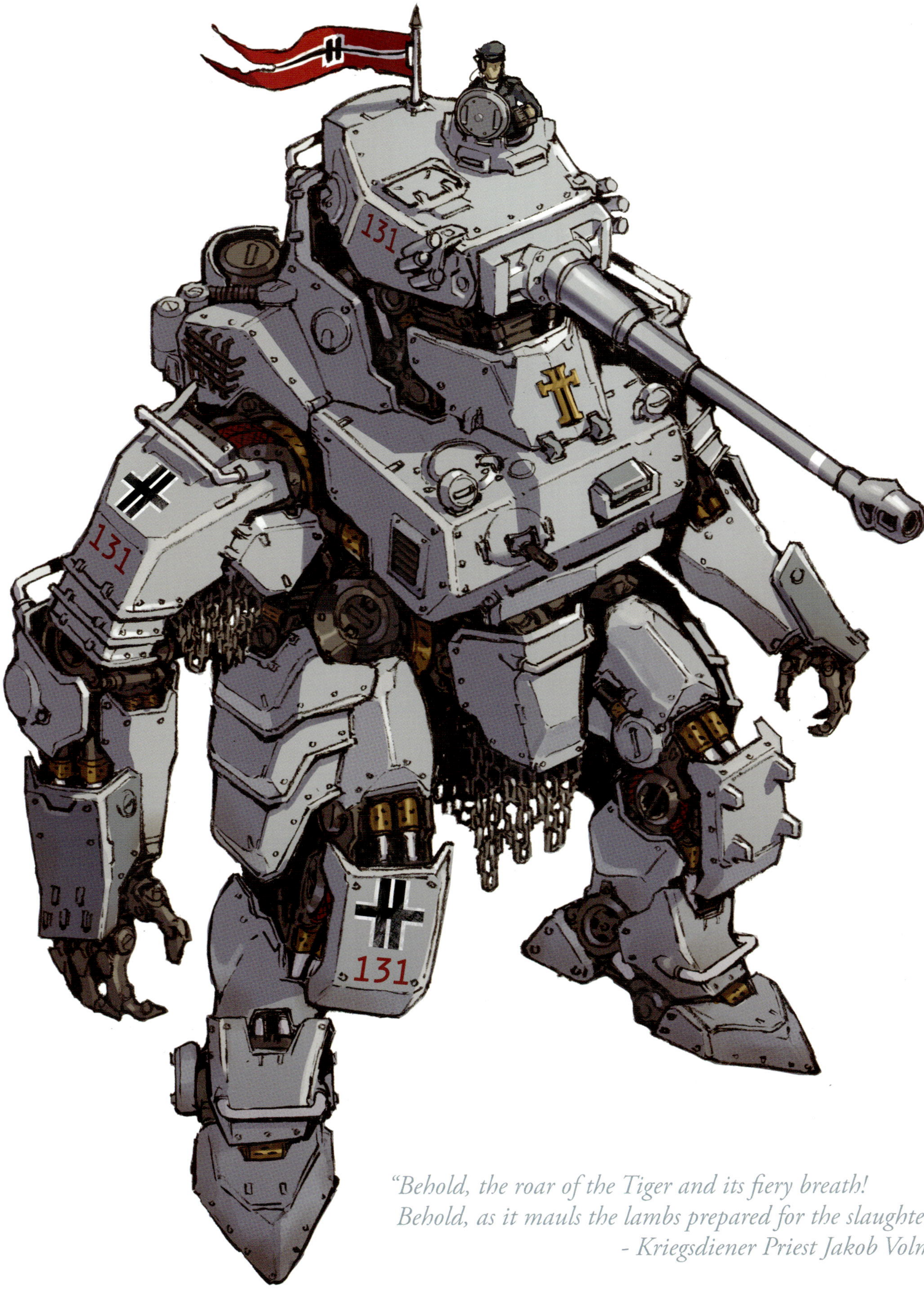

"Behold, the roar of the Tiger and its fiery breath!
Behold, as it mauls the lambs prepared for the slaughter!"
- Kriegsdiener Priest Jakob Volmar

PANZERRITTER TIGER AUSF. E

Much as the M4 Sherman became the iconic Alliance tankhead, the Panzerritter Tiger became that of the Kyzerian ground forces. In nearly every other aspect, however, the two were polar opposites, most notably in philosophy and design.

Most Kyzerian tankheads before the Tiger were utilitarian and modest in design. The Tiger was a noticeable deviation. The term "overengineered" has often been applied to the Tiger, but it should not be assumed that this characteristic was due to accident or oversight. In fact, the high-tech, complicated Tiger was always intended to be so, from the beginning, to create something truly cutting-edge.

The performance requirements for the Tiger were very high. It needed to be capable of destroying Alliance and Yamaichi tankheads while resisting enemy fire. The Kyzerians had always favored firepower over defense, but destroying a frontline tankhead required such tremendous weaponry that, in most armies, the job was left to dedicated tankhead hunters. Not so with the Tiger; it was expected to be the ultimate heavy tankhead, ruling the battlefield. The name chosen for this class of tankheads meant "armor knight." This tendency towards prestige projects would continue throughout the war, arguably to the detriment of the Kyzerian forces. Still, the Tiger can not easily be dismissed as a folly or distraction. The machine quickly garnered a reputation with the Alliance army as a devastating and nearly indestructible beast on the battlefield.

Wherever a monster arises, you will soon find monster slayers. Tactics for taking down the dreaded Tiger were quickly devised. While impressively fast for its size, the Tiger was still not nimble compared to any medium tankhead. Multiple Alliance tanks would often coordinate to destroy a Tiger unit. This was not easy, as its thick side armor, especially the overlapping shoulder and leg armor plating, was almost as tough as its frontal armor. The only reliable way to kill the Tiger was with close-range attacks to its weaker rear power plant. To this day, reports about these tankhead fights remain some of the most repeated war stories.

The reality of the Tiger's downfall is less exciting to report, being a consequence of logistics and maintenance. It was not designed with ease of maintenance in mind. The highly complex systems that made it such an advanced design left little room for accessibility, and what would have been a quick routine job that took hours on most contemporary tankheads could take days. Further compounding the issue, specialized tools and equipment could be required. Still, the Tiger was produced until the end of the war and would remain a figure of fear among New Kyzeria's enemies.

REAR VIEW

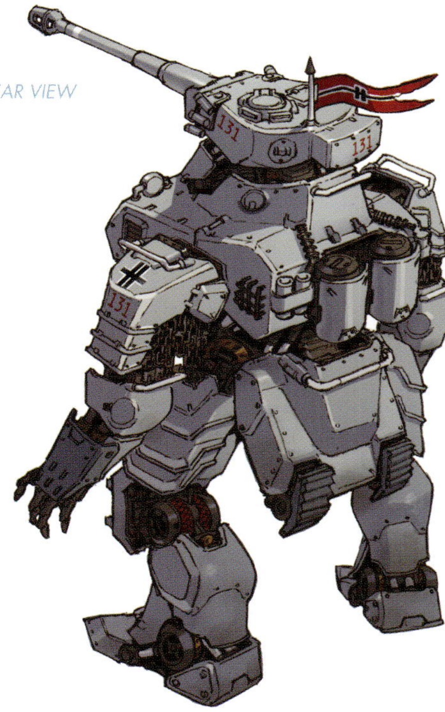

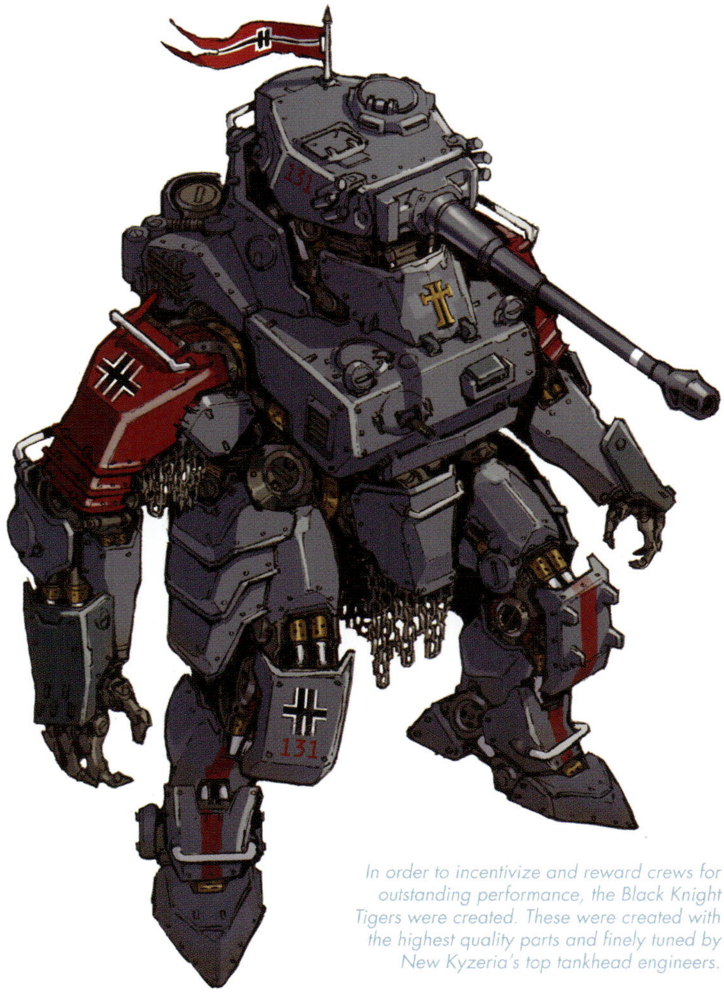

In order to incentivize and reward crews for outstanding performance, the Black Knight Tigers were created. These were created with the highest quality parts and finely tuned by New Kyzeria's top tankhead engineers.

SPECIFICATIONS

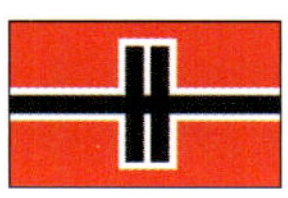

COUNTRY: New Kyzeria
UNITS BUILT: 2,347
NO. OF CREW: 5
MAIN ARMAMENT: 1 x 88 mm gun
SECONDARY ARMAMENT:
2 x 7.92 mm machine guns
ARMOR: 25–120 mm
MAX SPEED: 46 km/h

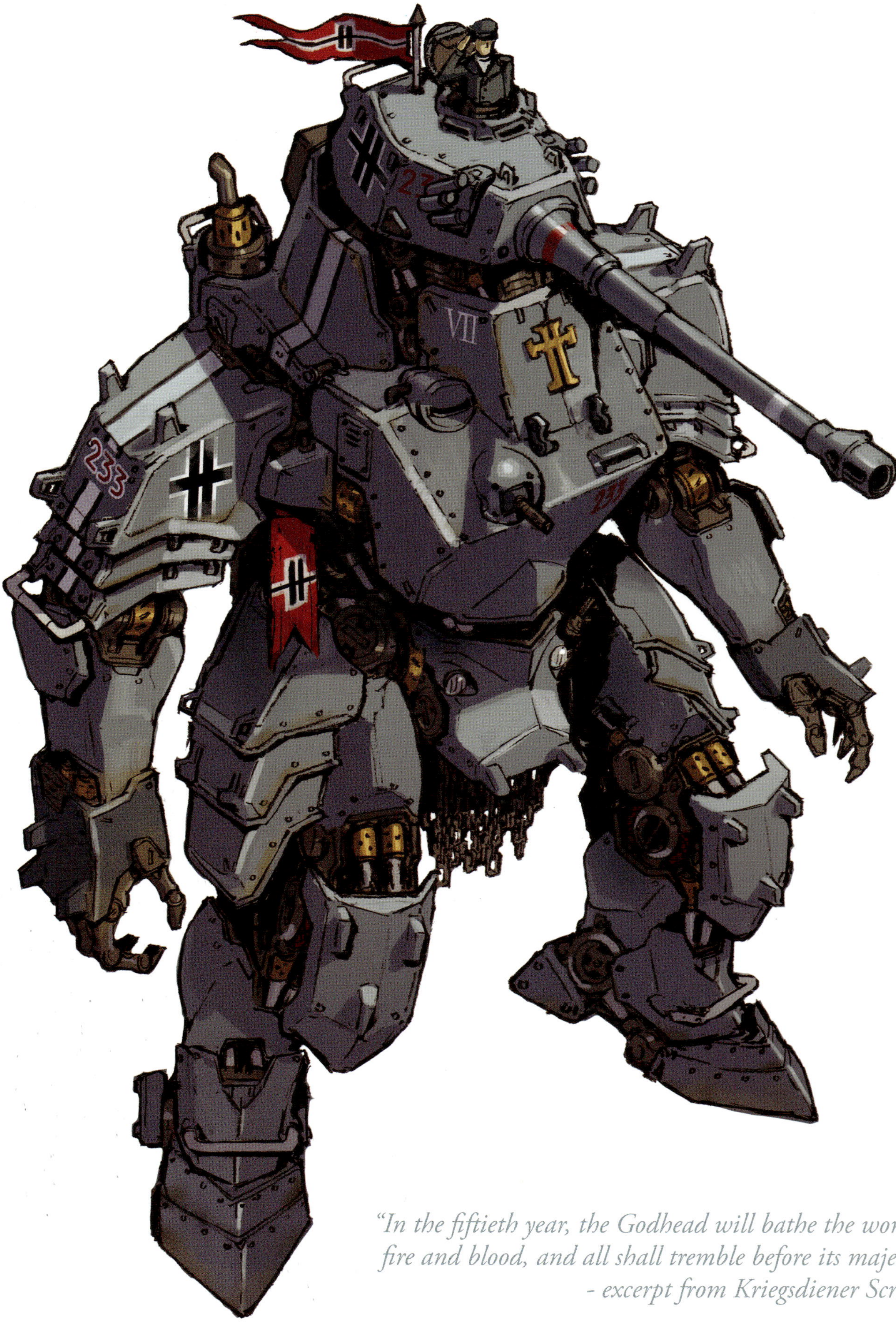

"In the fiftieth year, the Godhead will bathe the world in fire and blood, and all shall tremble before its majesty."
- excerpt from Kriegsdiener Scriptures

PANZERRITTER II KÖNIGSTIGER AUSF. B

A direct successor to the first Panzerritter heavy tankhead, the Königstiger, or King Tiger, was intended to be superior in every respect. It first entered the Neverwar in WC048, rather late in the war, as it turned out—although no one knew it at the time. It was, in fact, one of the last combat vehicles introduced by the Kyzerians to have made a significant impact on the war at all, though it was not necessarily a positive one.

Compared to its predecessor, the King sported thicker armor in all areas where it counted, especially in the front of the hull. Additionally it had more sloped armor on its chest and chassis regions—a significant upgrade over the original Tiger. Most notably, it stood taller, a true king towering over the battlefield.

Although Alliance troops at the time thought the head turret was comically small, even they knew that its weapon was far from a laughing matter. Its large size and its efficiency in almost all areas, combined with superb optics, made this one of the most deadly tankhead hunters of the war. However, overspecialization in armor penetration came at the expense of anti-infantry capabilities, which were limited to two machine guns, one of the few features that remained unchanged from the original Panzerritter design.

The leg units on the Panzerritter II were a mixed blessing. The actual mechanical configuration was ingenious and, if produced perfectly, would have been a marvel. However, at that point in the war, the Kyzerian economy was suffering; "perfect" was probably beyond its capacity. Shortages due to blockades, reduced productivity due to Alliance bombing of factories, and an outright lack of skilled laborers were just some of the many hurdles that prevented this tankhead from achieving true glory. Cost-cutting and simply making do with what was available had never been the Kyzerian way. The ego and pride of its leadership dealt poorly with any attempts to do so, no matter how necessary. A number of pneumatic interfaces in the legs—the knees and ankles, in particular—required valves that the Empire could no longer produce. At the same time, concessions were made in other areas where they should not have been. All this led to continual stumbles and breakdowns.

The first King Tigers saw combat during the infamous Operation Oberherr (WC047), when they made a fighting landing on the Liberterran coast. What was intended to be a glorious breakthrough ended in a stumbling slog. In the sand and surf, the legs of many King Tigers failed, though their weapons made more than enough of an impression on the Alliance defenders to secure their fearful reputation.

Recently-declassified documents from the Yamaichi Dominion archives also show that metallurgic testing on a pair of captured King Tigers indicated the use of steel of inferior quality, prone to fragmenting on impact. Wartime shortages were most likely to blame here as well. The study seems to answer the mystery of why King Tiger crews were sometimes killed by shrapnel, even though enemy shells had not penetrated its armor.

Altogether, the Panzerritter II King Tiger was a theoretically superb design that might have done so much more if produced to specifications. It instead revealed a fatal flaw in the Kyzerians' rigid mindset and inadaptability. It is estimated that, during its early days, only one in twelve King Tigers was combat ready when deployed. Still, the modern historian looking back should not undersell its offensive capability or its overall toughness. Its reputation was won in the field, despite its many flaws. In the big picture, however, it is reasonable to conclude that the energy, materials and time spent on the Königstiger's production could have been better invested in other, less ambitious units. As the saying goes, hindsight is 20/20, and the Kyzerian commanders did not conceive of the war ending in WC049. Instead, they still dreamt of baptizing the 50th year of war with the launch of the Godhead, and the kings of the battlefield—the King Tigers—confidently striding beside it.

SPECIFICATIONS

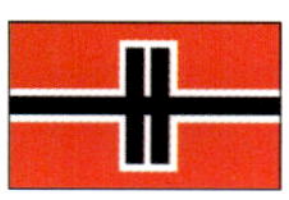

COUNTRY: New Kyzeria
UNITS BUILT: 692
NO. OF CREW: 5
MAIN ARMAMENT: 1 x 88 mm gun
SECONDARY ARMAMENT:
2 x 7.92 mm machine guns
ARMOR: 25-185 mm
MAX SPEED: 42 km/h

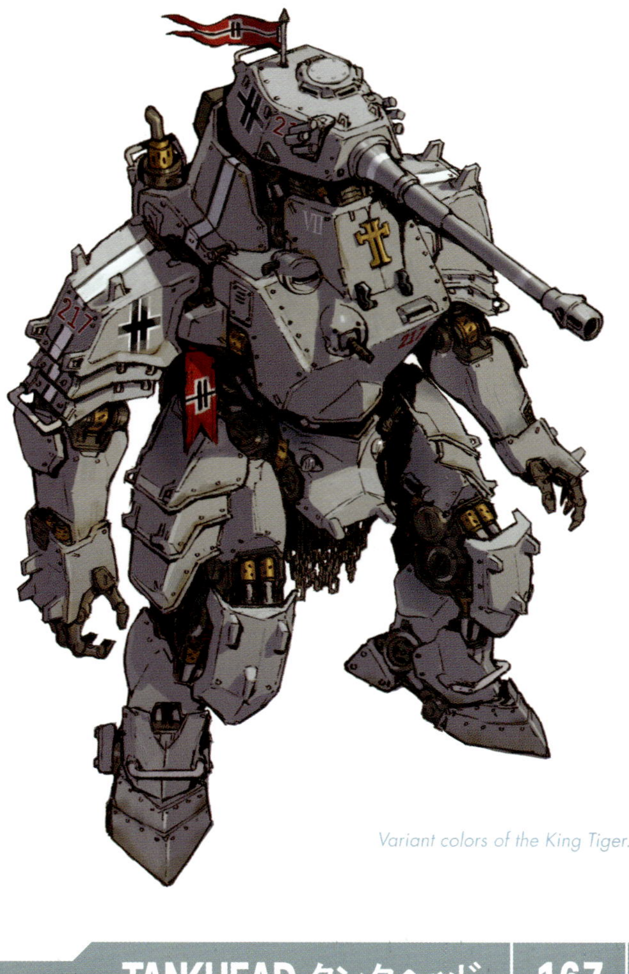

Variant colors of the King Tiger.

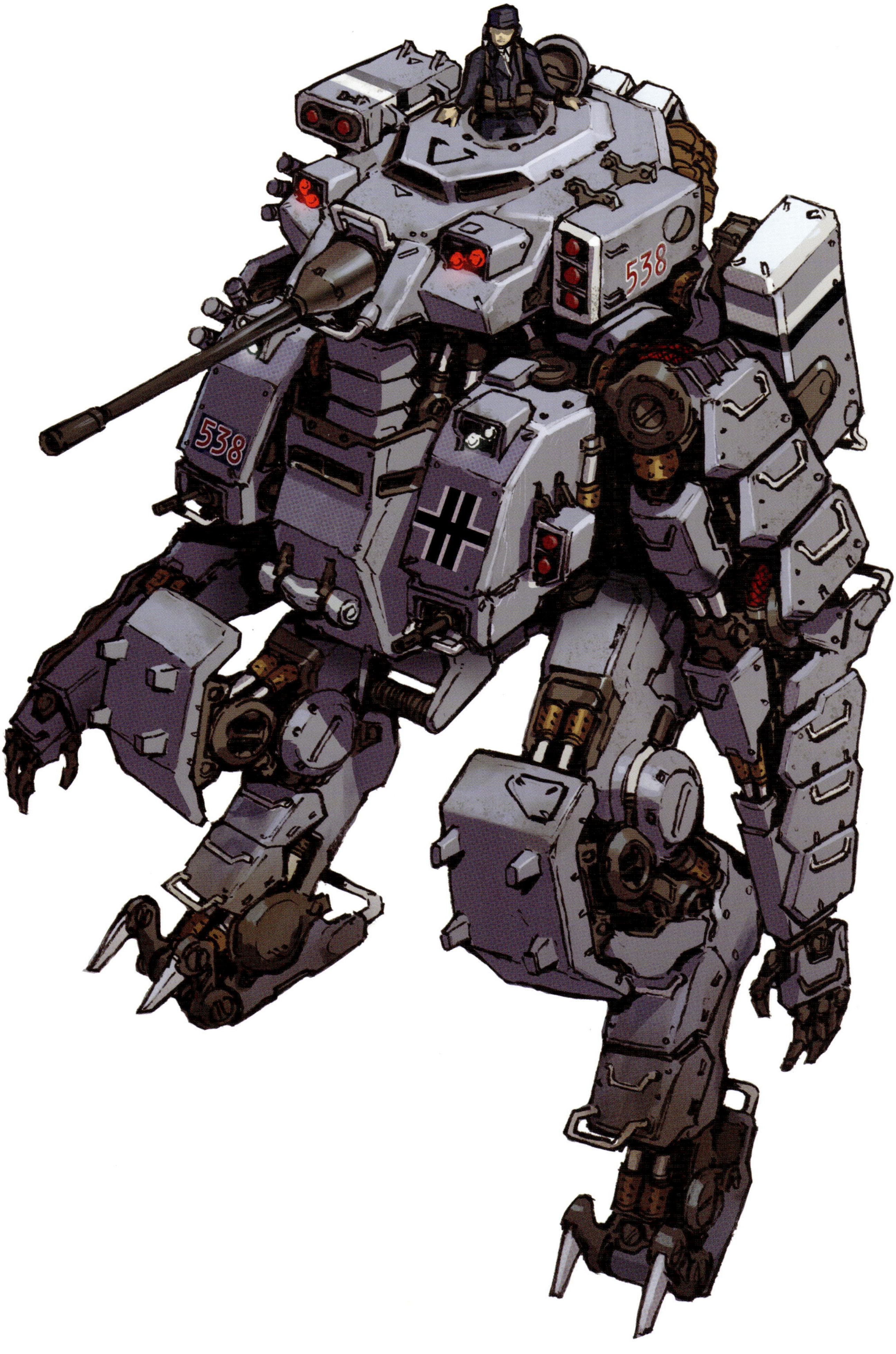
538
538

PANZERKOPF X DORN

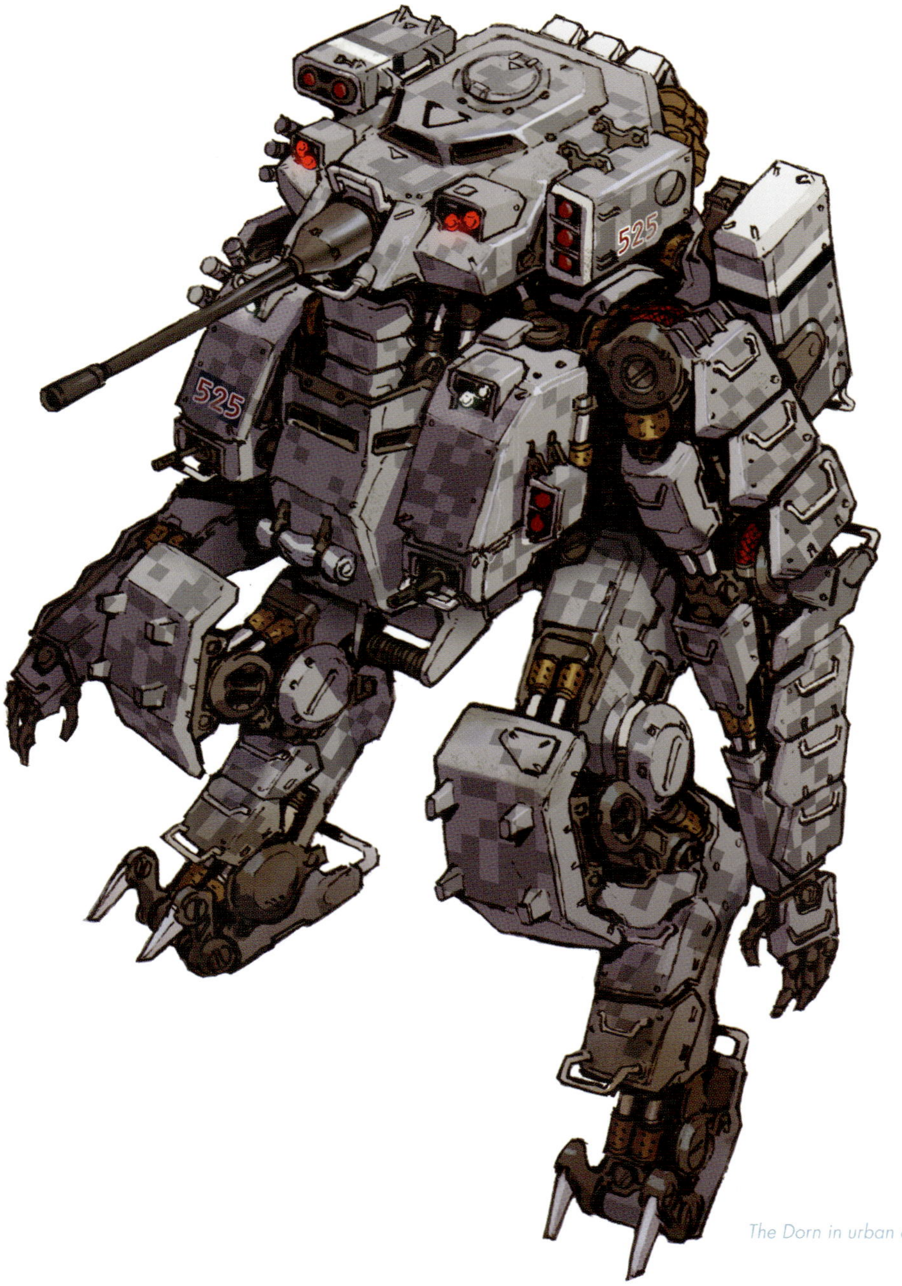

The Dorn in urban camouflage colors.

SPECIFICATIONS

COUNTRY: New Kyzeria

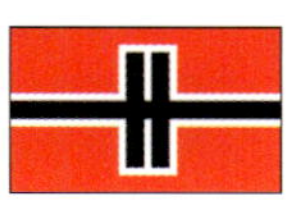

UNITS BUILT: 19
NO. OF CREW: 4
MAIN ARMAMENT:
5 x anti-tank missile launchers
SECONDARY ARMAMENT:
1 x 25 mm chain gun
2 x 7.92 mm machine guns
6 x smoke grenade launchers
ARMOR: 14–75 mm
MAX SPEED: 56 km/h

In WC043, Kyzerian forces were able to intercept and capture a Yamaichi convoy transporting tankhead armor plating. The Kyzerians retrofitted the captured parts with their own technology to create a terrifying hybrid, the Dorn. Yamaichi armor craftsmanship and Kyzerian material sciences and mechanical engineering proved a formidable combination in combat. Logistically speaking however, it was also an extremely demanding tankhead to build in both material and time investment. For this reason, modern historians generally agree that New Kyzeria could never have produced enough of them to change the outcome of the war even if it had had steady access to Yamaichi semi-ceramic armor.

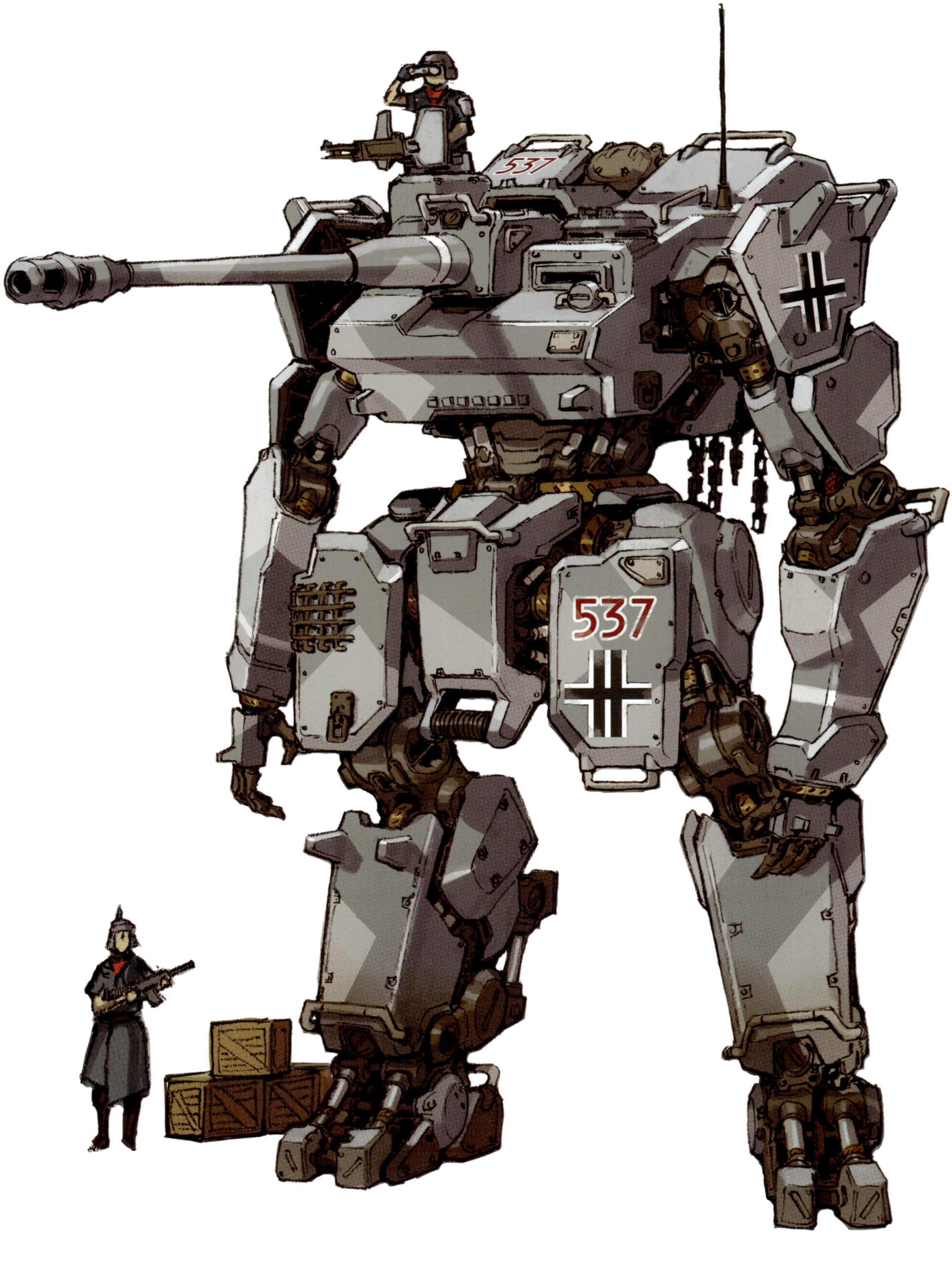
537
537

STURMGESCHÜTZ V AUSF. A

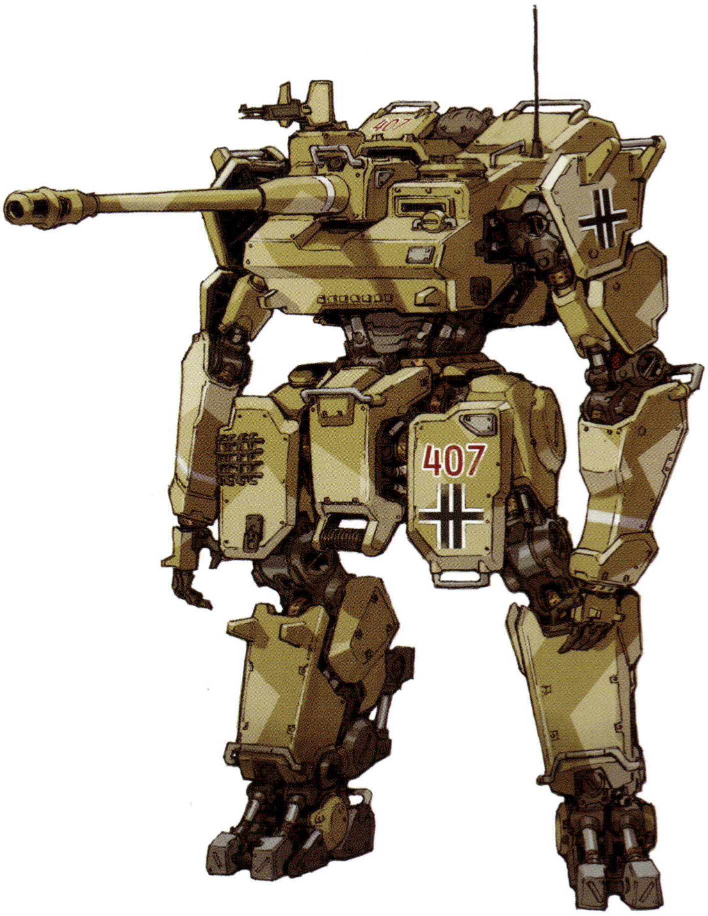

The StuG V in desert camouflage.

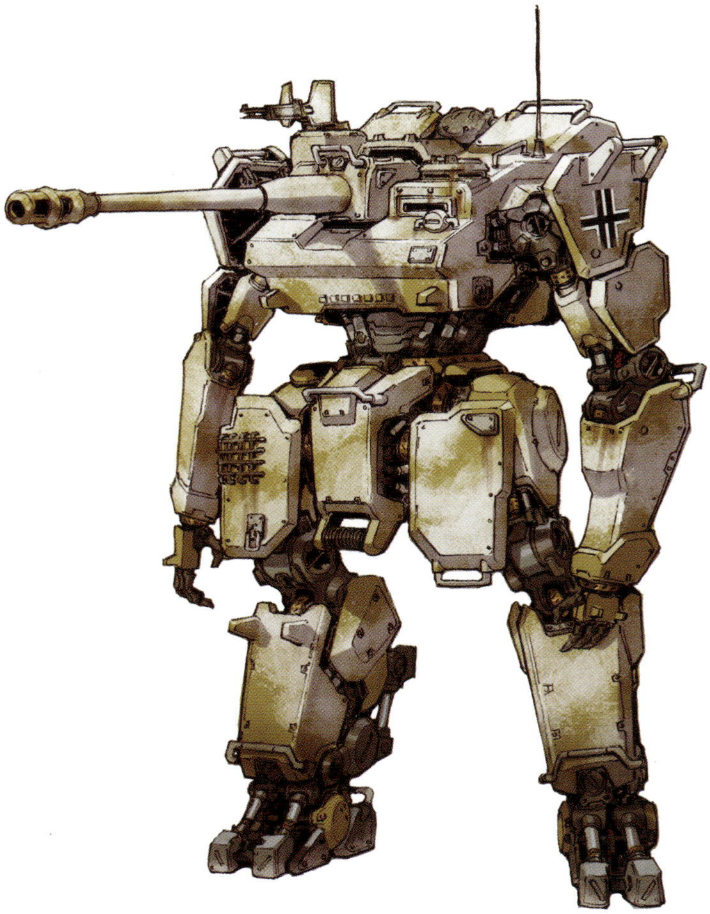

The StuG V in winter camouflage.

SPECIFICATIONS

COUNTRY: New Kyzeria
UNITS BUILT: 8,090
NO. OF CREW: 4
MAIN ARMAMENT: 1 x 100 mm gun
SECONDARY ARMAMENT:
1 x 7.92 mm machine gun
6 x smoke grenade launchers
ARMOR: 16–100 mm
MAX SPEED: 45 km/h

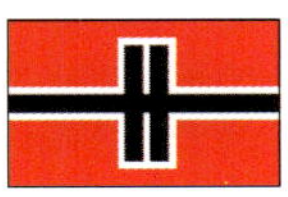

Halfway through development, Kyzerian Command decided to convert a portion of the Panzerkopf V into the Sturmgeschütz V. This was not only due to the successful performances of the StuG III and IV tank hunter variants out in the battlefield, but also as a cost- and material-saving measure. The production of a StuG type did not require a turret or the complex pneumatic components that the Panzerkopf did.

HAUBITZENSCHREITER WESPE

When the Panzerkopf II proved obsolete as a main battle tankhead, a number of conversion projects were initiated. These ran the gamut from stripped scout vehicles to recovery systems, but mainly focused on combining the agile frame with aging but still effective weapon systems. The first Wespe tests mounted a single 10.5 cm smoothbore gyrocannon, the mainline artillery field gun contemporaneous with the Panzerkopf II. The mobility of the frame remained mostly intact but it was found to be nearly impossible to fire the weapon without the recoil toppling the tankhead. A series of trials using more and more outlandish poses were undertaken, going so far as to experiment with supine and prone firing positions. Deeming these poses impractical or simply too undignified for a weapon of the Kyzerian army, the development team later settled on short stabilizer feet mounted to a new back skirt. Ever efficient, the Kyzerian engineers designed this area to double in function as ammunition storage.

Challenges remained, however, as any stabilizer leg long enough to support the machine in a standing position was too fragile to withstand the rigors of firing for long. Again, the engineers turned to posing and this time they found success. A squatting position, with the legs of the tankhead resting on specialized brackets on its haunches, interlocking with its lower leg armor braced by a pair of short thick stabilizers in the skirt finally solved the issue.

Assuming this posture was a slow process in the field and, in production, required production of a number of special parts, rather in excess of the design brief. In an effort to appease the design committee, the engineers took a brute force approach and mounted twin cannons instead, effectively doubling the firepower mobilized with each conversion kit and halving the cost.

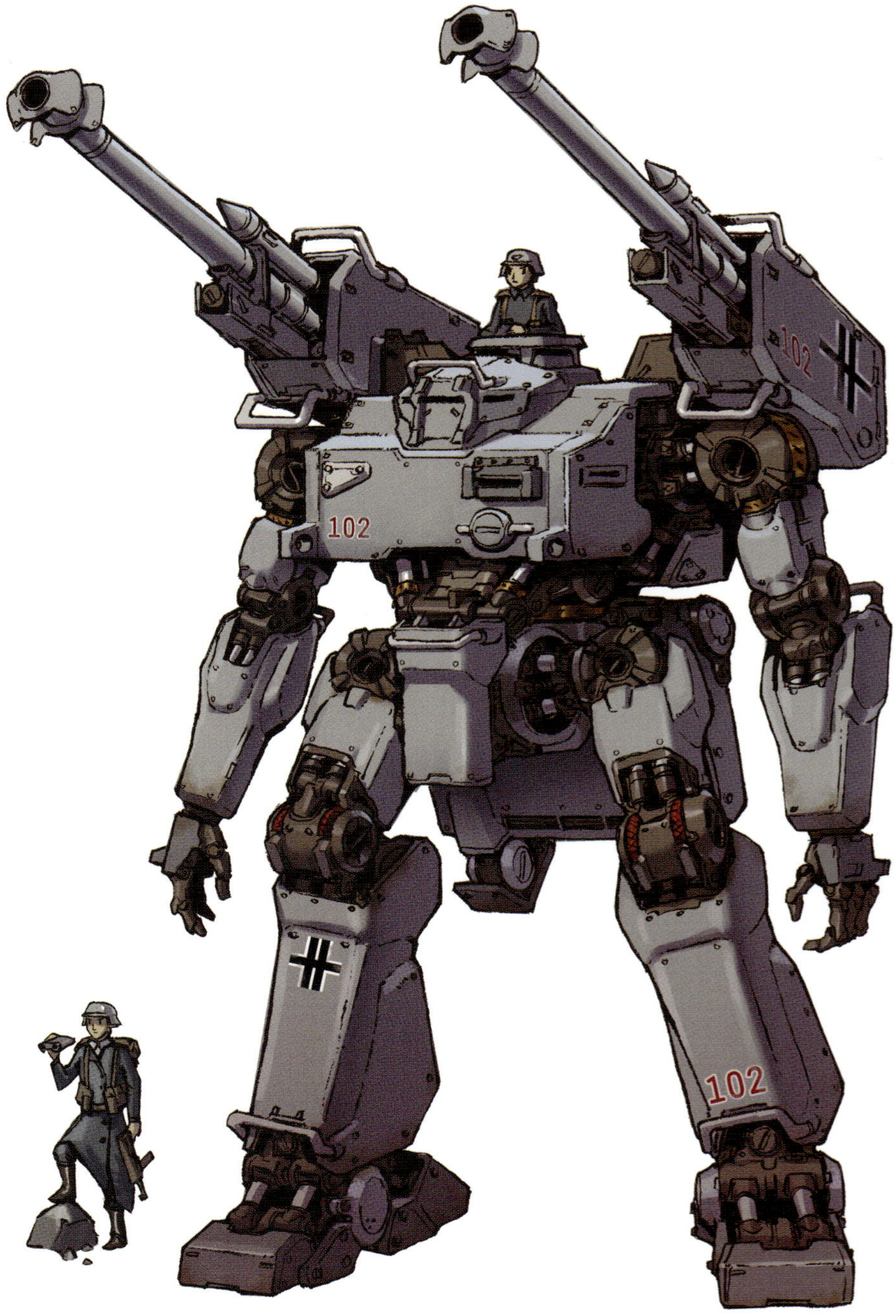

The new unit was named Wespe and it proved well-suited for its intended role. Soon after, all Panzerkopf II production was earmarked for the sole purpose of producing further Wespe units. The vast majority of the Wespes served in the land war against the Yamaichi, who left remarkably few records concerning their encounters with the machines. Most battle reports designated the Wespes as mechanized artillery, without calling out any peculiarities or properties. Against Alliance forces, however, the Wespe was more notorious. Its dual-cannon system initially acquired it the moniker "double trouble," despite it being simply another average gun among many more impressive and overengineered weapon systems.

SPECIFICATIONS

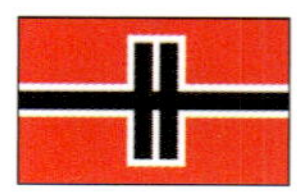

COUNTRY: New Kyzeria
UNITS BUILT: 1,276
NO. OF CREW: 6
MAIN ARMAMENT:
2 x 105 mm light howitzers
ARMOR: 5–30 mm
MAX SPEED: 34 km/h

MINENRÄUMER PANZERKOPF

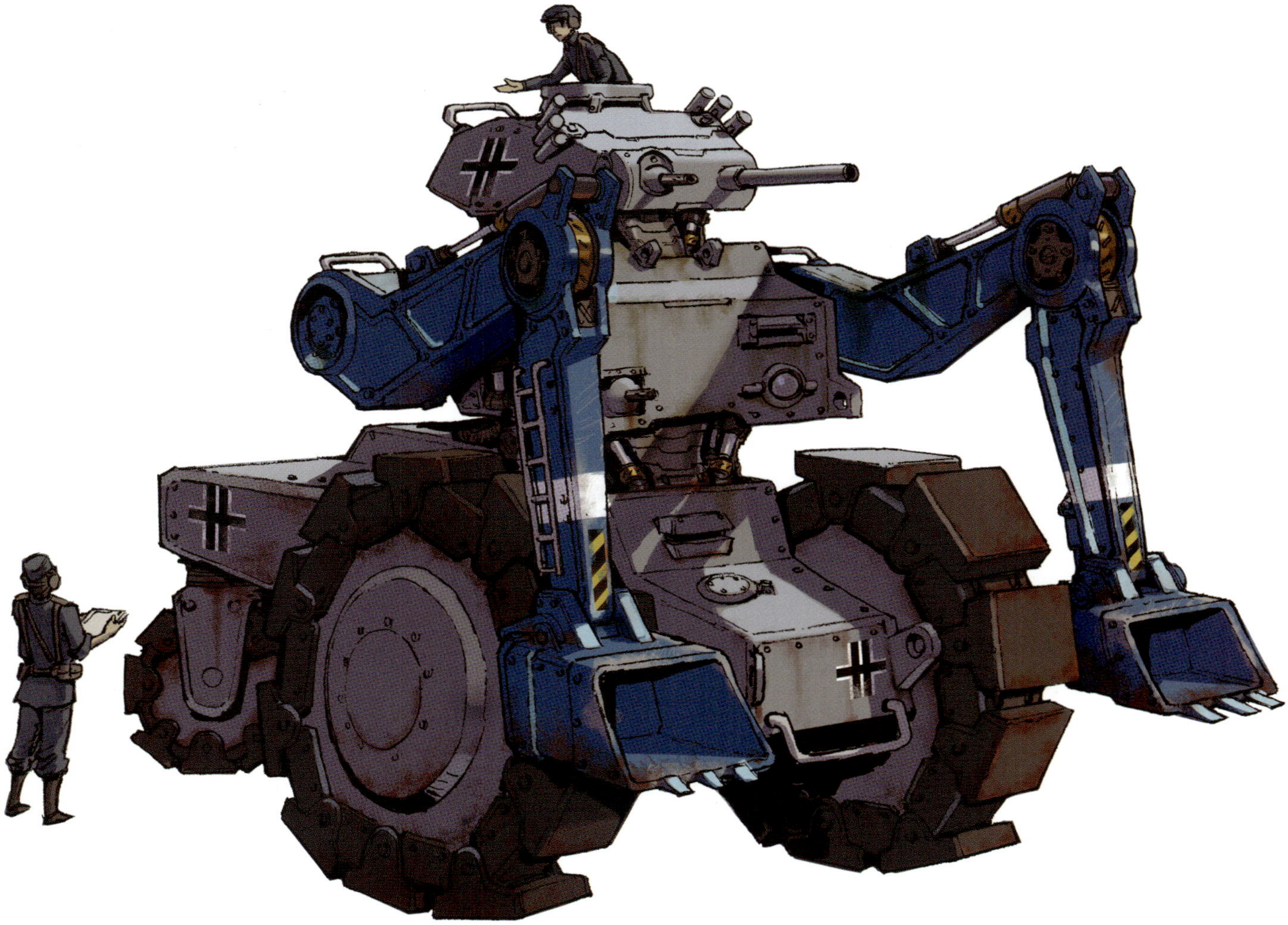

The Minenräumer was a very early example of tankhead technology being adapted for engineering purposes. Note the padded wheels designed to spread weight on soft ground in the one shown here. Such wheels were also designed to take the brunt of any mines the Minenräumer might miss while clearing the way.

"Ask to drive a tankhead, you get a Minenräumer instead."
-From the Kyzerian Army Marching Song 'To Triesburg and back'

SPECIFICATIONS

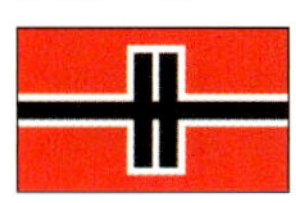

COUNTRY: New Kyzeria
UNITS BUILT: 959
NO. OF CREW: 4
MAIN ARMAMENT: 1 x 37 mm gun
SECONDARY ARMAMENT:
2 x 7.92 mm machine guns
6 x smoke grenade launchers
ARMOR: 7–25 mm
MAX SPEED: 10 km/h

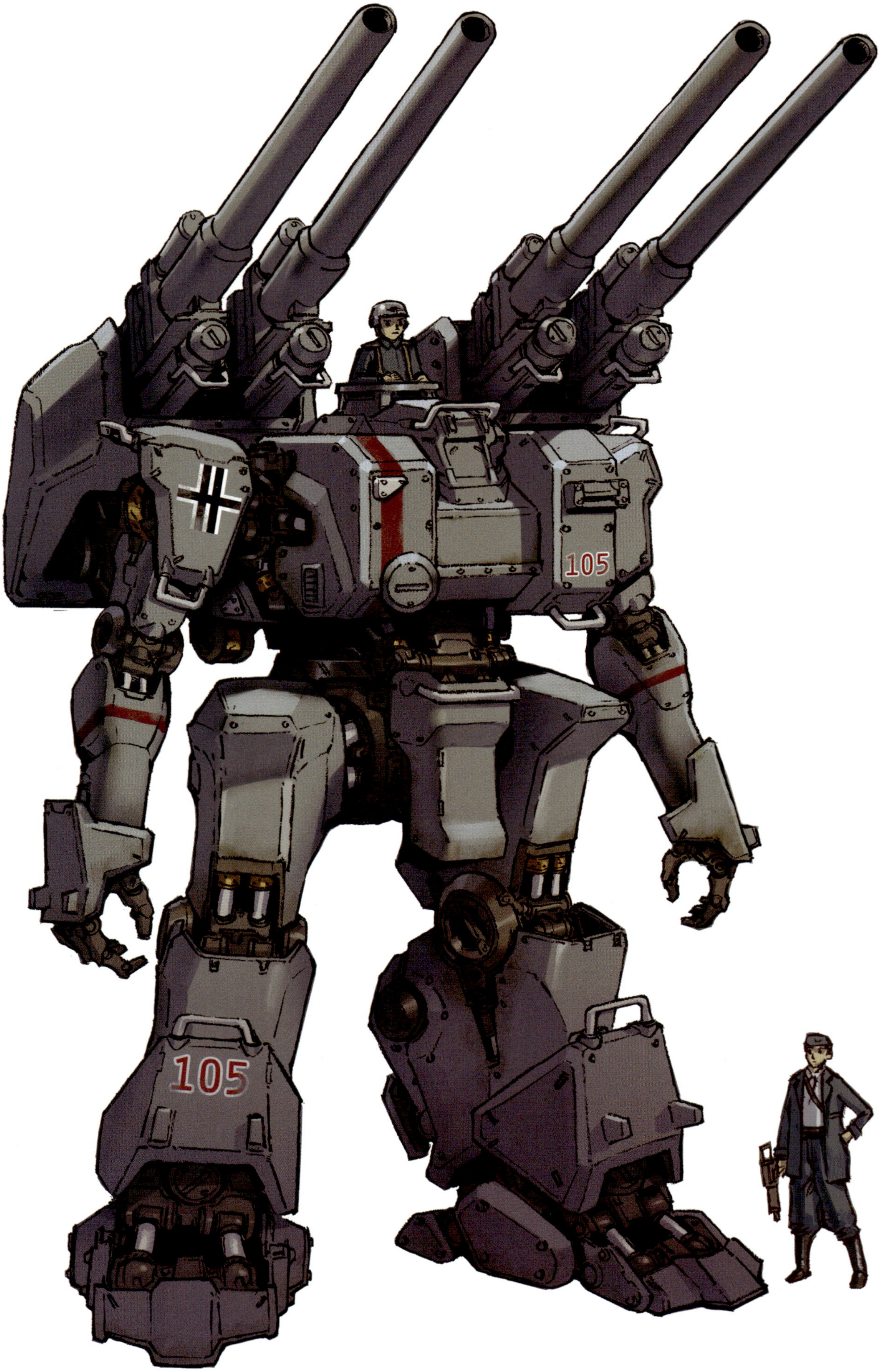
105
105

SCHWERER HAUBITZENSCHREITER HUMMEL

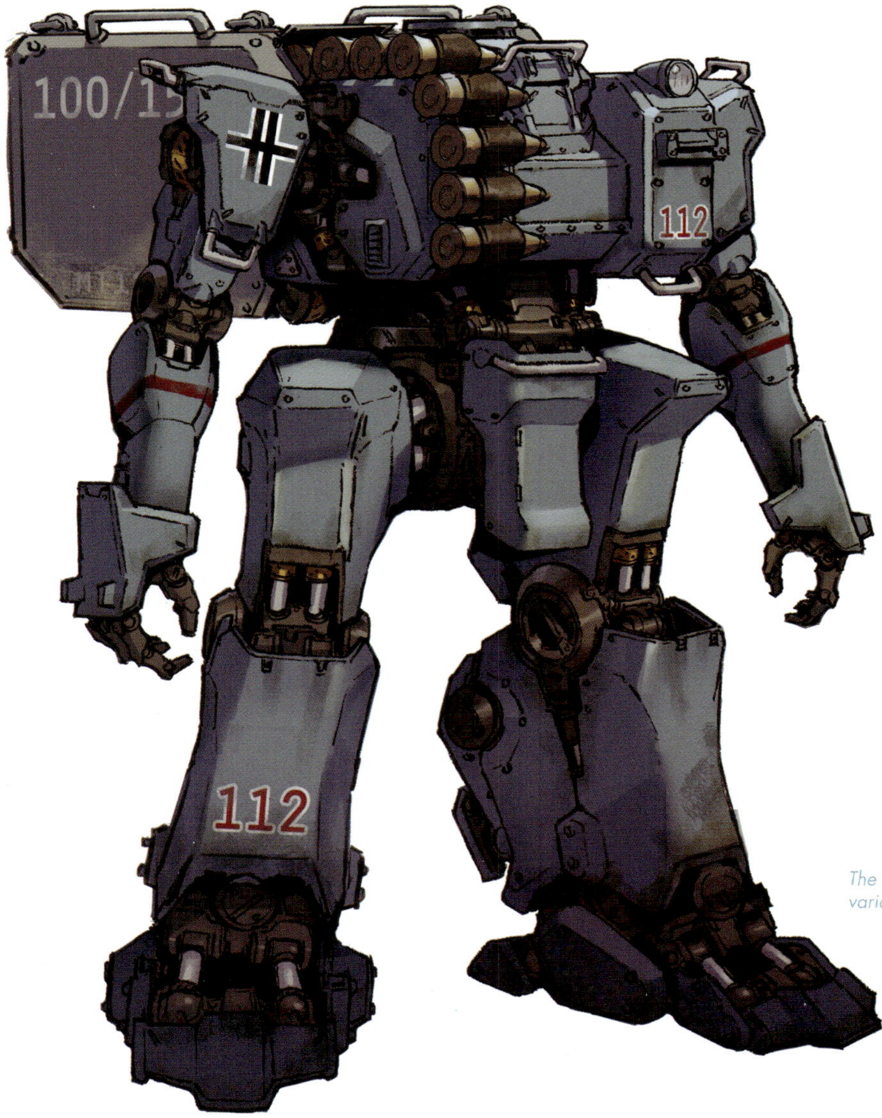

The ammo transport variant of the Hummel.

SPECIFICATIONS

COUNTRY: New Kyzeria
UNITS BUILT: 1,414
NO. OF CREW: 7
MAIN ARMAMENT: 4 x 150 mm howitzers
ARMOR: 10–30 mm
MAX SPEED: 40 km/h

The Hummel was a hybrid heavy-artillery piece based on the Panzerkopf III and IV chassis. In many regards, it was a larger and more heavily armed counterpart of the Wespe. Its quad artillery guns were effective and the Hummel would prove an adequate, if unremarkable, design. One of its main flaws was that the machine could not transport enough ammunition to keep its four guns firing for long. To combat this problem, an ammo transport variant was developed, though it was designed to be convertible to the fully-armed model with minimal tools.

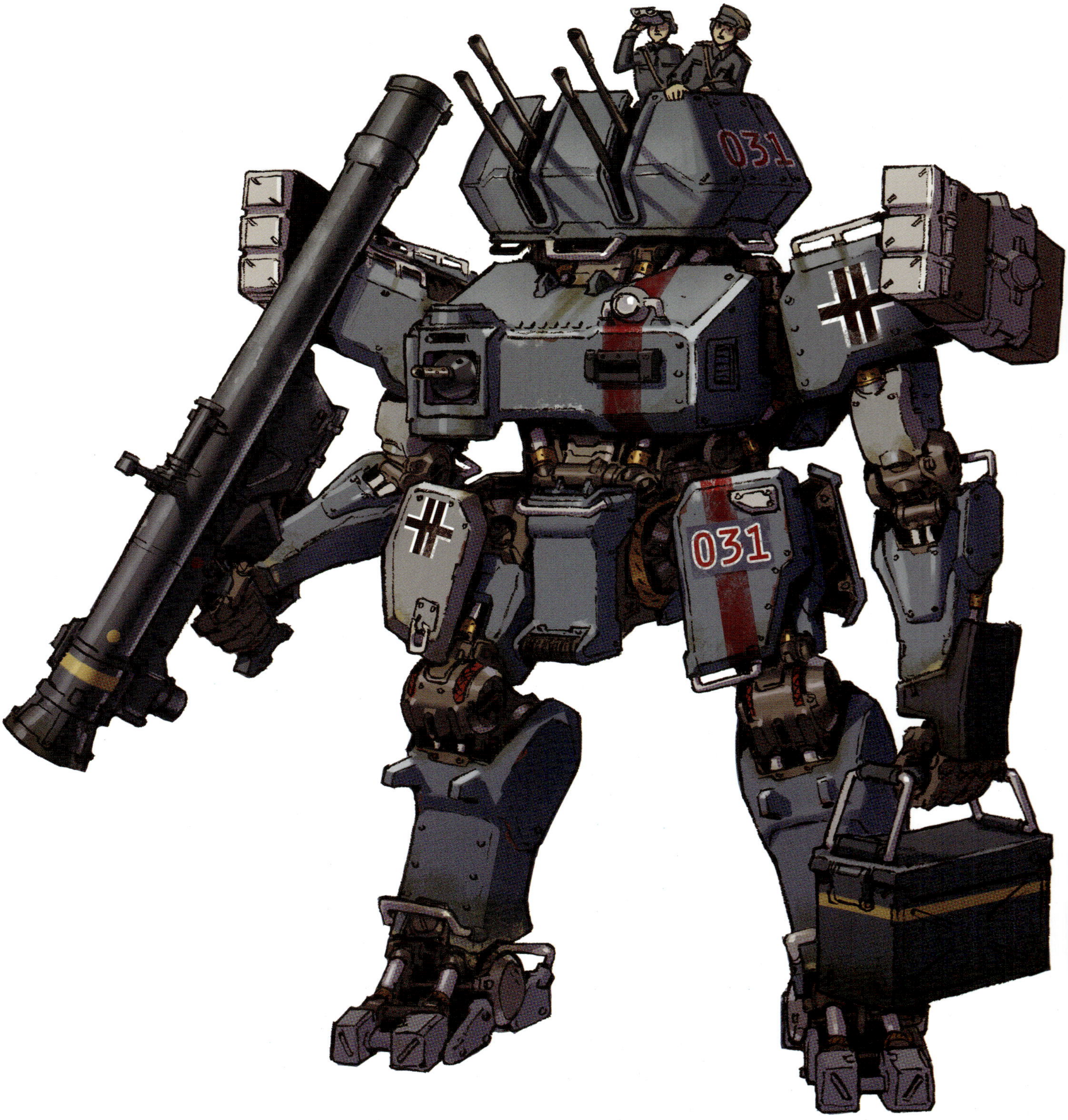
031
031

FLAKPANZERKOPF IV WIRBELWIND

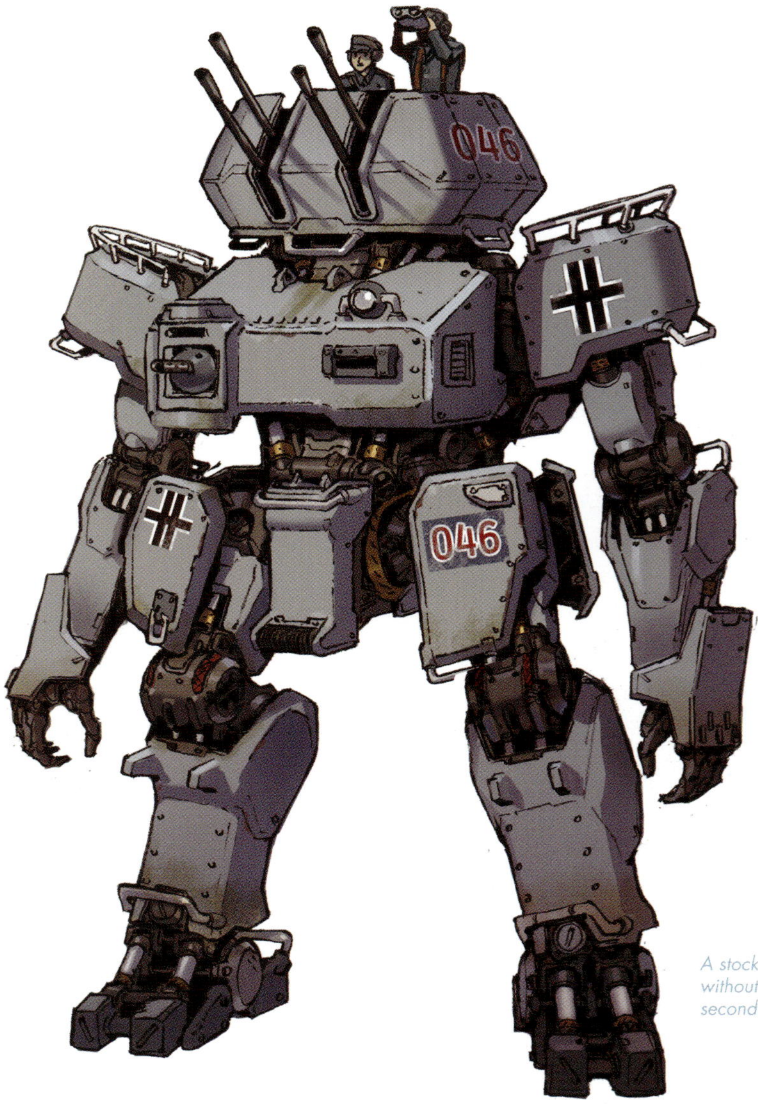

A stock Wirbelwind without its additional secondary armaments.

SPECIFICATIONS

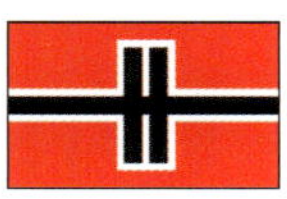

COUNTRY: New Kyzeria
UNITS BUILT: 445
NO. OF CREW: 5
MAIN ARMAMENT:
1 x quad 20 mm anti-aircraft guns
SECONDARY ARMAMENT:
1 x 7.92 mm machine gun
1 x handheld wire-guided surface-to-air missile launcher
6 x barrage air-deterrent rockets
ARMOR: 10–80 mm
MAX SPEED: 40 km/h

While capable of a blistering rate of fire with its quadruple anti-aircraft guns, the Wirbelwind had trouble taking down the fast and well-armored jetheads of the late war. Unlike most tankheads, its turret head was open-topped to vent gun smoke. The saved weight allowed for supplementary handheld weaponry to compensate for its lackluster main armament. The shoulder-mounted launchers seen here were a very late addition. These launchers fired rockets which trailed metal cables. At their zenith, the rockets would deploy small parachutes, allowing them to stay in the air for a limited amount of time. If a jethead were to pass through the dangling cables, the cables would inevitably get tangled in the propellers or air intakes, dragging in the rest of the cable and the rocket with it.

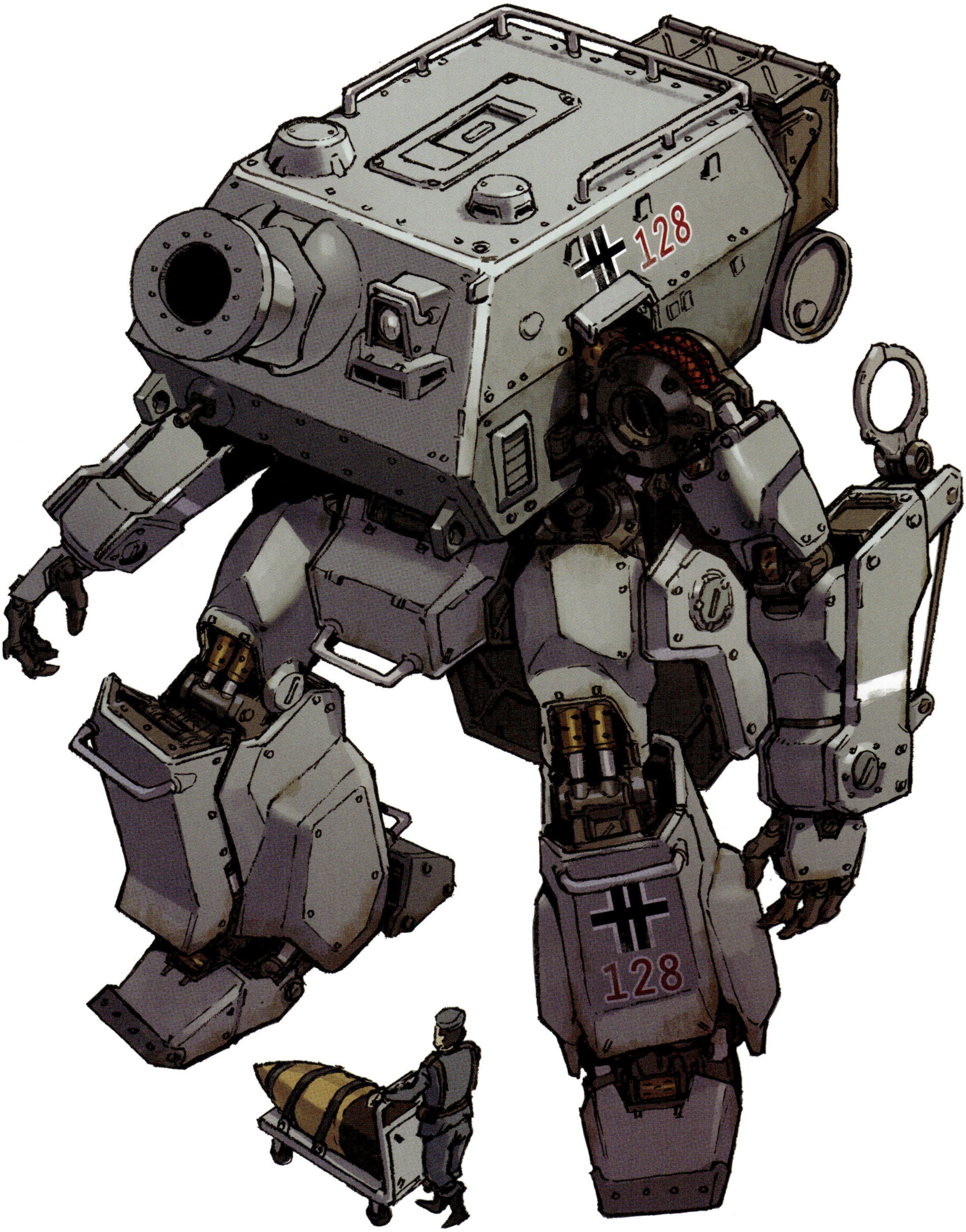
128
128

MÖRSERKOPF STURMTIGER

In the second half of the Neverwar, following a period of relative quiet, combat shifted away from traditional battlefields and entrenched fronts towards running battles across the countryside, culminating in close-quarters urban combat. While the Kyzerian tankheads of this era were strong in dynamic, mobile engagements, they performed poorly in the cramped environments of cities and towns. Despite being well armored, the leg joints on advanced weapons platforms like the Panzeritter Tiger were susceptible to infantry or other tankheads using short range anti-tankhead weapons from cover behind buildings.

By this point, the need for a weapon capable of demolishing a fortified building in a single shot had become apparent to the Kyzerian army's leadership. In response to this need, the Sturmtiger was born. Though based on the chassis of the Panzerritter Tiger, its armor was redesigned from the ground up. Its designers took particular care to reinforce defense against attacks from low-trajectory origins, resulting in thick plated legs with spaced anterior armor on the shins, as well as side and back skirts to protect the vulnerable hip joints. The lower arms had solid plating, but the shoulders were left entirely unarmored, as an attack from above was considered unlikely. Topping all of this, a solid plated box formed the torso, giving this mortarhead its iconic look.

The Sturmtiger was originally intended to be equipped with the 210 mm gyrohowitzer that was in development at the time. However, that process had encountered delays, so Kyzerian engineers sought an alternative. In an uncharacteristically flexible and creative move, the Kyzerians looked to naval weaponry development. The naval department had just developed a depth charge launcher, intended for use against subheads at greater ranges, that used a conventional (non-gyro) rocket engine to propel the charge over five kilometers. Almost by chance, this weapon was tested on dry land against fortifications and was found to be tremendously effective.

Adapting this weapon platform to its new role proved remarkably simple. The large grenades were equipped with either the largest explosive charges they could carry or similarly-sized shaped charges for bunker busting. The peculiarities of such a weapon did, however, impose a number of problems for the design of the mortarhead itself. Gyrojets shot relatively fast-accelerating shells that cleared the barrel quickly enough for the smoke and gas that built up to be extracted by a fan. The Sturmtiger's new weapon, due to its impressive weight, took much longer to get moving and its heavy rocket engine blasted tremendous amounts of fumes into the hull compartment. A system for evacuating these fumes was fitted around the barrel, producing the iconic pepperbox-style muzzle.

The weapon system that emerged from all of this was so short, with such a long shell, that you could actually see the grenade sitting in the barrel when it was loaded—an intimidating sight for those unlucky souls who encountered it up close. Only half a dozen shells could be stowed in the hull of the mortarhead, so the rest were stored in its backpack and lower leg compartments. The shot was much too heavy to handle by hand, so the tankhead was equipped with a specialized crane on its left forearm for hoisting shells up to its ammo hatch on the left back plate.

For all of its devastating firepower and outside-the-box thinking, this mortarhead failed to have the impact its tremendous weapon would suggest. The development process took too long and personal conflict between Vaimund II and Vailinde IV, the ruling twins, resulted in further delays. The first Sturmtigers ended up serving in rural conflicts at first, and then in urban defense rather than assault. Alliance troops that faced or captured one of the rare Sturmtigers were impressed by its unique weapon, ensuring the design's fame to this day.

SPECIFICATIONS

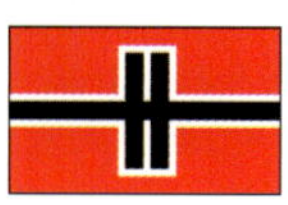

COUNTRY: New Kyzeria
UNITS BUILT: 33
NO. OF CREW: 5
MAIN ARMAMENT: 1 x 380 mm rocket launcher
SECONDARY ARMAMENT:
1 x 7.92 mm machine gun
ARMOR: 62–150 mm
MAX SPEED: 40 km/h

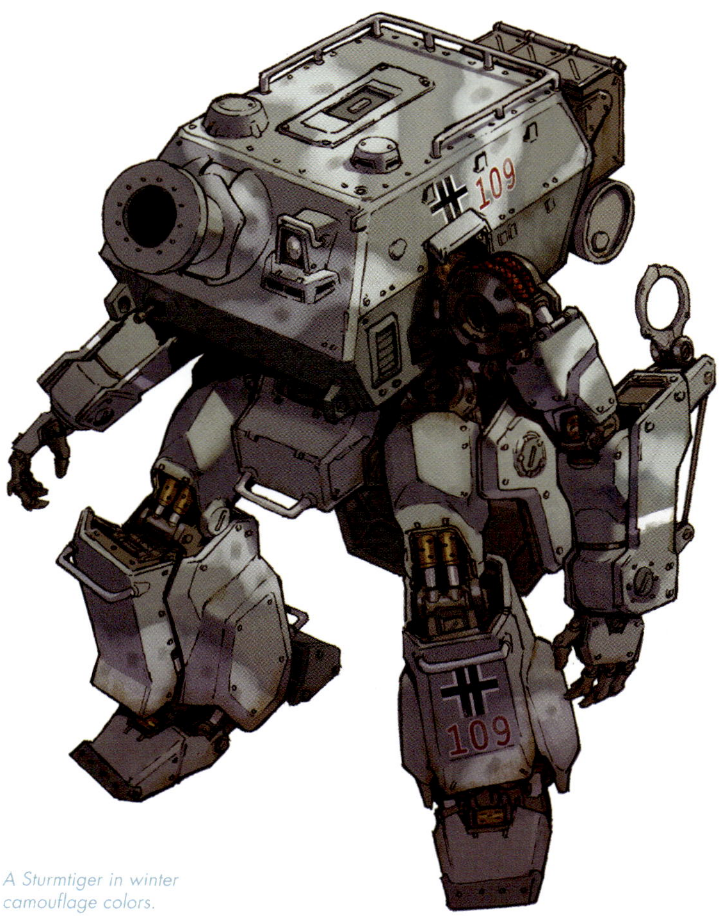

A Sturmtiger in winter camouflage colors.

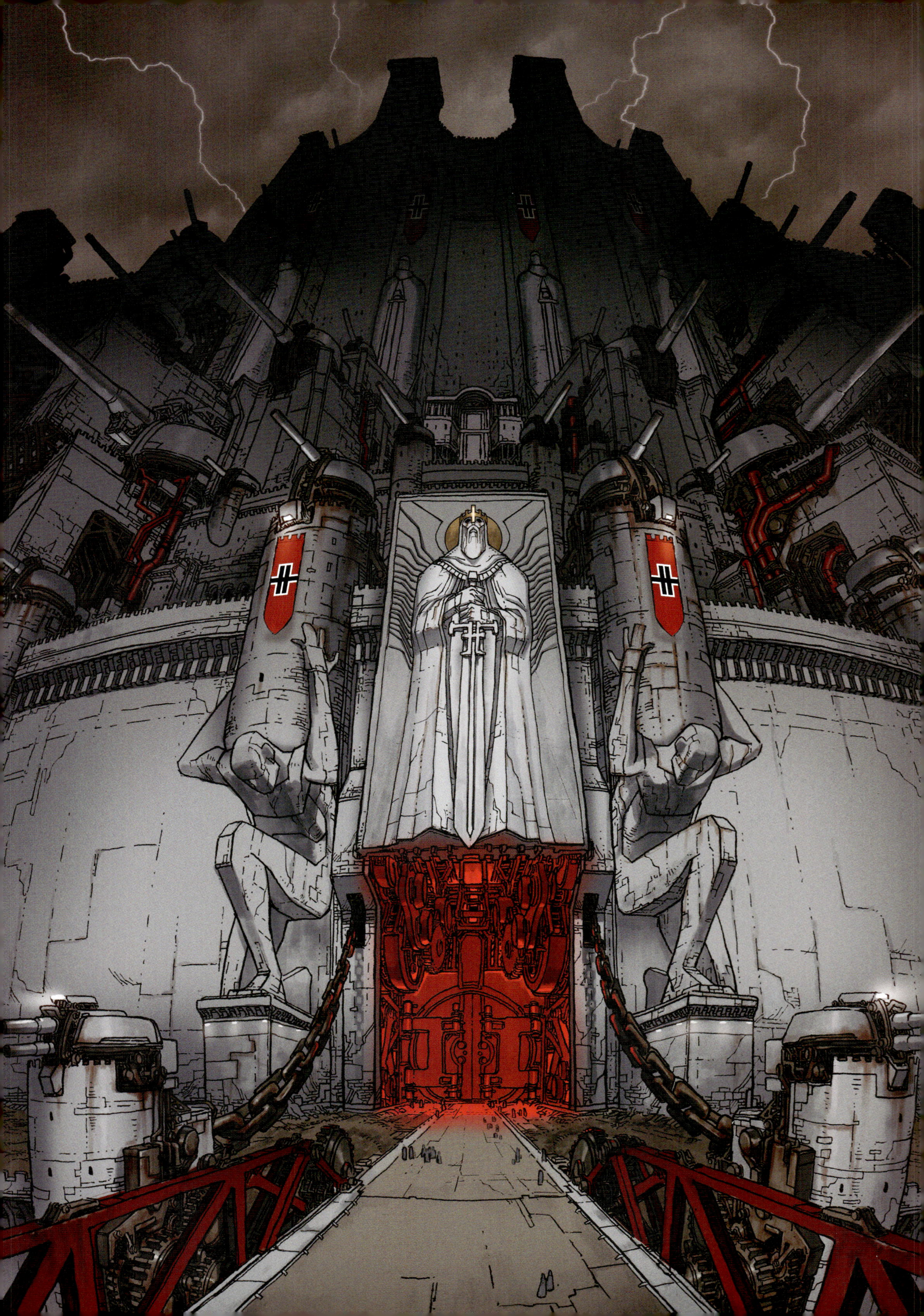

TRIESBURG WEST GATE

*"The young recruits complain that we have too many guns on the West Gate.
They do not know it's for that reason we have no need to fire a single one of them."*
- Kyzerian General Gustav Hausen

The fortification known as the "Triesburg West Gate" has always been famous as one of the oldest defenses still in use in New Kyzeria. Triesburg is located between a river crossing and the largest pass through the Stahlbirge Mountains, which allowed its gigantic gate to control traffic through the area for centuries. A large part of the current defenses were built by the Triesburgers during the Kyzerian conquest era in the hope of keeping the Triesburg city-state independent of the Kyzerian Empire. Triesburg was eventually conquered in BWC 183 by the Kyzerians and became Kaiser Bogomar III's headquarters for a while. The original gate was broken during the siege and was replaced by the current one, depicting Ansas, father of Ennike the Conqueror. The Kyzerians have always taken great pride in the continual upgrades and overhauls of their gate's tremendous firepower, which was long held up as the unbreakable western gate to the Kyzeria heartlands, forming an impenetrable wall along with the Stahlbirge peaks and the Taubensee Lake to the north.

During the early years of the Neverwar Triesburg was a vital staging ground for forces deploying to the Southern Front, being the only reliable railway route north of the Ramziegel line. The infamous Triesburg defenses were also heavily relied upon as they prevented any Alliance breakthrough into Kyzeria by the Triesburg pass. This meant that any invaders would need to go the long way round by the Ramziegel route, allowing the Kyz to aggressively deploy their forces to the Southern Front, confident in their defenses.

This exact overconfidence would eventually play a significant factor in Operation Grandmaster Gamble and the fall of the Kyzerian Empire. While the Stahlbirge remain a daunting obstacle, technology had advanced so far by the end of the war that the latest generation of tanks and tankheads were in fact capable of traversing other more precarious passes through the mountains. The Kyzerian War Ministry had received reports on this fact, but they deemed it a passable weakness. They reasoned that while a force might penetrate through the passes, they would not support any logistical network to speak of. As the report stated, any strike force using such a route would effectively be on a suicide mission, cut off from any supply lines the moment they crossed the mountains. This was exactly what the Alliance Southern Special Strike Force ended up doing during the Grandmaster Gamble. Outrunning the Kyzerian defenders once they broke through, they bypassed Triesburg entirely through the Barkas Pass in the East and rapidly advanced north to assault the Godhead.

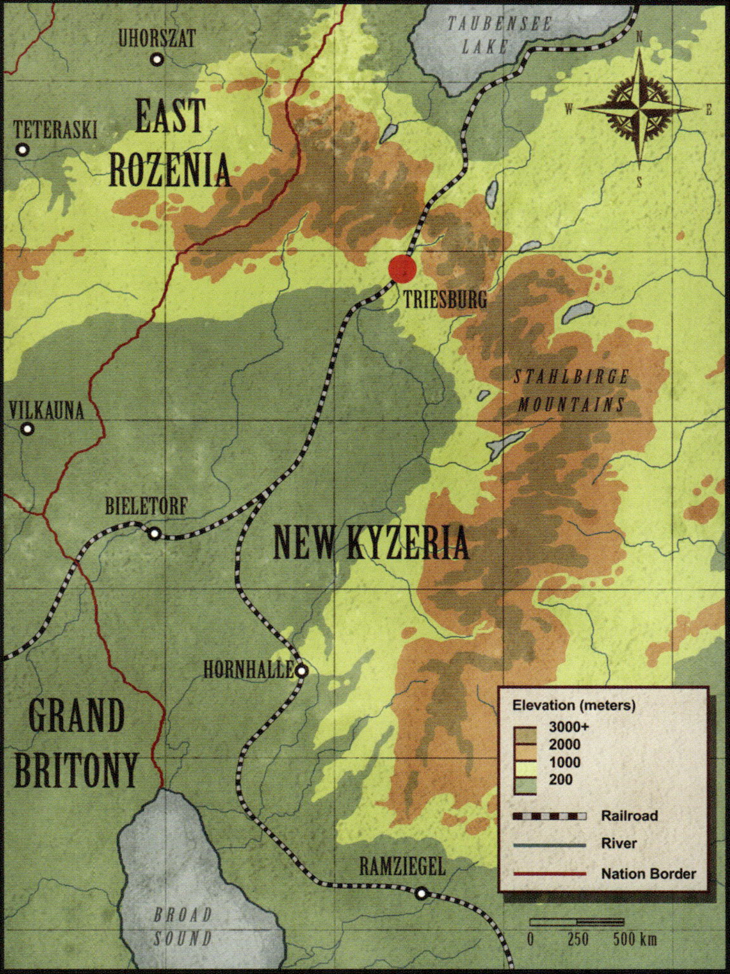

The topographic map of the Stahlbirge Mountains shows the natural defenses they form, and the strategic importance of the Triesburg pass.

Modern archaeological excavations suggest that the gate's past goes back much further than previously thought. Early carbon dating records show that part of the wall's lower foundation vaults dates to the same era as the excavated Isle à Von artifacts. Unfortunately, further research has been progressing at a snail's pace as the mass of the current walls and the many layered foundations are so tremendously heavy that digging requires extreme caution and specialized equipment to prevent collapse. However, each foot of progress made in these excavations seems to reveal new clues to the ancient architecture hidden below the walls.

KARL-GERÄT

SPECIFICATIONS

COUNTRY: New Kyzeria
UNITS BUILT: 12
NO. OF CREW: 21
MAIN ARMAMENT: 1 x 600 mm siege mortar
SECONDARY ARMAMENT:
1 x 30 mm multi-barreled rotary cannon
3 x 50 mm guns
ARMOR: 50-120 mm
MAX SPEED: 10 km/h

The Karl-Gerät heavy mortarhead is often posited as the earliest example of the Kyzerian obsession with giant weapons. Unlike its successors, however, it was a reasonably practical design. The gigantic siege mortars of the twelve heavy mortarheads that were built proved effective throughout the latter half of the Neverwar, notably participating in the Siege of Grigomos (WC036–039) and the assault on the Lumpe Line (WC045). In order to assist with reloading, three dozen Panzerkopf IVs were converted to dedicated ammo carriers and loaders.

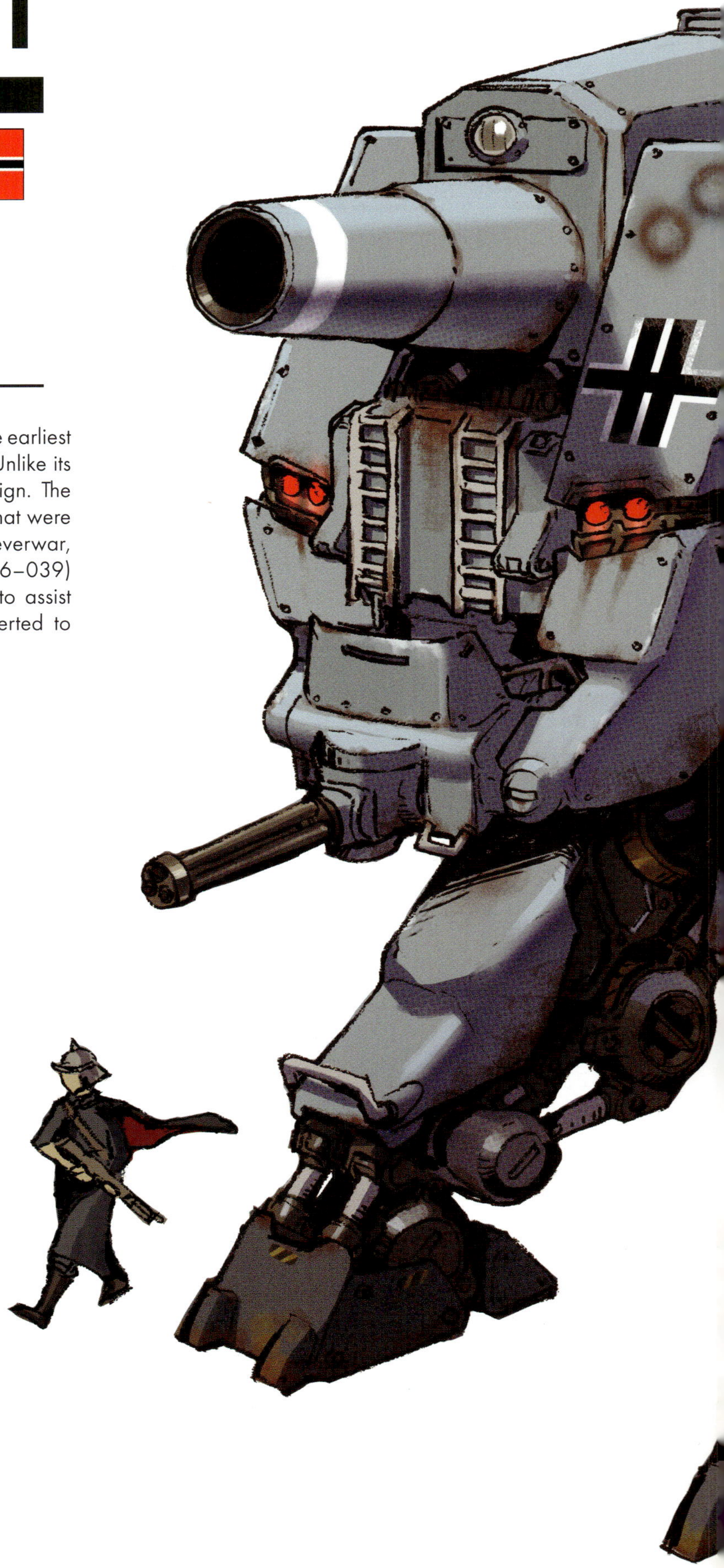

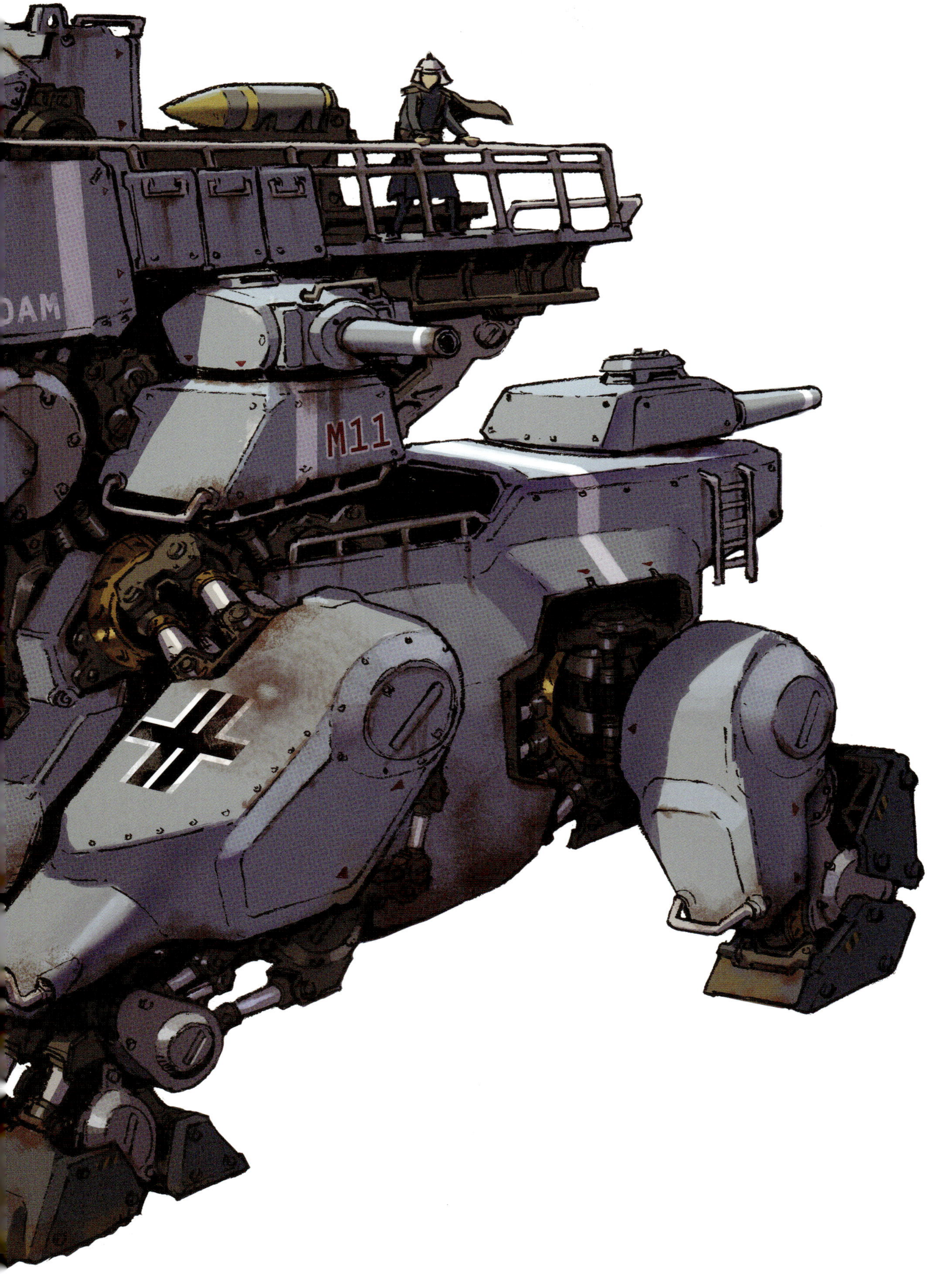
OAM
M11

TK-2F TANK KONG 2 (FLAK TYPE)

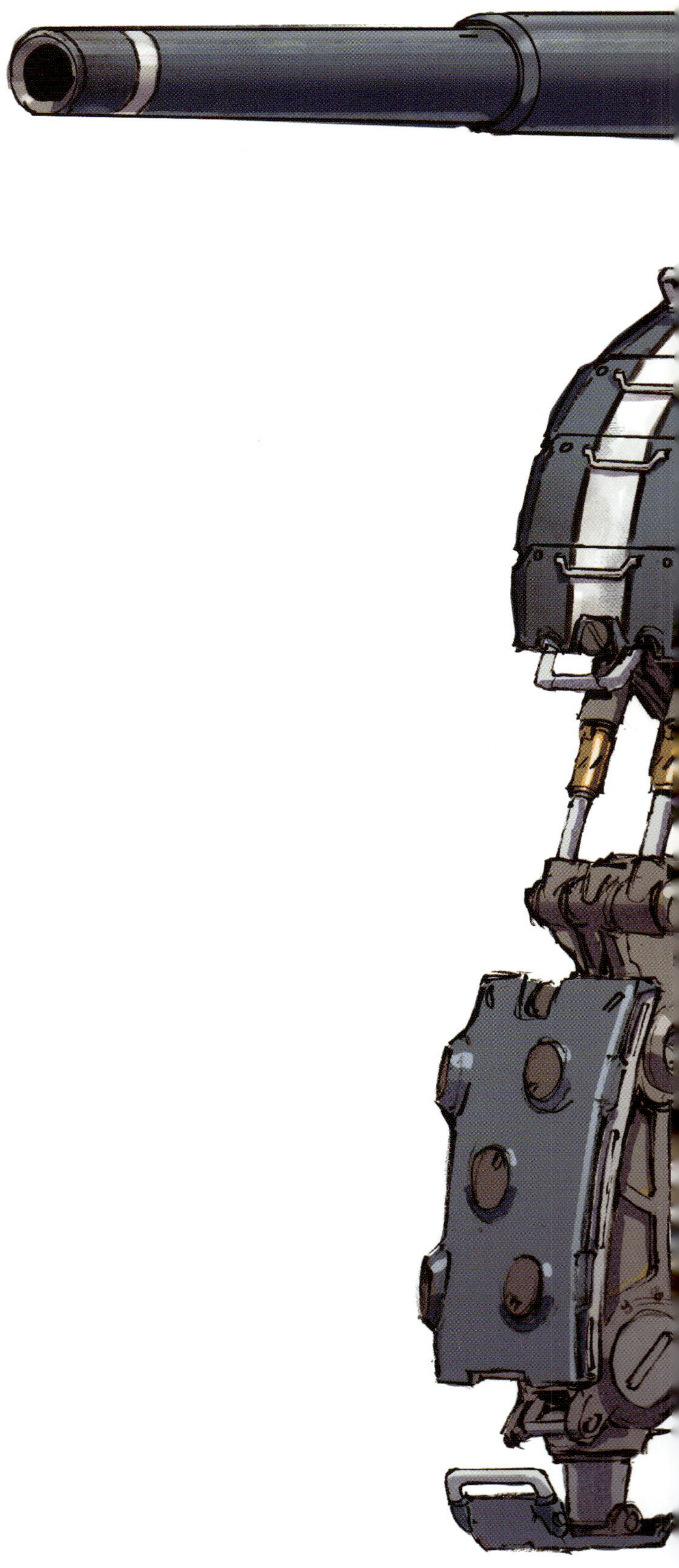

SPECIFICATIONS

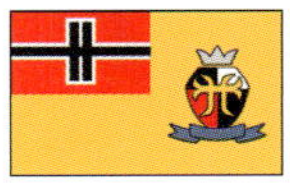

COUNTRY: East Rozenia
UNITS BUILT: 356
NO. OF CREW: 4
MAIN ARMAMENT: 1 x 128 mm gun
SECONDARY ARMAMENT:
2 x 12.7 mm machine guns
ARMOR: 16–70 mm
MAX SPEED: 75 km/h

The Bol-Obez factory, located in the West Rozenian province of Novgryov, provided the Alliance with a steady stream of Tank Kongs and Tank Kong variants. All that changed in WC047, when Novgryov changed its allegiance to East Rozenia. Despite this drastic shift in politics, Bol-Obez did not cease operation and business went on as usual, with the exception of the color of paint applied to the units.

Alliance high command deemed recapture of the Kong production facility non-essential. Much to the contrary, in fact, it designated Bol-Obez as a strategic target. The Alliance commenced bombing runs on the factory within a week of the Novgryov defection. To counter the constant threat of air raids, most of the new Kyzerian Kongs were equipped with the infamous Flak 88, which by then were also well known as highly effective anti-tank weapons.

The Kong's highly recognizable profile and lack of communication concerning the captured factory led to confusion on the battlefield. Multiple instances of friendly fire within both the Alliance and Kyzerian ranks were reported in the following years, as the Alliance mistook friendly Kong units for Kyzerian Kongs and vice versa.

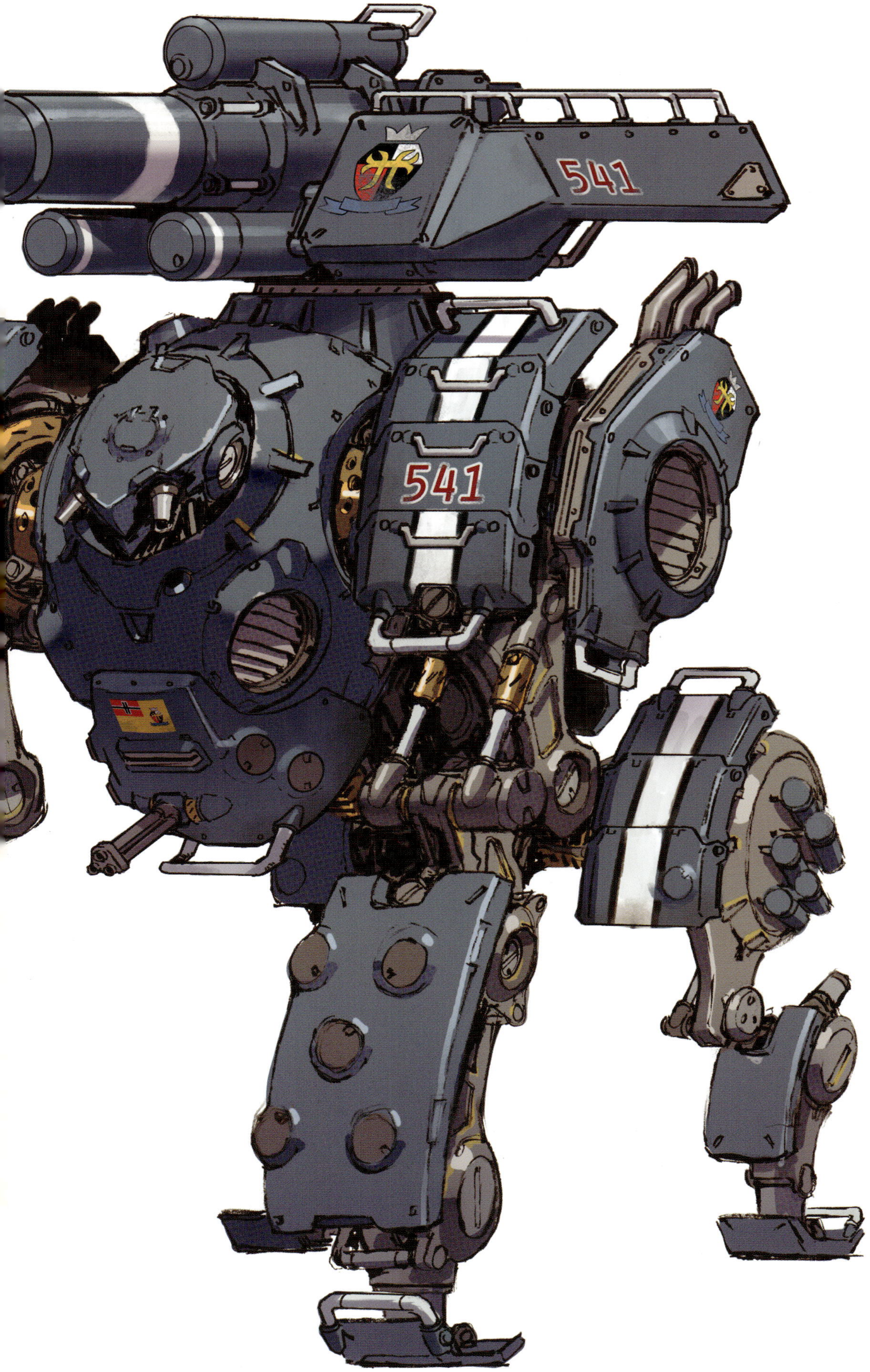
541
541

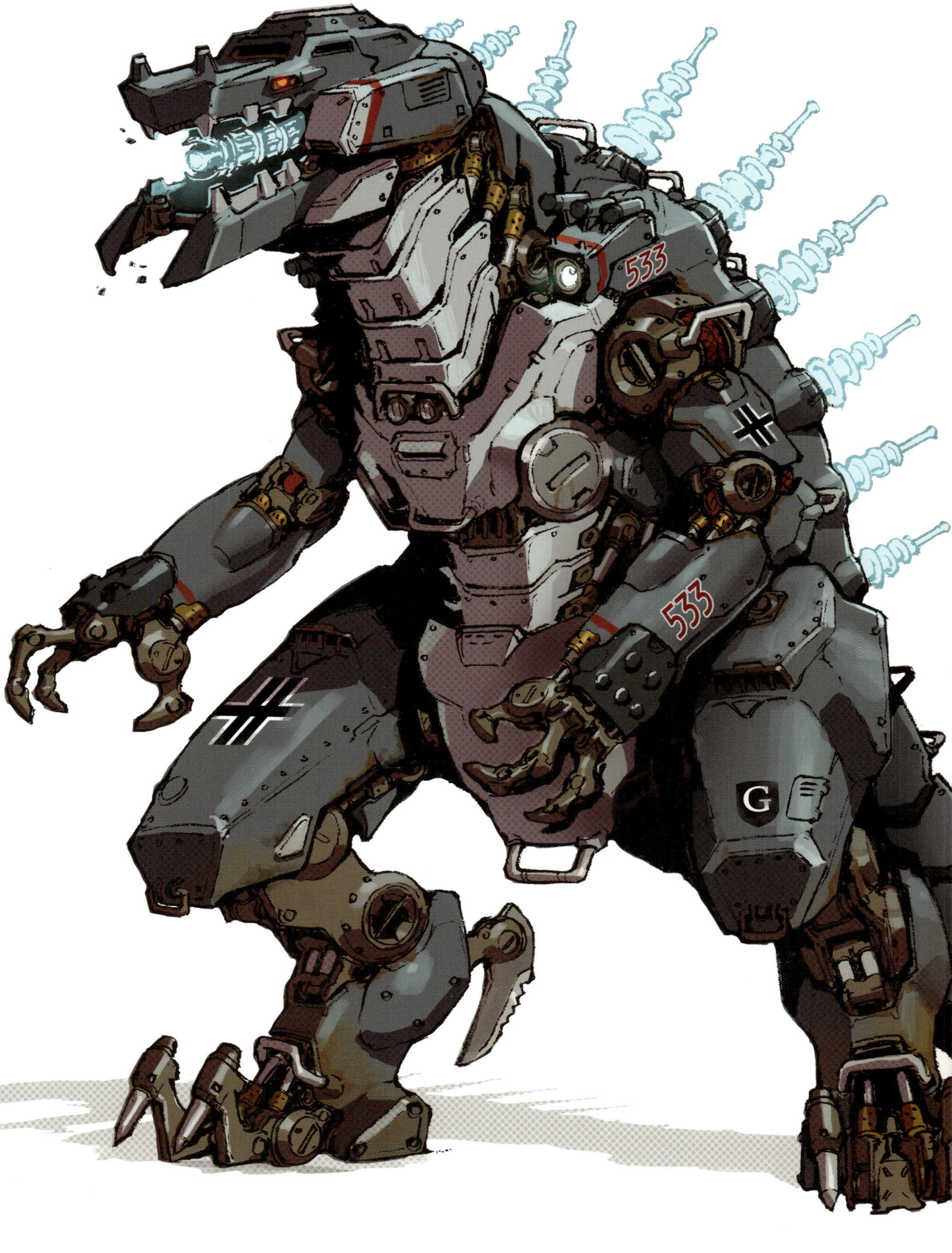

533
533
G

PANZERBLITZECHSE AUSF. E

During the final years of the Neverwar, rumors and fiction about the bizarre technology developed by the Kyzerian Empire ran rampant through Alliance media. Liberterran pulp magazines could not print these stories fast enough to keep up with demand.

As far-fetched as many of those stories seem today, there were in fact some few Kyzerian projects which did, indeed, stray into the fantastical. The Panzerblitzechse was a testing platform for voltaic weapon systems, a weapon reverse-engineered from early data garnered from the Operation Zeehond research. At first glance, its dragon-like appearance would appear to be purely a cosmetic choice. However, this unconventional design came about due to a series of complications necessitated by the particular operation of its voltaic discharge cannon. First, the system required a number of pylons angled up and away from the weapon to create a charged field from which the weapon drew from while firing. Secondly, it needed a pair of polarizing grounding rods, which could be articulated in a specific direction, to control the weapon discharge. This led to the addition of the tail and its supporting bogie. Finally, in order to balance the heavy weapon system, the machine's torso needed to lean forward, requiring powerful double-articulated legs to move it all along at a ponderous pace.

The main weapon of the Panzerblitzechse was devastating but difficult to aim, as the electric beam would be attracted to nearby objects. Another quirk had to do with an effect of the field projected by the back pylons; their operation led to the continual arcing of powerful static electricity charges. In an urban environment, this would inevitably blow out the local power grid even before the tankhead fired its cannon. The heavy stomping gait of the Panzerblitzechse made it a challenge to steer, but it was more than capable of ramming through most obstacles and buildings.

"In days of darkness, the Thunder Leviathan shall rise from the ocean and bring down its wrath on the nations of the East."
- excerpt from Kriegsdiener Scriptures

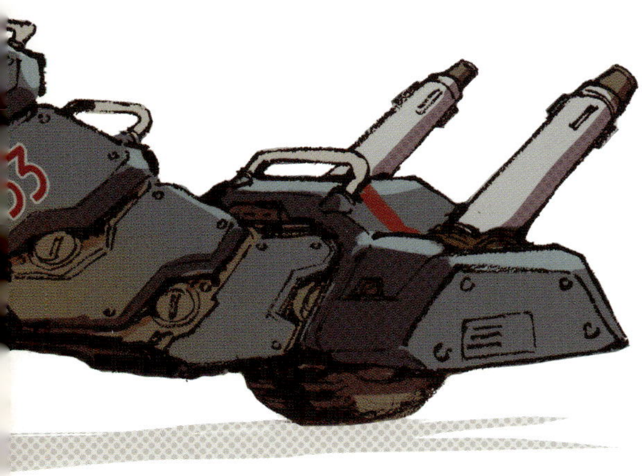

With all these particularities, this unique tankhead became something of a solution in search of a problem. As part of the Godhead revival initiative, field testing of the Panzerblitzechse's systems was considered imperative. However, Kyzerian High Command felt that simply marching the machine to the frontline would be a waste of its potential. A number of solutions to this problem were proposed. Many were seriously pursued, among them a drop pod which could be deployed behind enemy lines and, even more unorthodox in character, a proposal to modify the Panzerblitzechse so that it could be buried entirely, its crew living inside the hull for days or weeks until an unsuspecting enemy had overtaken their position. The machine would then emerge and rain down terror in a surprise attack. This proposal gained significant traction at first, but it was quickly discovered that creating a hole deep enough to hide the machine would inevitably draw the attention of enemy scouting reports. Additionally, the crews of these machines had always complained of severe static shock while operating and were ill-inclined to the notion of bivouacking in these conditions for extended periods of time.

Eventually, researchers realized that the Panzerblitzechse design was very well suited to scaling rocky inclines. This led to the final winning proposal: the electric menace would be used to assault targets of opportunity on the rocky Yamaichi coastlines. Many factories and other industrial and infrastructure targets had always been considered unconquerable by ground forces due to the difficult approach and terrain, but that very defensive strength also meant they were left relatively unguarded. Consequently, the powerful Panzerblitzechses made easy work of them more often than not.

The result of the final efforts of this uniquely creative series of Kyzerian development projects was the Ausf. E, a variant of the Panzerblitzechse specifically designed to be amphibious and semi-submersible to approach its targets covertly, under water. Unfortunately, its weaponry would short out if operated when submerged. Additionally, it lacked operational range and had to be deployed from a transport unit. The records of these operations were lost at the end of the war, so speculation as to the exact details of the vehicle used to deliver the Ausf. E on its notorious raids became a popular pastime. A modified cruiser or transport ship seemed, and still seems, to be the most likely candidate, possibly camouflaged as a civilian cargo ship. The notions of a special submarine that towed the dragon-like tankhead and an airship have also been put forward, though serious historians generally pay no heed to such theories.

SPECIFICATIONS

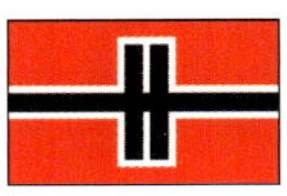

COUNTRY: New Kyzeria
UNITS BUILT: 3
NO. OF CREW: 5
MAIN ARMAMENT: 1 x voltaic discharge array
SECONDARY ARMAMENT: 6 x smoke grenades
ARMOR: 16–110 mm
MAX SPEED: 55 km/h

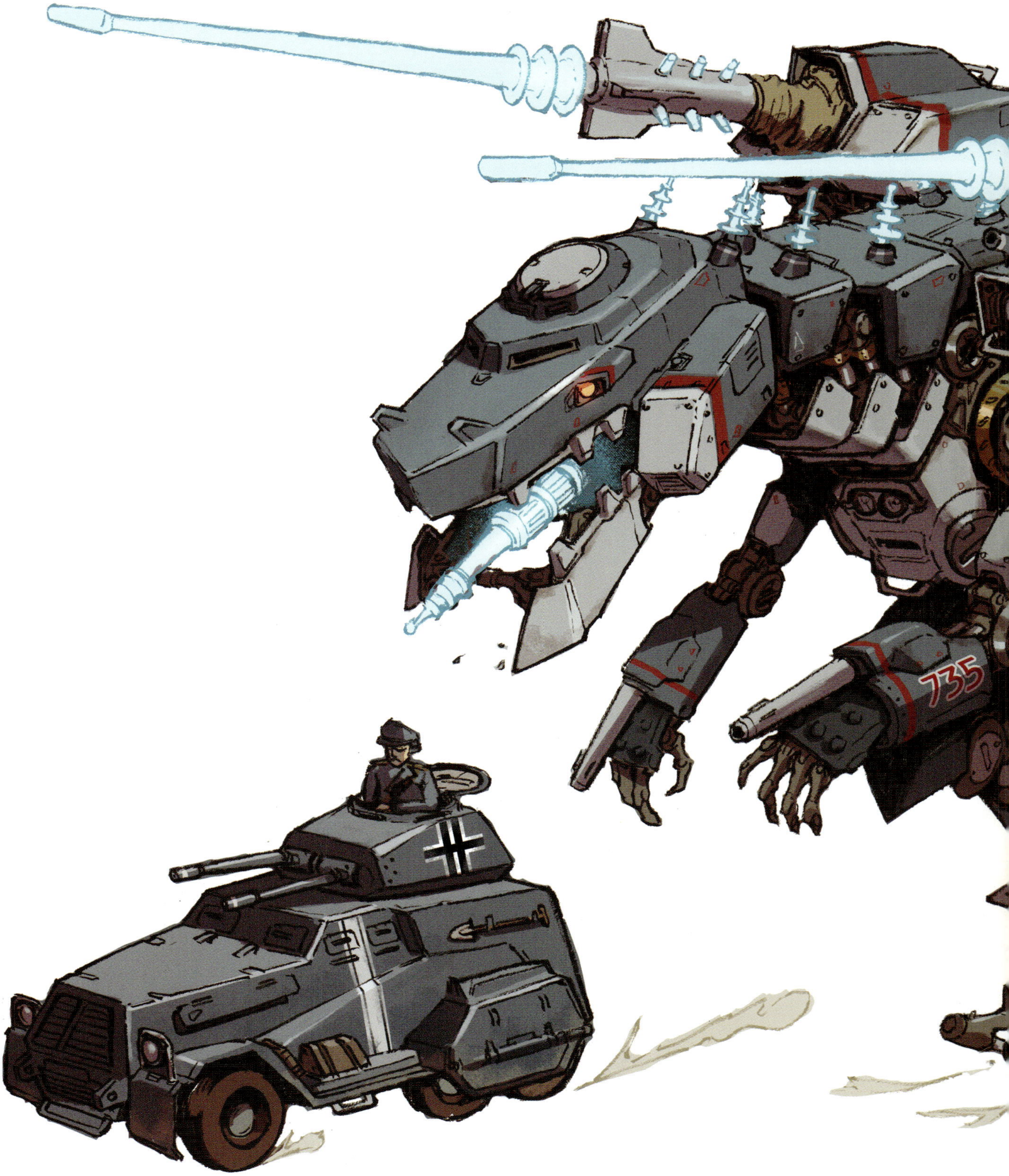

PANZERBLITZECHSE II AUSF. E

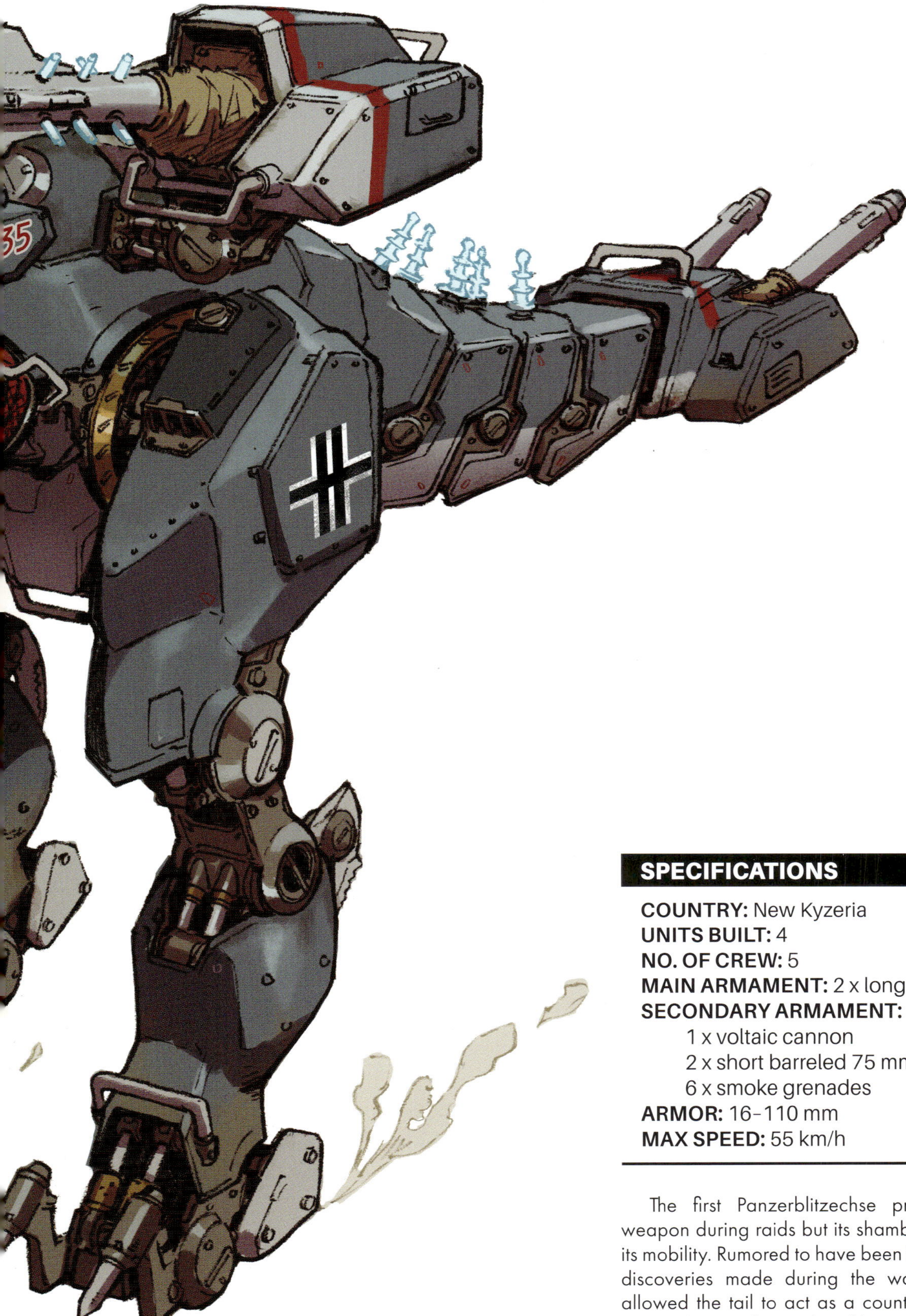

SPECIFICATIONS

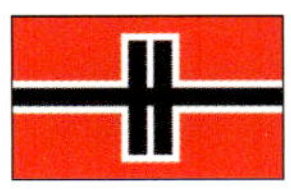

COUNTRY: New Kyzeria
UNITS BUILT: 4
NO. OF CREW: 5
MAIN ARMAMENT: 2 x long-range voltaic cannons
SECONDARY ARMAMENT:
1 x voltaic cannon
2 x short barreled 75 mm guns
6 x smoke grenades
ARMOR: 16–110 mm
MAX SPEED: 55 km/h

The first Panzerblitzechse proved an effective terror weapon during raids but its shambling gait greatly hampered its mobility. Rumored to have been inspired by paleontological discoveries made during the war, clever linkage systems allowed the tail to act as a counterbalance. Combined with the much more focused and longer-ranged energy weapons, the Mark II was capable of more directed attacks on hardened military targets but, like most experimental weapons, only had a limited impact on the war.

PANZERHAUBITZE MAMMUT AUSF. B

The Mammut was a continuation of the Panzerblitzechse line and was used to further develop voltaic weapon systems. While the Panzerblitzechse II was capable of projecting its energy at a range of roughly 200 meters, it was easily drawn off-target, its energy being dissipated prematurely by other grounded objects along the line of trajectory.

The main armament of the Mammut operated by way of a two-step system. The main gun would first fire a sabot round which, in effect, was barely more than a stabilized gyroharpoon. The intent of this first shot was not to destroy or even damage the target. Rather, it would trail a very fine line of wire behind it, reeling off a coil as the projectile flew. The exact composition of this wire seems to have varied according to constant experimentation. Some versions were spun from miniscule shavings of relic alloy armor from the Isle à Von artifacts. It was very unlikely that New Kyzeria was capable of producing this alloy at the time, meaning it must have been extracted from the relics they already possessed. Battlefield finds of this wire are still incredibly valuable, the daydreams of many metal detectorists. They are incredibly rare, though, for reasons that shall become apparent below.

When the projectile reached its target, or otherwise ceased its flight, the second stage of the weapon system would trigger automatically. Systems located in the lower half of the barrel complex would release a voltaic charge through the barrel, gaining speed and increased focal density as it traveled along the wire. Arriving at the harpoon, the charge would atomize the projectile and continue into the target, causing tremendous damage. Interestingly, it has been speculated that this accelerative effect was quite gradual but constant, not even topping out before reaching the projectile. Theoretically, this meant that the weapon was at its most powerful when fired from the maximum possible distance.

SPECIFICATIONS

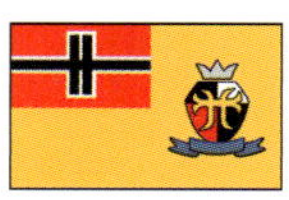

COUNTRY: East Rozenia
UNITS BUILT: 30
NO. OF CREW: 22
MAIN ARMAMENT:
1 x gyroharpoon-guided voltaic cannon
SECONDARY ARMAMENT:
1 x 210 mm mortar
2 x 128 mm anti-tank guns
2 x 37 mm anti-tank guns
1 x 30 mm multi-barreled rotary cannon
3 x 7.92 mm machine guns
14 x smoke grenade launchers
ARMOR: 60–120 mm
MAX SPEED: 15 km/h

It is worth pointing out that voltaic weapons are not simply high-voltage electrical systems. An unguided or unfocused voltaic release will indeed soon diffuse with a lightning-like effect, as seen in the case of the Panzerblitzesche's weapon. A focused effect, on the other hand, like what the Mammut endeavored to create, becomes something far more devastating—closer to a dense beam of pure light than electricity. Witnessing the firing off the Mammut's voltaic coil cannon would often be described, by those unfortunate enough to know, as being like looking directly at the sun or a camera flash, burning an afterimage onto the retina.

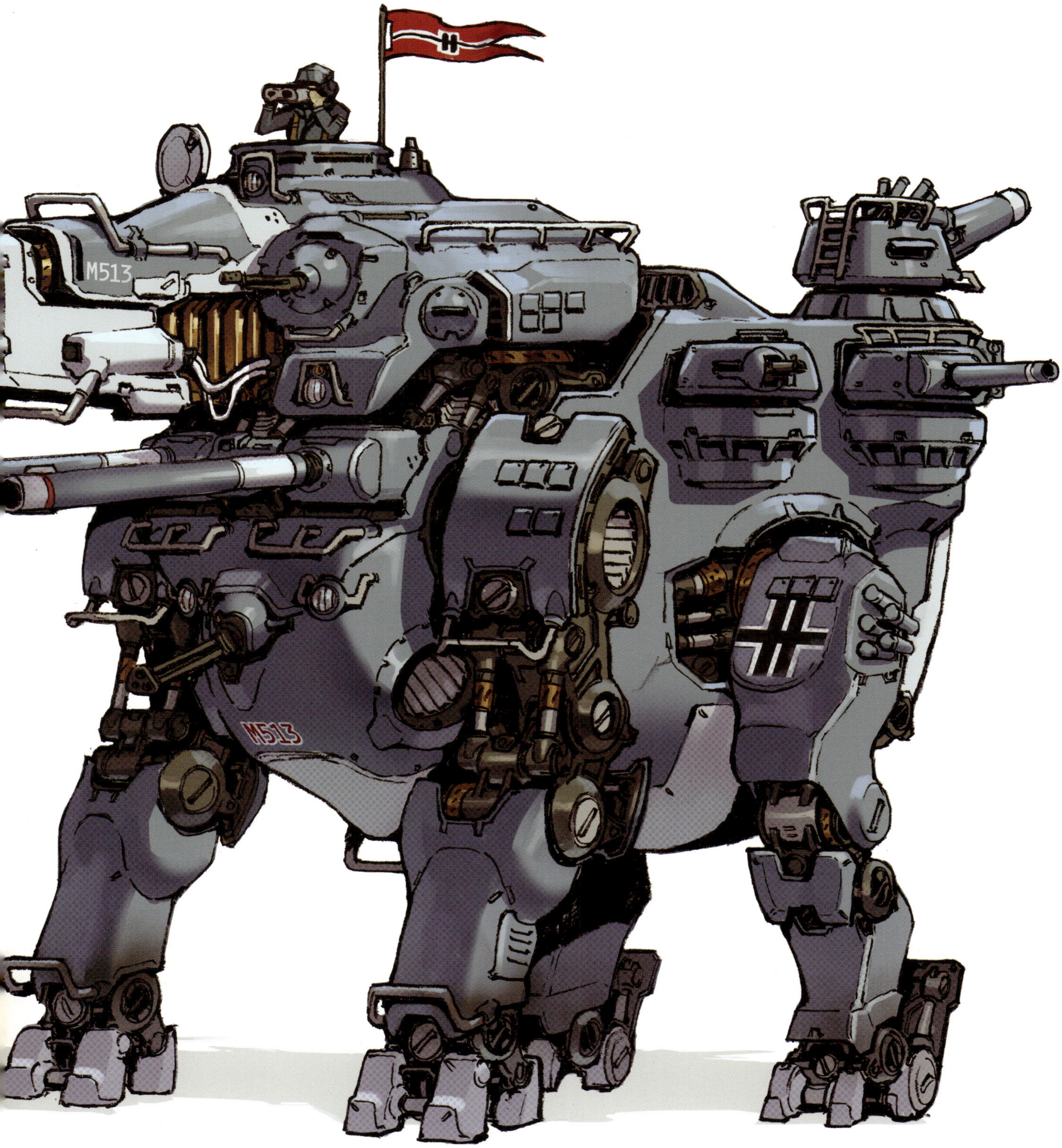
M513
M513

BLOHM UND VOSS BV-141

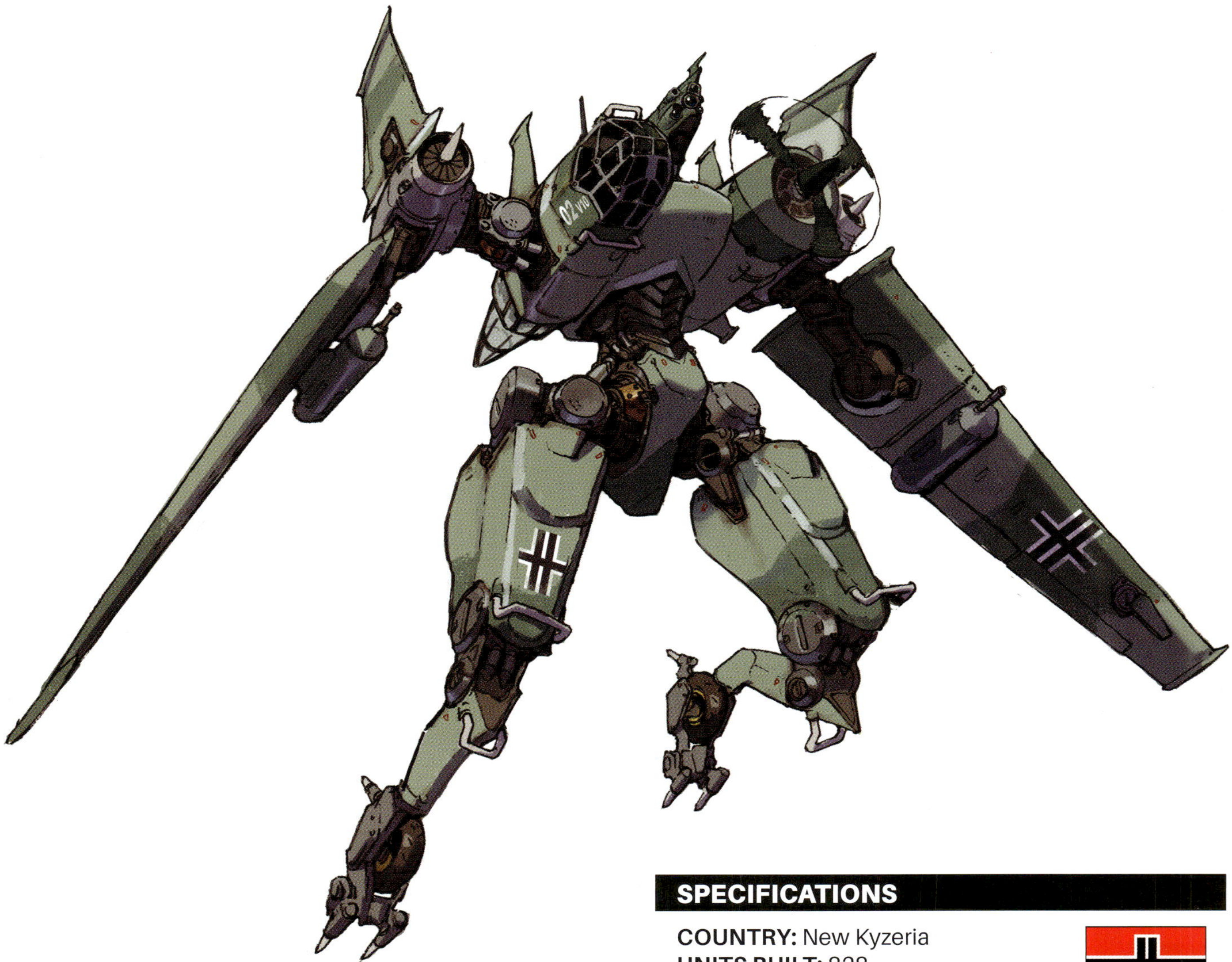

SPECIFICATIONS

COUNTRY: New Kyzeria
UNITS BUILT: 828
NO. OF CREW: 3
MAIN ARMAMENT: 2 x 20 mm gun pods
PAYLOAD: 200 kg
MAX SPEED: 368 km/h

When jetheads thundered onto the Neverwar scene, they were even more experimental and even less understood than the tankheads that preceded them. The engine power required to get these metal machines into the air was tremendous, leading to novel engineering challenges that tankheads had never had to deal with. A number of different design philosophies developed during this early jet-powered era. The Yamaichi jetheads were powered by multiple small jet systems, carefully balanced by expert engineering and the willingness of their pilots to manually compensate for any peculiarities in handling. The Alliance remained fairly conventional and balanced, relying on paired units of turboprop engines on most of their machines. So common, these days, is their approach that it is easy to forget that they were not, necessarily, the perfect solution in every situation.

Jetheads were usually built with paired engine units mainly to counter gyroscopic roll and irregular slipstreaming. In short, unless an aircraft mirrored its engines on both sides, it was likely to create an asymmetrical flight character, constantly pulling or rolling towards one side while in flight due to the imbalance. Some daring designers from New Kyzeria set out to utilize this characteristic, rather than fight against it. The BV-141 was a highly asymmetrical jethead, offsetting its single large turboprop to one side with a large cockpit window on the other side. This allowed the pilot superior range of vision, and the carefully calculated shape of the body along with its, again, asymmetrical control surfaces meant that the BV-141 actually had a very gentle and stable flight.

While the Kyzerian war ministry was skeptical about the odd-looking machine, it proved itself a highly-effective reconnaissance jethead with bombing capabilities beyond those originally required of it. With the air war evolving at a blistering pace during those final years of the Neverwar, there was a pressing need to produce effective machines of every type, as quickly as possible, or risk being overtaken by the enemy. Therefore, an order was issued to design and produce a fighter-bomber based on the already successful reconnaissance model.

BLOHM UND VOSS BV-237

The engineers working on the BV-237 project realized they could utilize many of the principles and mechanics of the BV-141 in the new design. For, while the new turbojet engine did not create a corkscrew slipstream like the earlier propellor system, it did still exert a powerful gyroscopic torque on the airframe. By relying on calculations done for the earlier model, the designers managed to compensate for these similar effects, once again creating the blueprint for an unconventional asymmetrical jethead frame.

SPECIFICATIONS

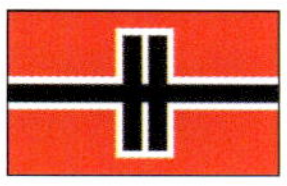

COUNTRY: New Kyzeria
UNITS BUILT: 4,330
NO. OF CREW: 3
MAIN ARMAMENT: 2 x 13 mm machine guns
SECONDARY ARMAMENT:
2 x 20 mm gun pods
ARMOR: 800 kg
MAX SPEED: 805 km/h

Besides its much-increased speed and improved armor, the most notable difference between the 237 and 141 was to be found in the new bomb delivery system. Whereas the 141 was found to be powerful enough to mount four light bombs onto its landing-hips, the 237 was designed with bombing capabilities in mind from the start. Unlike the Alliance, which relied on knee chutes, the 237 used a secondary joint on its knee assembly to rotate the lower landing-leg out of the way of the upper hip-bays entirely. This allowed for a more vertical dead drop of its payload and the simpler external mechanism left more room for additional ordinance. While the cockpit and observation blister of the 237 was much smaller than its reconnaissance predecessor, it still provided a wide field of vision—including an expanded range of vision toward the nadir—to allow for fairly accurate bombing runs. A notable downside of the bay configuration was apparent during landings. The 141 had already proven to be weak-kneed and frequently toppled over upon touchdown. The additional joint on the 237 created another point of failure that made it more prone to these catastrophic "Schwalbe dives," as the pilots came to refer to them.

All in all, the BV 237 was an above-average strike bomber with adept handling that made it capable of fighting off any fighters trying to intercept it.

*"Then Man said unto God,
'Graft onto us wings so we may
fly like the birds of the air
and angels in heaven.'"*
*- excerpt from
Kriegsdiener Scriptures*

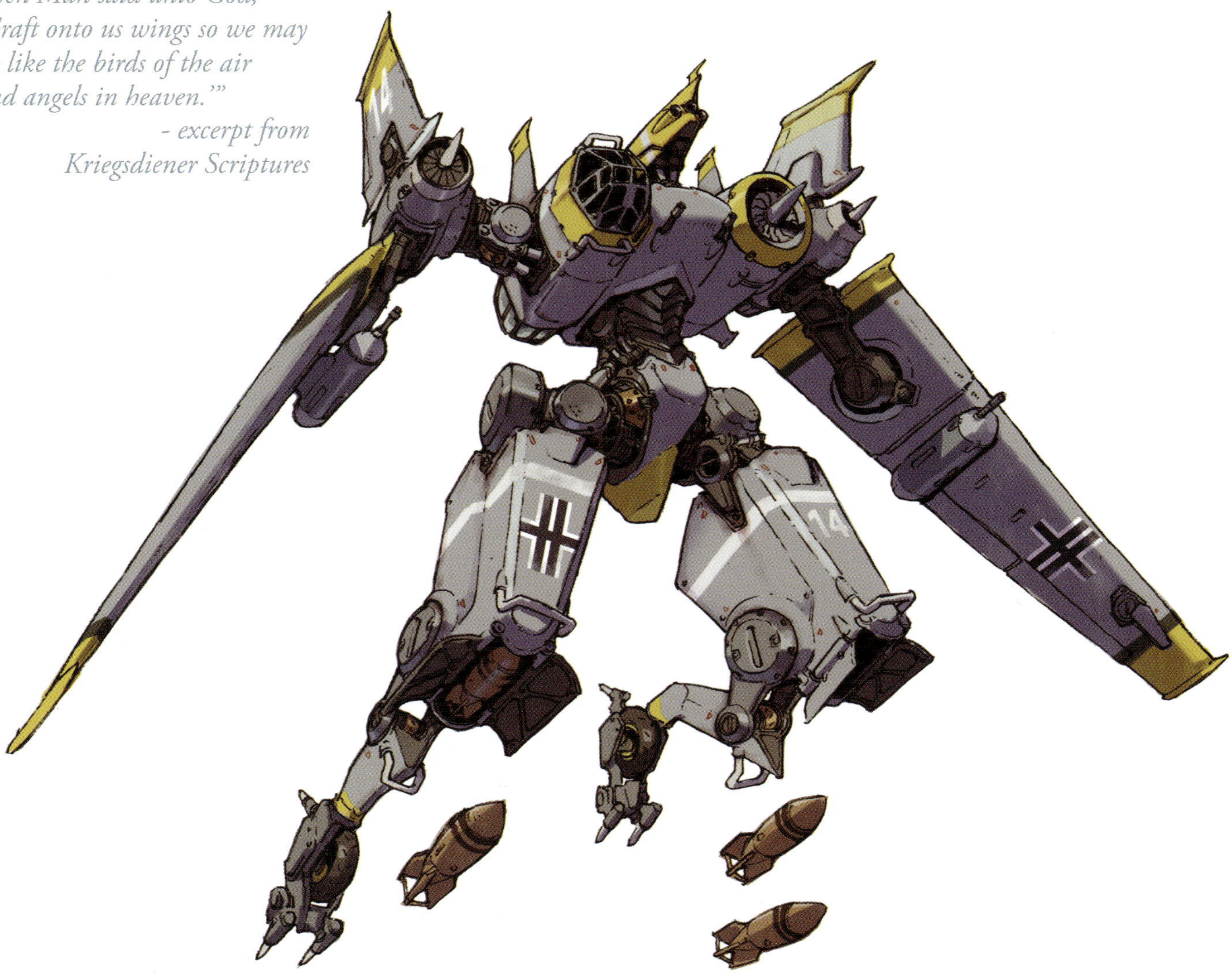

FOCKE-ACHGELIS FA 223 DRACHE

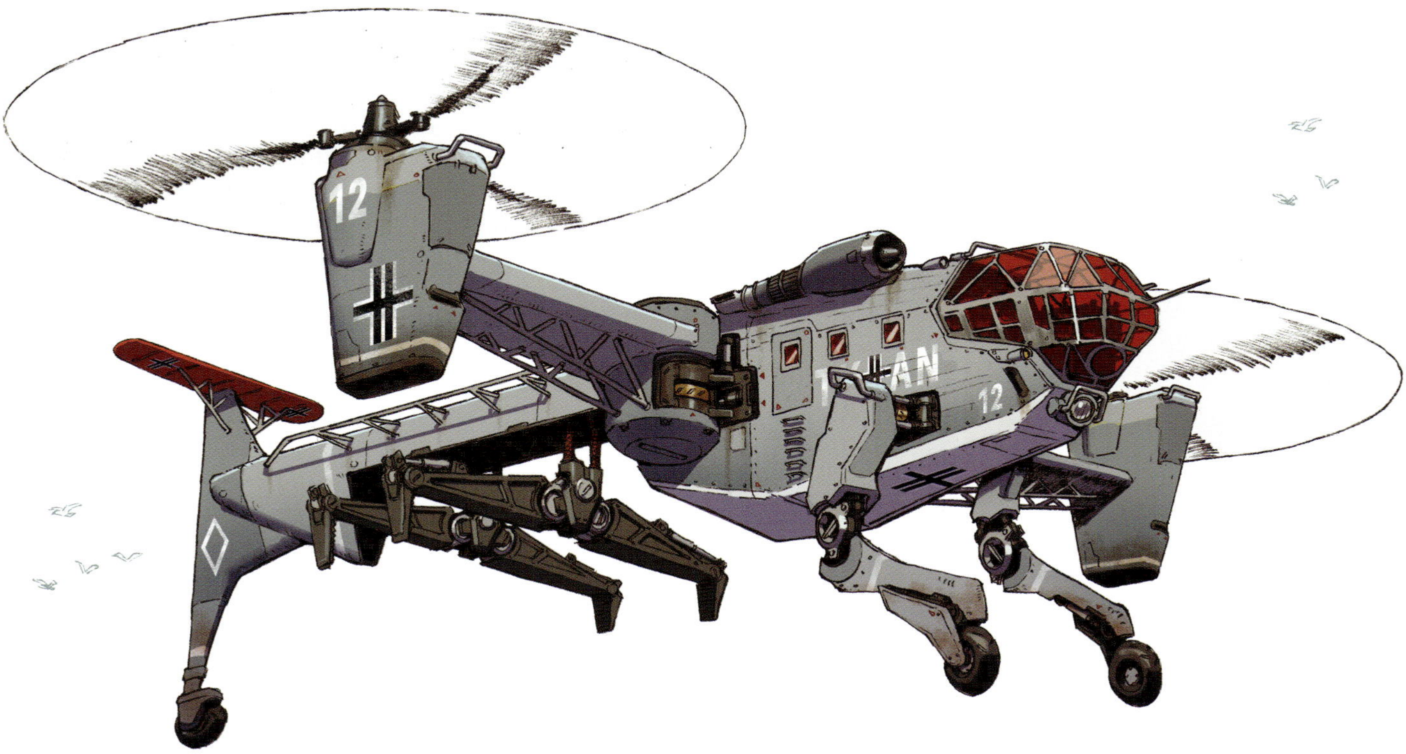

SPECIFICATIONS

COUNTRY: New Kyzeria
UNITS BUILT: 175
NO. OF CREW: 3
INFANTRY CARRIED: 12
CAPACITY: 29,000 kg
MAX SPEED: 286 km/h

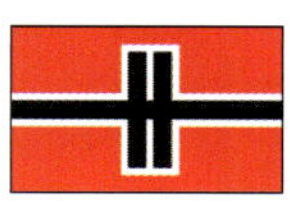

A design with great potential which, nonetheless, was shut down by Alliance operations at every turn. When the prototype made its first test flight in WC041, it proved itself the fastest helicopter in the world. Unfortunately, the factory was destroyed by Alliance bombers before the second prototype could be completed. Following that, the production plant of the Drache would be bombed or otherwise destroyed another five times between WC042 and WC048 before full mass production finally began.

A key feature of the Drache was its folding wing arms, which allowed for easy transportation and storage. Some say that this feature inspired the tiltrotor design of modern-day helicopters.

HORTEN HO 229

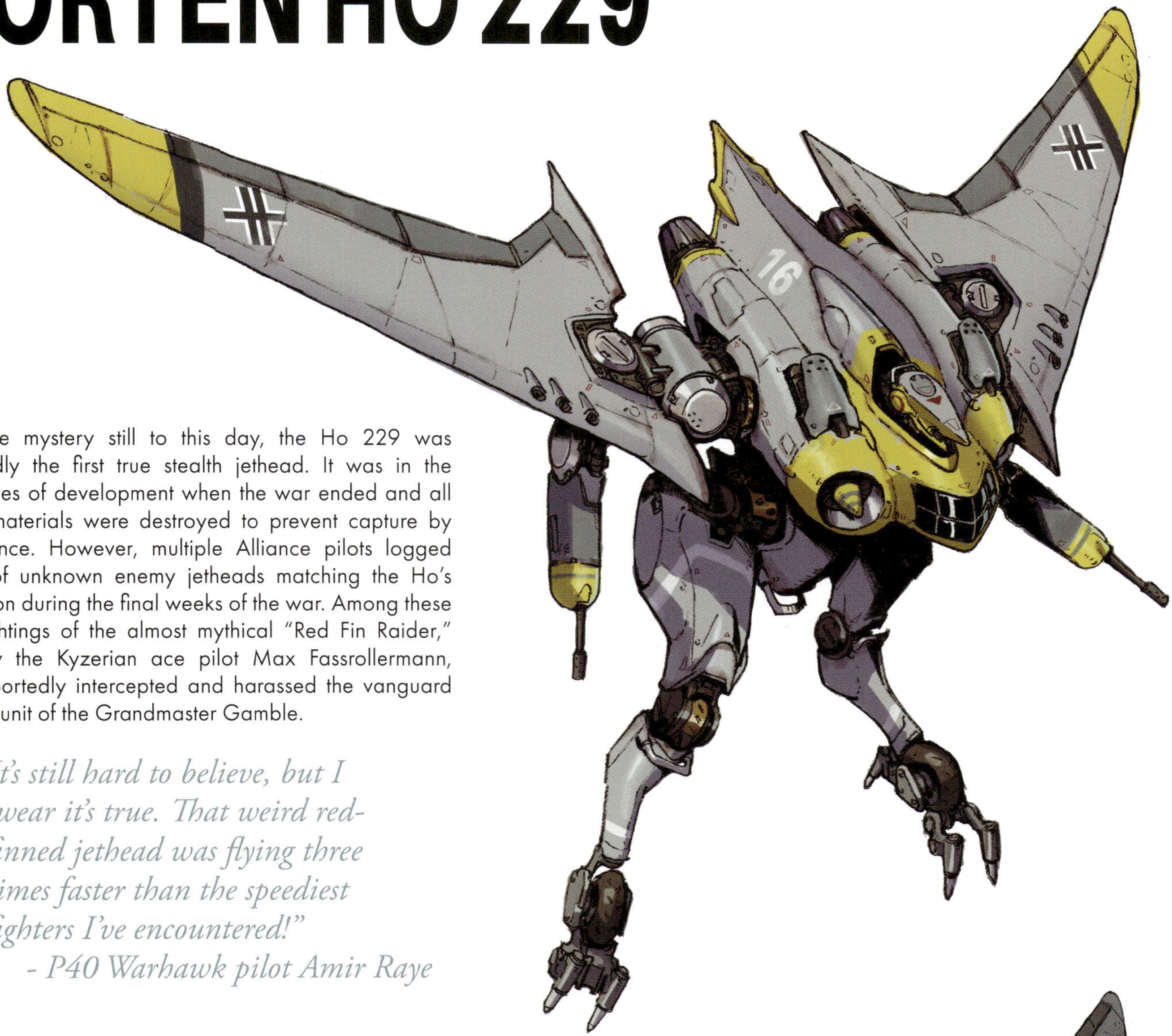

A true mystery still to this day, the Ho 229 was supposedly the first true stealth jethead. It was in the final stages of development when the war ended and all related materials were destroyed to prevent capture by the Alliance. However, multiple Alliance pilots logged reports of unknown enemy jetheads matching the Ho's description during the final weeks of the war. Among these were sightings of the almost mythical "Red Fin Raider," flown by the Kyzerian ace pilot Max Fassrollermann, that purportedly intercepted and harassed the vanguard airborne unit of the Grandmaster Gamble.

"It's still hard to believe, but I swear it's true. That weird red-finned jethead was flying three times faster than the speediest fighters I've encountered!"
- P40 Warhawk pilot Amir Raye

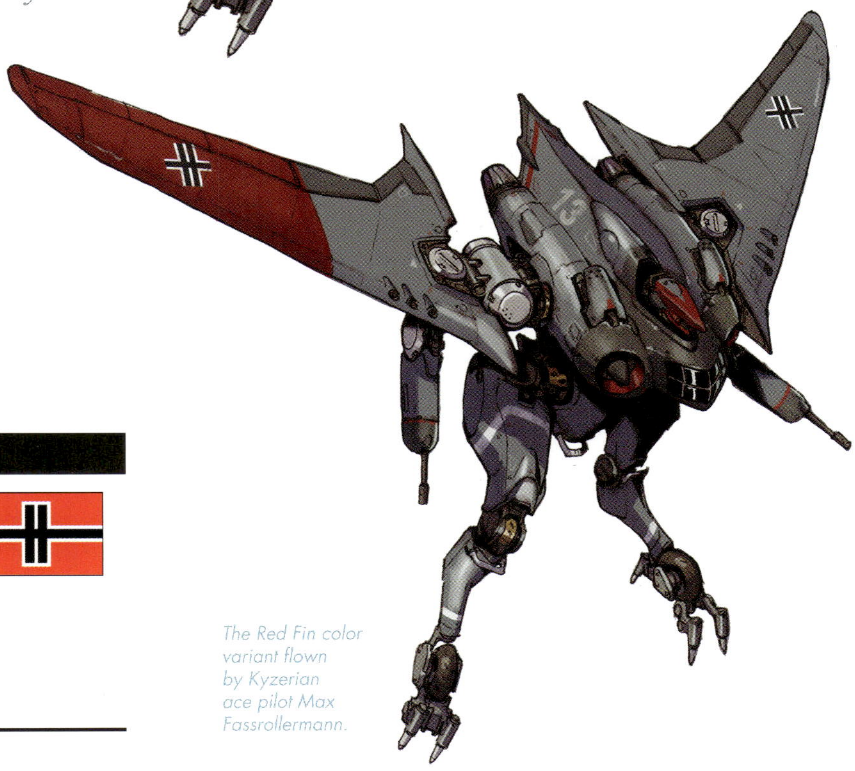

The Red Fin color variant flown by Kyzerian ace pilot Max Fassrollermann.

SPECIFICATIONS

COUNTRY: New Kyzeria
UNITS BUILT: 7
NO. OF CREW: 1
ARMAMENT:
2 x 20 mm cannons
6 x 7.62 mm machine guns
MAX SPEED: 960 km/h

TYPE XXVII SEEHUND

The Kyzerians were much more aggressive with their micro-subheads than even the Dominion. The Yamaichi would deploy their micro-subheads in advance strikes against harbors and other coastal targets in preparation for followup naval assaults. Similarly, the Alliance used them to strike at coastbound targets, but mainly in covert operations meant to harry and hamper the enemy more than as a prologue to another attack. The Kyzerians, on the other hand, were of the opinion that a micro-subhead should be used for striking at enemy naval elements at sea. The operational limitations of these smaller vessels meant that this could only be done around contested waters where there was a base to return to close by.

The previous subheads used by the Kyzerians were limited to little more than the notorious "manned torpedo" designs which had been developed by all combatants since the middle years of the war. Innovation came late in the war, when a pair of Alliance mini-subheads were captured during a failed raid on the Kyzerian Imperial Battleship *Tirpitz* in WC046.

The Type XXVII Seehund was arguably the first successful mini-subhead in the Kyzerian navy. Taking inspiration from the captured Alliance vessels, the Kyzerians equipped the Seehund with a cable-cutting apparatus to allow it to penetrate harbors protected by anti-subhead nets. To further aid in these operations, the number of fins and rudders were kept to a minimum to prevent snagging. However, this led to the machine having to rely on internally-shifted ballast to adjust its angle underwater, a ponderous and clumsy process.

In the area of armaments, the Seehund is notably better designed than the subheads of the other factions. As previously mentioned, the Kyzmarine had a doctrine of using these miniature subheads in anti-ship operations, much like their larger cousins, and they were equipped appropriately. A pair of torpedoes were attached to its small hull and, while this limited its arc of fire, it was capable of sinking small naval vessels and threatening any civilian or cargo ship within 5 km. It is fortunate for the Alliance that the Kyzerians never reverse-engineered the superior Yamaichi long lance torpedoes, as those would have turned the Seehund into a ferocious strike craft.

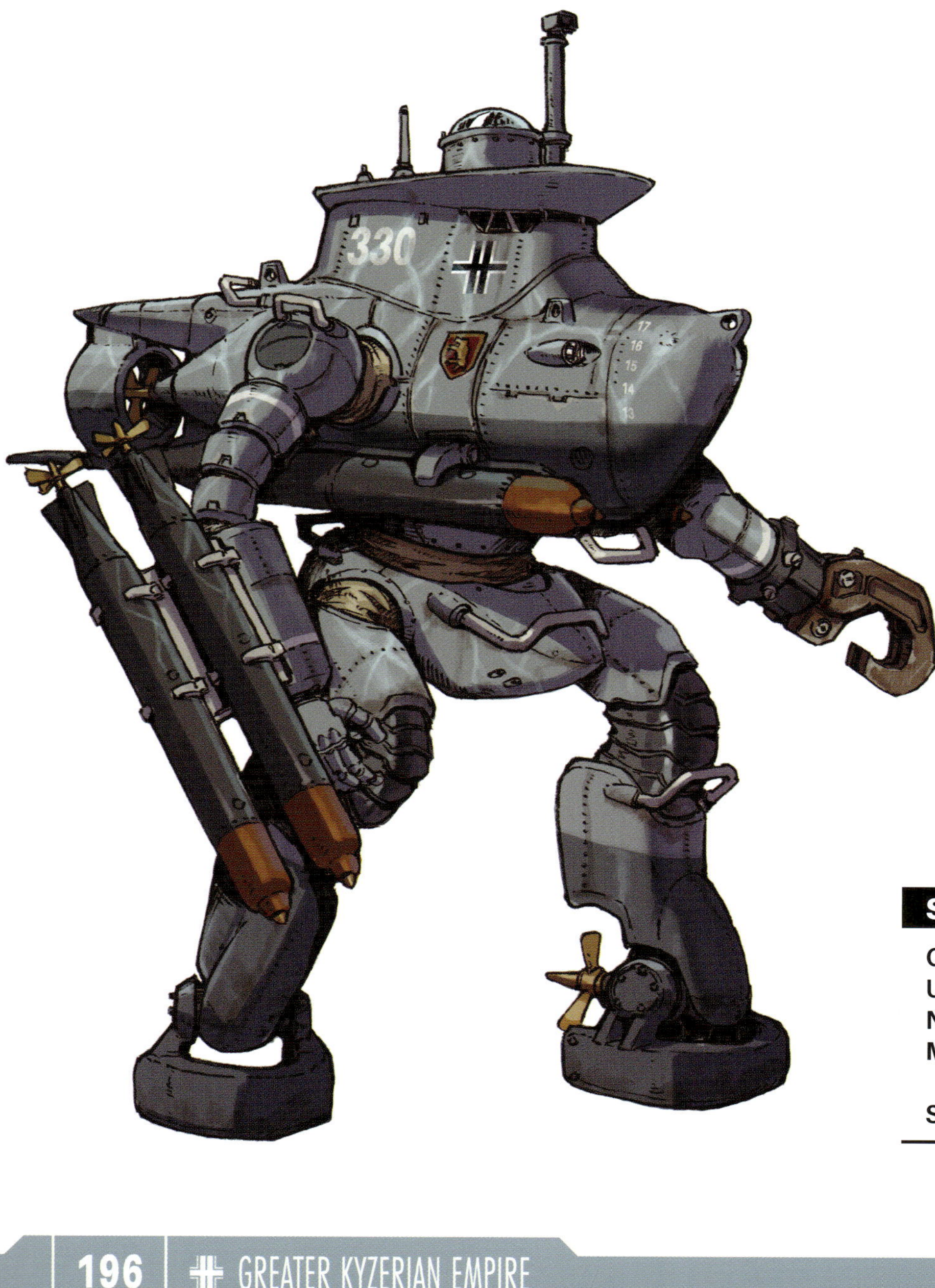

During the closing months of the war, the armaments were further extended by another pair of torpedoes on the starboard arm. The armor on the arms was significantly reduced in order to accommodate them. While this doubled the available firepower on these later models, no Seehund ever managed more than two successful hits. The first torpedoes launched by the Seehund would still have taken an enemy convoy by surprise but, before the additional weapons could be brought to bear, the next target would have outrun and outmaneuvered the slow subhead.

The Seehund was considered an integral part of the defense of the coastal areas between the Alliance- and Kyzerian-held territories, justifying a number of planned improvements to the design. Most of these never saw development, let alone adoption in the field, as the destruction of the Godhead (WC049) brought a spectacular and unexpected end to the war. Many Seehund remain in museums and private ownership today, and every other year or so the wreck of one is discovered on the Rozenian coast.

SPECIFICATIONS

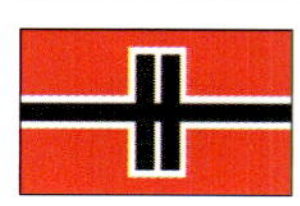

COUNTRY: New Kyzeria
UNITS BUILT: 385
NO. OF CREW: 2
MAIN ARMAMENT:
6 x 534 mm electric torpedoes
SPEED: 3 kn

GROSSES KRIEGSSCHIFF BISMARCK

The construction and launch of the Kyzerian Imperial Battleship *Bismarck* in WC042 would be such a tremendous event that it would only ever be eclipsed by the advent of the Godhead itself at the end of the war. While many details about the *Bismarck* remained hidden, Alliance Command was well aware of its development. They knew that she was painted in a new black waterline paint (because the Yamaichi had blocked New Kyzeria's access to the material used for the previous red paints) and they knew, for a fact, that she would be the largest shiphead in existence when she launched. Alliance naval command, therefore, kept a weather eye out for her imminent deployment. She was finally spotted by an East Rozenian resistance cell which managed to get a report out to its western compatriots, who quickly informed the Alliance. Soon after, a Mistraland report announcing a sighting was leaked to the Alliance.

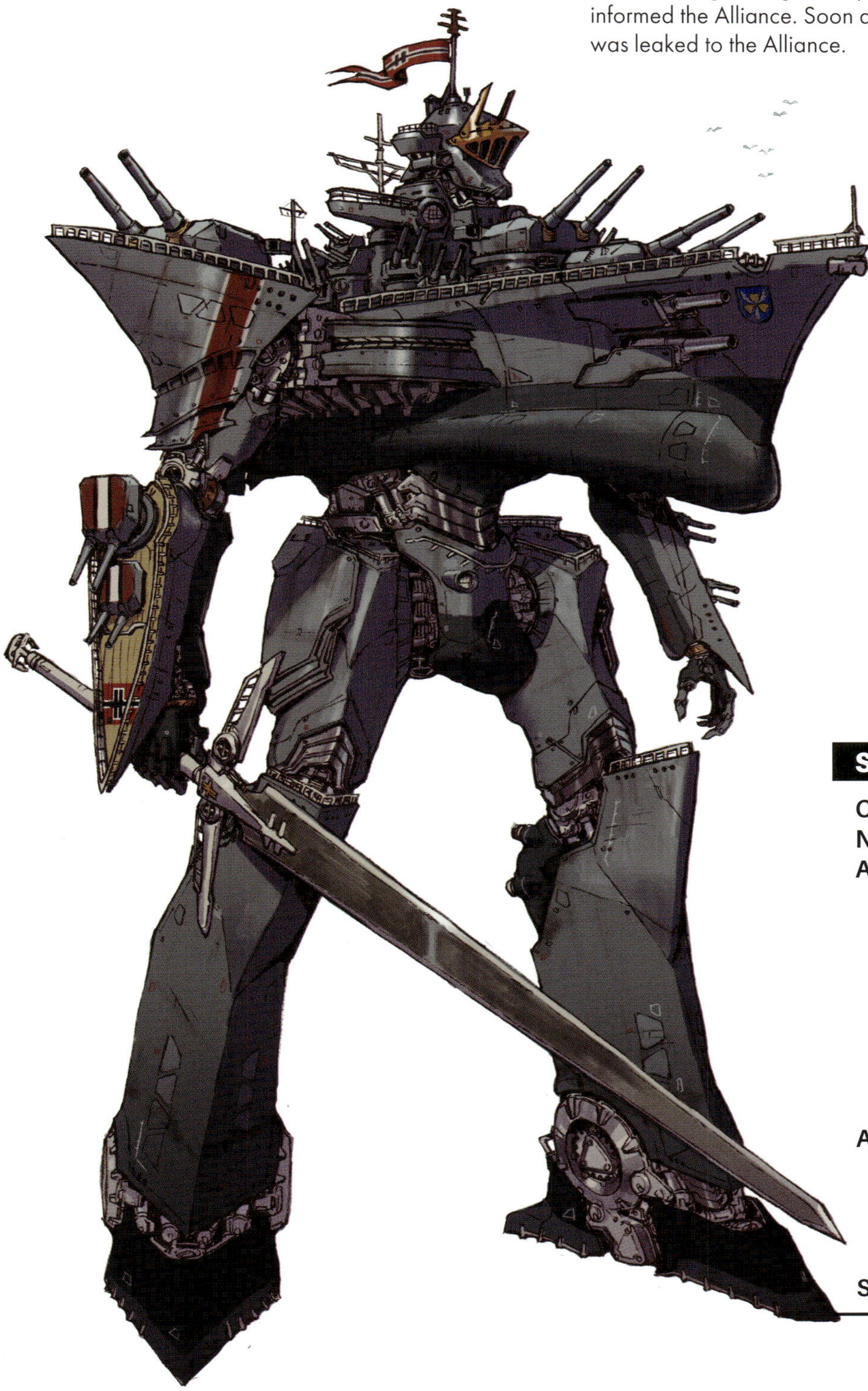

From this moment on, nearly the entire Alliance Tramonic fleet was issued new orders to search for the titanic *Bismarck*. The Alliance shiphead *Hood*, pride of the Britonican Navy, would manage to intercept her in extremely poor winter weather, en route to the Eardenaug Sea. During the encounter, the *Bismarck* closed the distance to the *Hood* and bisected it with a single blow of her mighty blade.

Alliance Command then put out a fleetwide directive to "Sink the *Bismarck*." Evading and fighting small skirmishes, the *Bismarck* tried to reach docks in Eastern Rozenia to repair the damage sustained in her battle with the *Hood*, but the Alliance fleet managed to catch up with her. It would take four fully armed Alliance vessels over an hour of firing to fulfill their mission.

SPECIFICATIONS

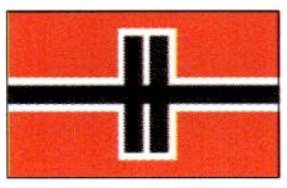

COUNTRY: New Kyzeria
NO. OF CREW: 3,162
ARMAMENT:
- 2 x twin 406 mm guns
- 5 x twin 380 mm guns
- 2 x twin 305 mm guns
- 2 x twin 150 mm guns
- 16 x 105 mm anti-aircraft guns
- 12 x 20 mm anti-aircraft guns
- 22 x 534 mm torpedo tubes
- 1 x handheld shipbreaker claymore

ARMOR:
- 320 mm (belt)
- 360 mm (turrets)
- 120 mm (main deck)
- 50 mm (upper deck)

SPEED: 30 kn

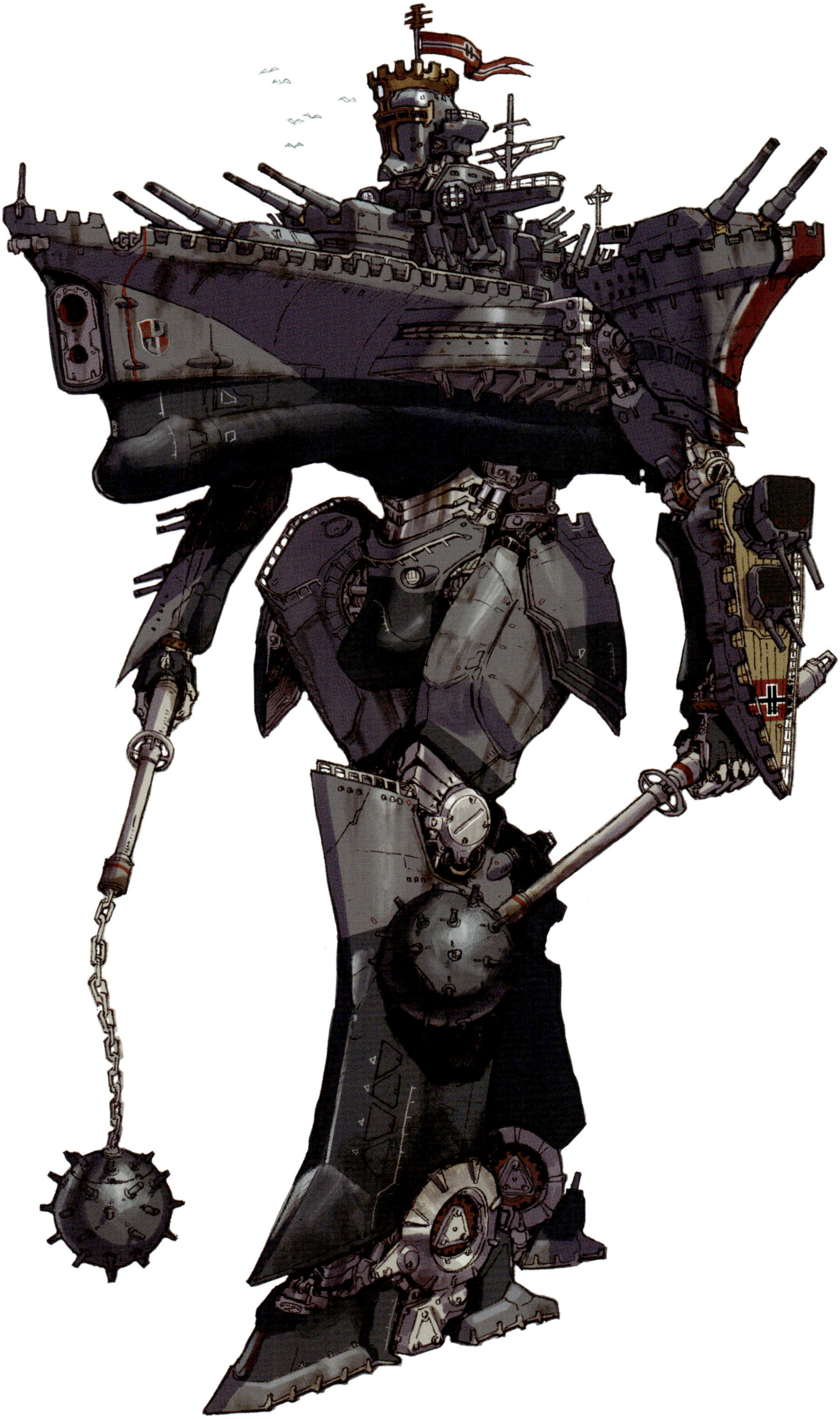

GROSSES KRIEGSSCHIFF TIRPITZ

The sister ship to the *Bismarck* and the largest shiphead constructed by the Kyzerians during the war, the *Tirpitz* was mainly deployed defensively against the Yamaichi Dominion navy in the Gregale Sea. Her presence there was such a deterrent to Dominion fleets that she saw very little combat. In WC046, Kyzerian Command recalled her to the Szynburg dry dock on the Tramonic Ocean for upgrades. Alliance Welman subheads managed to catch her en route and severely damaged her legs.

Repairs took nearly half a year, during which time an experimental railgun was installed in her bow. A week later she would use this new weapon to bombard the Dominion's North-Gregale weather monitoring and refueling station, to devastating effect. After this, multiple attempts to destroy the *Tirpitz* were launched, but failed. She was eventually sunk by jethead bombers from twelve Yamaichi carriers a week before Operation Grandmaster Gamble (WC049). If she had been present in the Nehalean Channel as planned, history might have told a different story; the southern amphibious force might well have been repelled and the attack by the Alliance would likely have stalled.

SPECIFICATIONS

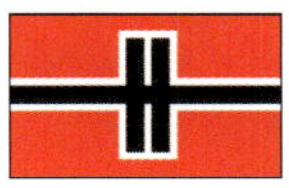

COUNTRY: New Kyzeria
NO. OF CREW: 3,162
ARMAMENT:
- 1 x 1,140 mm railgun
- 10 x 380 mm guns
- 4 x 305 mm guns
- 4 x 150 mm guns
- 16 x 105 mm anti-aircraft guns
- 12 x 20 mm anti-aircraft guns
- 22 x 534 mm torpedo tubes
- 2 x hand-held mineflails

ARMOR:
- 320 mm (belt)
- 360 mm (turrets)
- 120 mm (main deck)
- 50 mm (upper deck)

SPEED: 30 kn

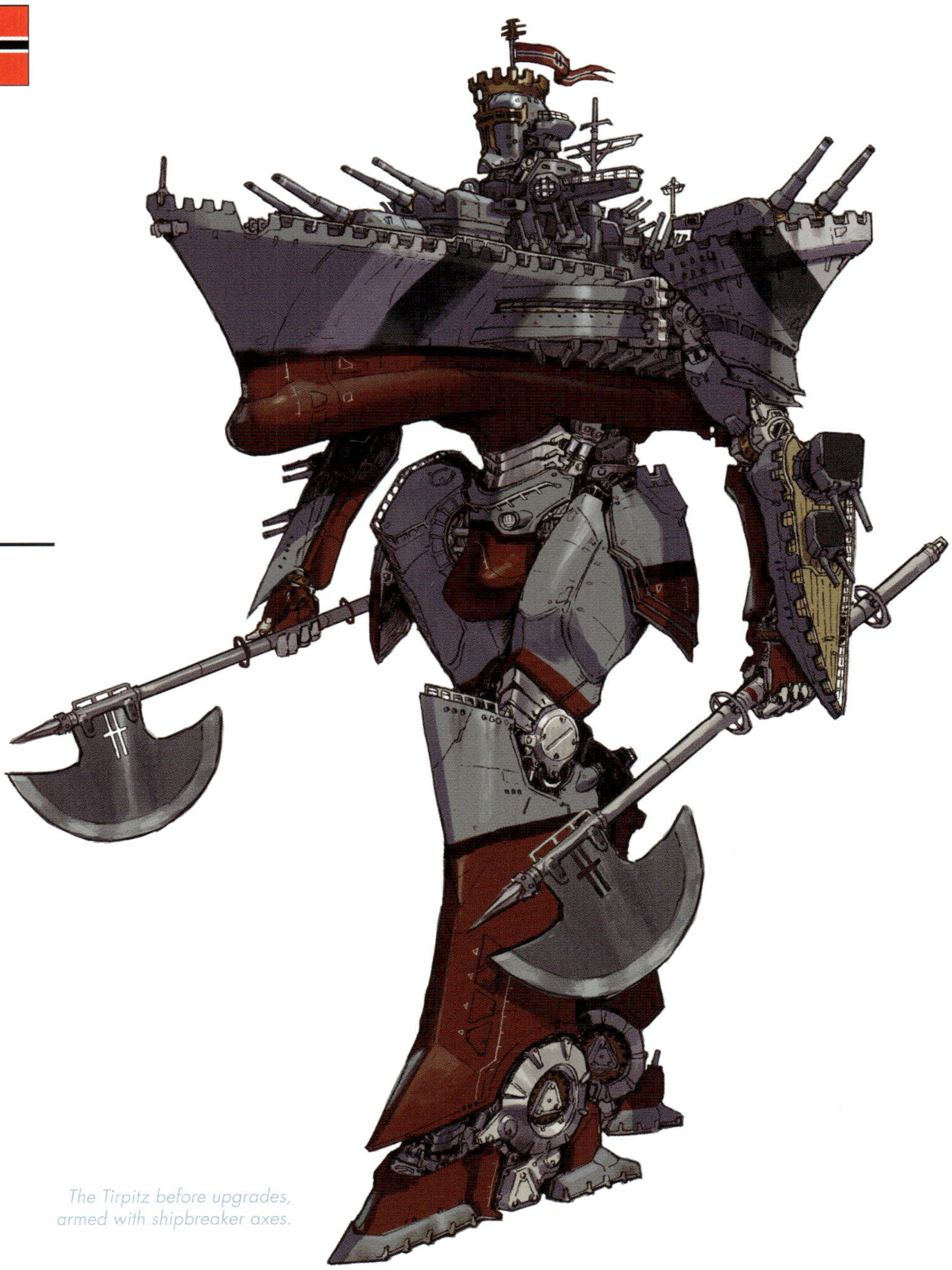

The Tirpitz before upgrades, armed with shipbreaker axes.

GROSSER FLUGZEUGTRÄGER GRAF ZEPPELIN

While most shipheads stand out as targets due to their immense height, they utilized this height as an advantage in terms of increased range and visibility. The *Graf Zeppelin*, being a dedicated aircraft carrier, was designed with a stooping semi-quadruped configuration to reduce the tall profile usually seen in other shipheads. Mainly relying on the fighters it carried for offense, it would gain little advantage from increased sight ranges as spotter planes were already capable of providing superior recon without exposing the vulnerable carrier.

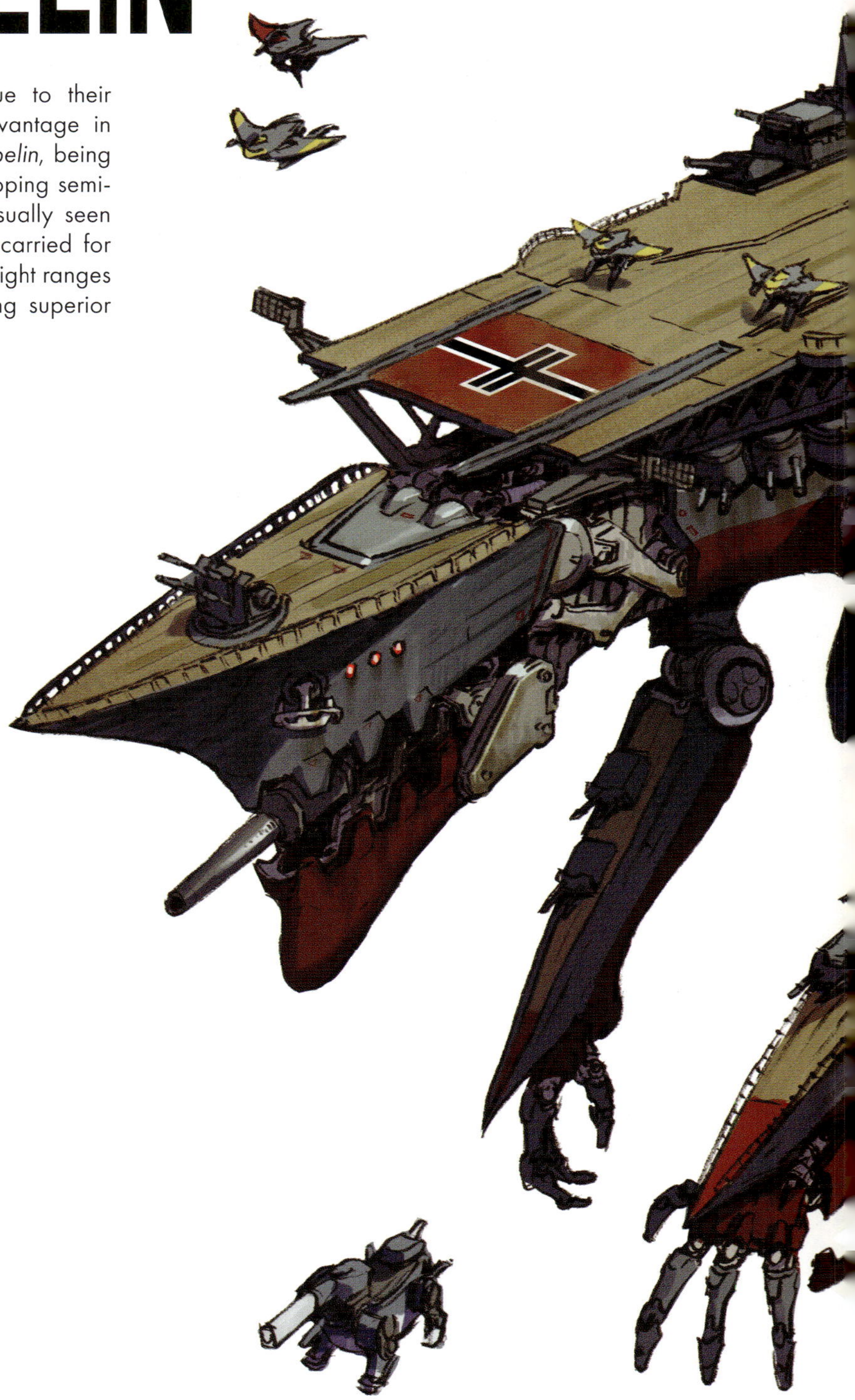

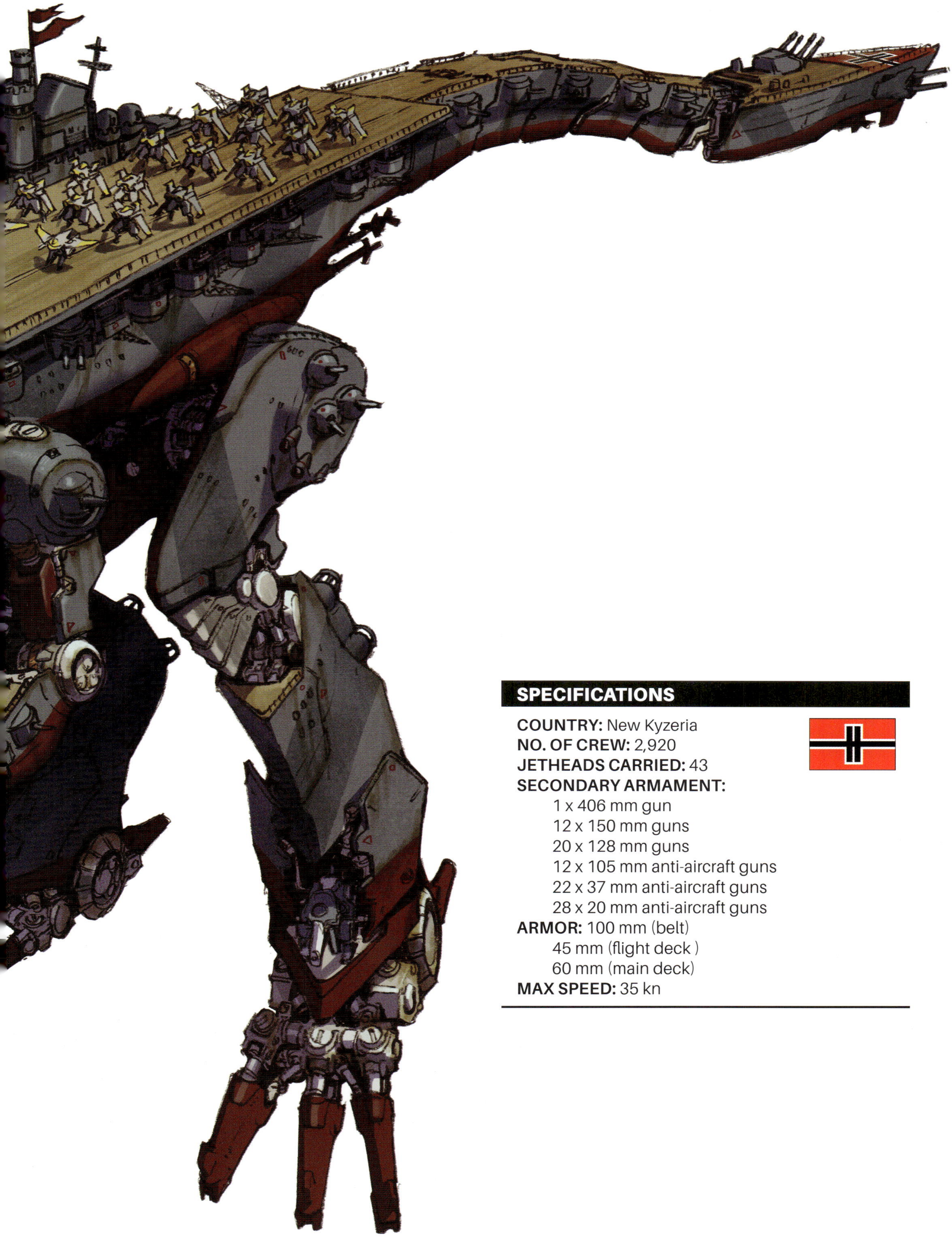

SPECIFICATIONS

COUNTRY: New Kyzeria
NO. OF CREW: 2,920
JETHEADS CARRIED: 43
SECONDARY ARMAMENT:
1 x 406 mm gun
12 x 150 mm guns
20 x 128 mm guns
12 x 105 mm anti-aircraft guns
22 x 37 mm anti-aircraft guns
28 x 20 mm anti-aircraft guns
ARMOR: 100 mm (belt)
45 mm (flight deck)
60 mm (main deck)
MAX SPEED: 35 kn

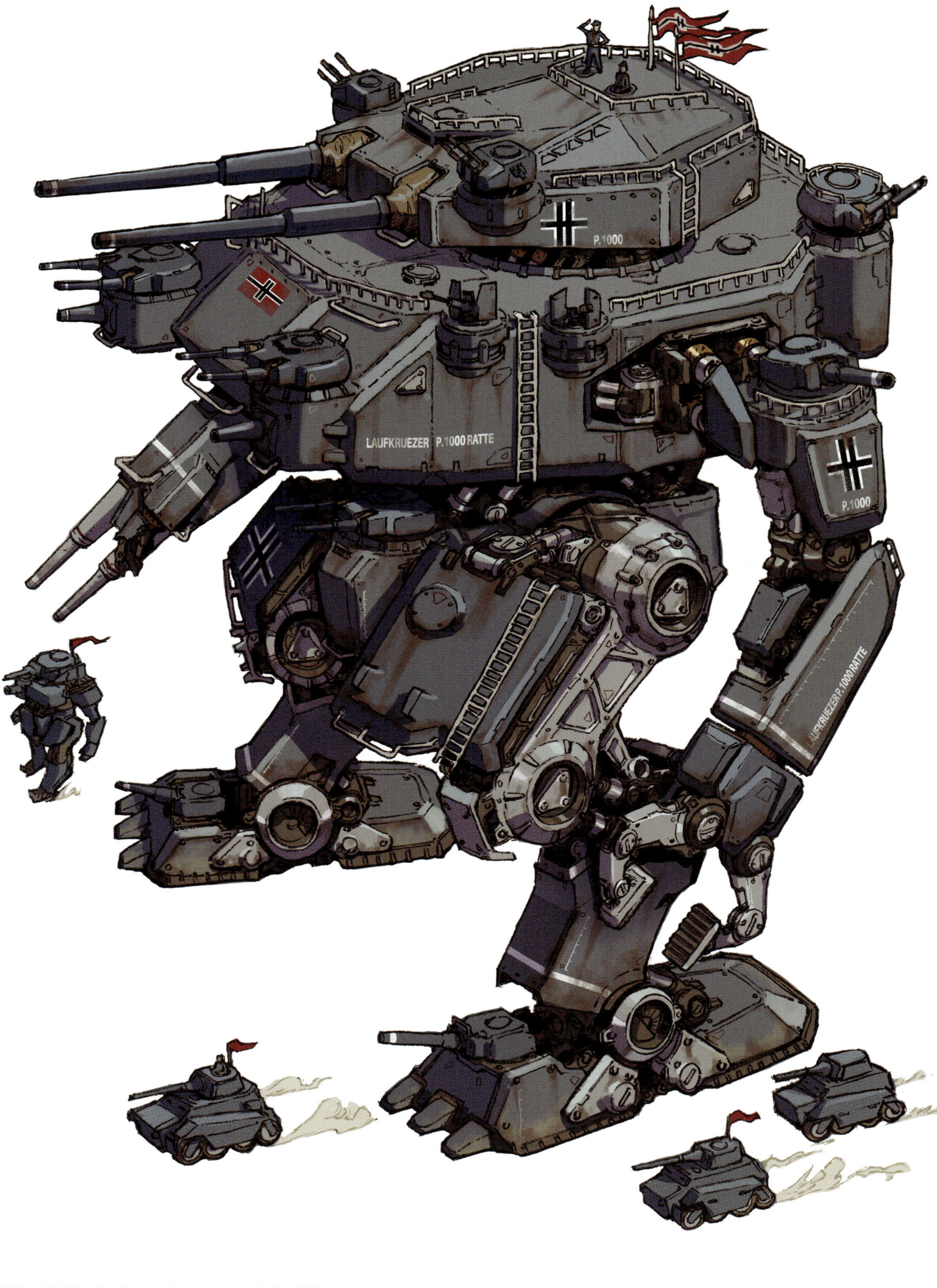
P.1000
LAUFKRUEZER P.1000 RATTE
P.1000
LAUFKRUEZER P.1000 RATTE

LAUFKREUZER P.1000 RATTE

The original intent of the walker-cruiser Ratte was, as its name might suggest, to create a vehicle that could deploy naval-level weaponry on dry land. The notion was less outlandish than it might at first seem if one considers the fact that New Kyzeria was expecting to have a number of leftover naval turrets from a large-scale refit of its cruisers. Given the enormity of the project, it gained the attention of Vaimund II and Vailinde IV, the Kyzerian twin rulers, who had gained quite a taste for titanic military projects. They wanted to see the Ratte as a frontline combat vehicle, rather than a simple carriage for leftover guns.

The project would start to lose favor as the twins' interest shifted to the Godhead Project, an even more grandiose endeavor. Therefore, it came as a surprise to the Ratte development team when, on New Year's Day WC047, they received the order to prepare their unfinished prototype for deployment. The "final" Ratte, shown here, was somewhat of a hodgepodge of parts. Its left arm, for example, was actually an early prototype claw manipulator, since construction on the cannon-armed version had previously halted due to lack of funding.

The reason for this sudden deployment would later become evident. Kyzerian forces were preparing for the ill-fated Oberherr invasion of Liberterra later that year. The Ratte had been deemed too large to transport for the invasion, but it was thought that it would be useful as a diversion unit. By deploying the bodged-together but still massively impressive Ratte on the Southern Front against Grand Britony, the Kyzerians were hoping to create the impression that they were making their big push on that front, drawing the Alliance forces there. In reality the Kyzerian army was boarding ships for an invasion thousands of miles away.

The Alliance did not anticipate any movement on the Southern Front at this time, so the Ratte was more successful in its mission than even its commanders had anticipated. Easily breaking through the Alliance defensive lines the Ratte managed to capture a fuel depot, extending its operation time tremendously. Only through clever deception and heroic action by the shiphead *Dreadnought*, converted into a training vessel at that time, was the Ratte eventually stopped. The Ratte was lured into the oncoming ocean tide by Alliance ground and naval forces in hopes of sinking the Kyzerian titan. In an act of ultimate sacrifice, the *Dreadnought* rammed into the Ratte while it was stuck in the coastal muck, finally bringing it down for good. Only 200 crew members from the *Dreadnought* survived this attack and the shiphead was destroyed, along with the Ratte, off the coast of Reeds.

It is commonly believed today that, had the Kyzerians put more stock in the Ratte and properly supplemented its assault with supporting forces, its offensive might well have changed the course of the war. Instead, those forces were lost in the catastrophic failure that was the Oberherr invasion.

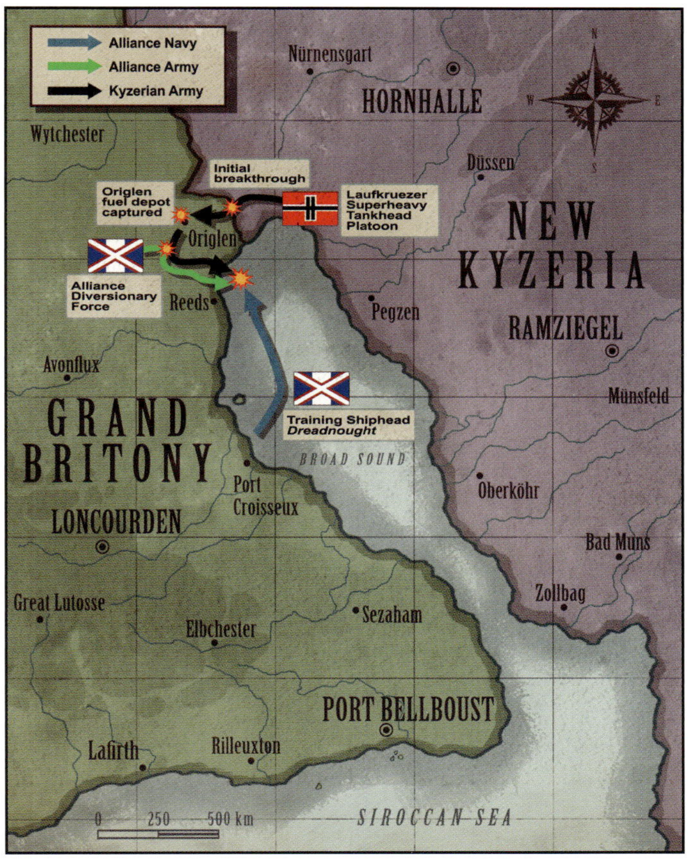

The path of the Ratte's rampage and its eventual destruction.

SPECIFICATIONS

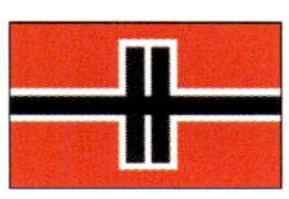

COUNTRY: New Kyzeria
UNITS BUILT: 1
NO. OF CREW: 550
MAIN ARMAMENT:
2 x 280 mm guns
SECONDARY ARMAMENT:
2 x 230 mm guns
4 x 150 mm howitzers
10 x 128 mm anti-tank guns
4 x 37 mm anti-aircraft guns
2 x quad 20 mm anti-aircraft guns
ARMOR: 150–260 mm
MAX SPEED: 21 km/h

"Then Man said, 'Let us make war vessels in our own image, after our likeness, to rule over air, land, and sea, and over every nation and over all of the world.' So man created them, colossi of steel and fire, to wage war against one another. For it was mankind's destiny to shed the blood of his brothers."

- excerpt from Kriegsdiener Scripture

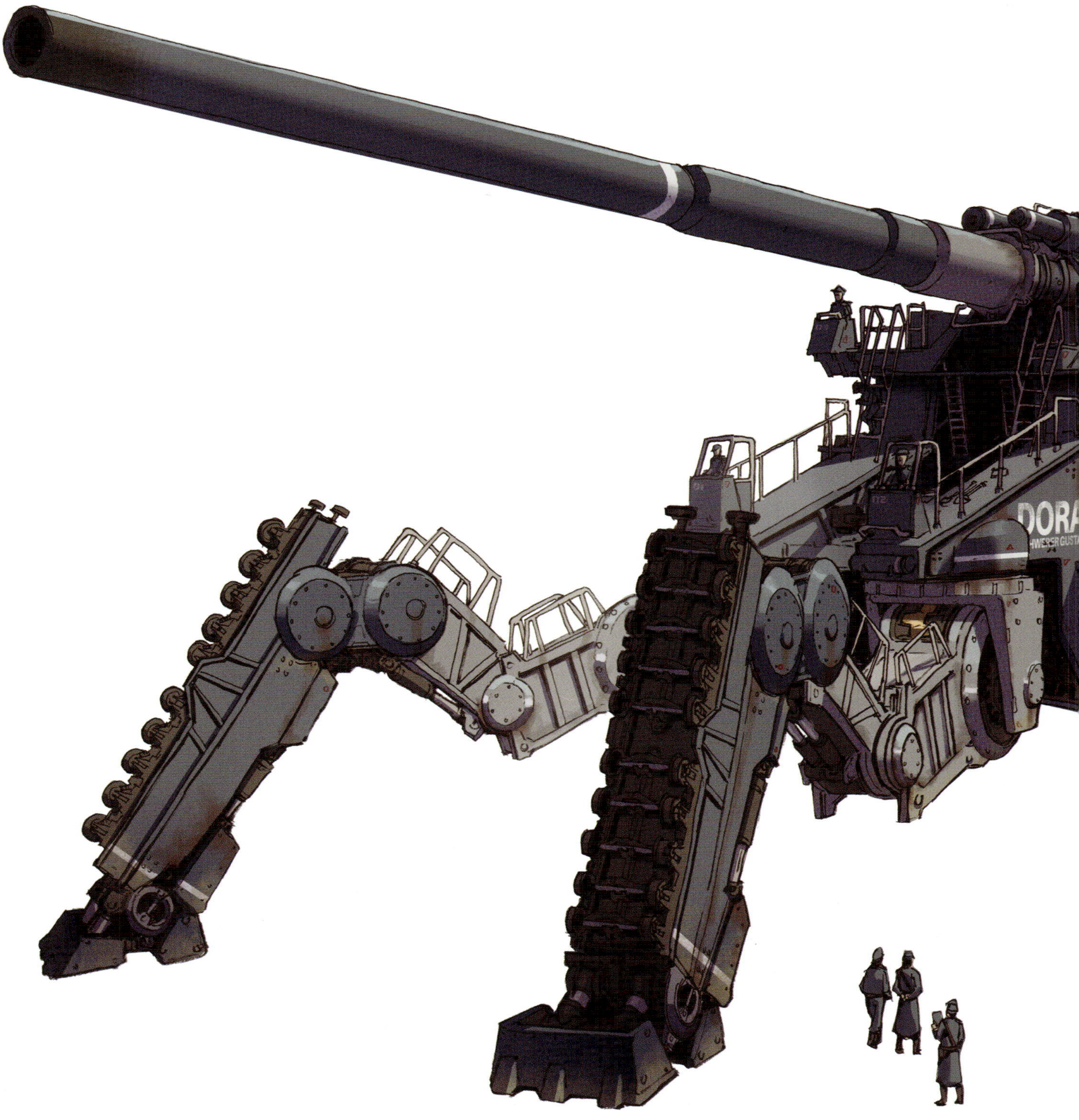

When the Lumpe Line was further fortified by the Alliance during the ceasefire period, the Kyzerians speculated on what sort of gun it would take to penetrate the tremendously thick concrete and steel armor of its fortresses. The requirements they arrived at were large enough to likely have dissuaded most development teams. The Kyzerian engineers were Kyzerian engineers, though, and they developed this idea to the point that tests, at least, were feasible. The cannon alone was too heavy to construct under the terms of the ceasefire treaty, which limited the weight of military materiel produced. Even were they able to assemble such a weapon, getting it to the front would still be a monumental achievement.

The already gigantic K5 Krupp 280 mm railroad gun was in full development and that style of railroad cradle seemed the only possible way to move the new gun, which its developers had named the Gustav.

SCHWERER GUSTAV

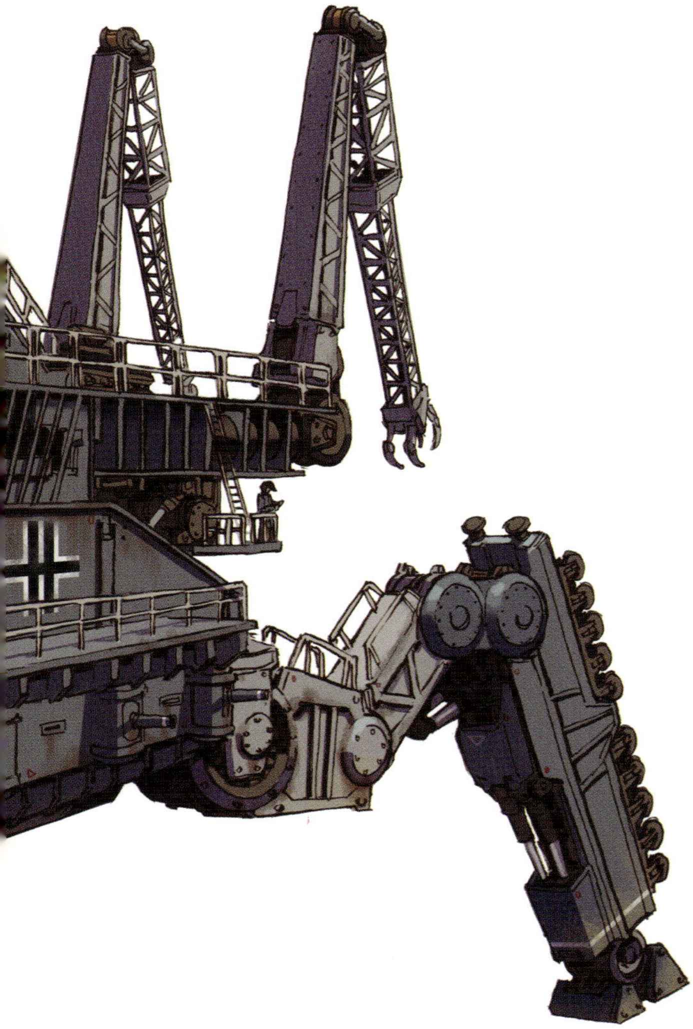

SPECIFICATIONS

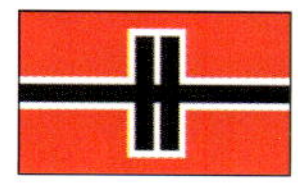

COUNTRY: New Kyzeria
UNITS BUILT: 5
NO. OF CREW:
250 gun crew + 2000 rail crew
MAIN ARMAMENT: 1 x 800 mm gun
SECONDARY ARMAMENT: 6 x 37 mm guns
ARMOR: 8–30 mm
MAX SPEED: 25 km/h

Improbable events and thoughts lead to big things, and few were bigger than the Schwerer Gustav.

Whereas the Krupp only needed one railroad track to transport it, though, the Gustav gun would be so large and heavy that only double carriages on specifically-spaced twin tracks could support it. And still, even then, more specialized track intersections would be needed to aim the gun in any lateral direction. Such a weapon would have required over a thousand crew, and would have constituted a diverting of time and resources that was considered unreasonable at the time. Hence, the project was scrapped.

That might have been the end for the design, had the war not started up again in the spring of WC015. By this point, Vaimund and Vailinde had taken a solid hold of the entire Kyzerian Empire, and they were not ones to shy away from ambitious projects. When they discovered the designs for the unrealized giant cannon, they demanded that the design be made ready for production, revised and updated with the most modern tankhead technology. Only two years later, in WC017, the first real functional prototype materialized, at about the same time Grand Britony launched their walking battleship, the *Dreadnought*. Next to that titan of steel, the Gustav gun, ultimately a giant gun on legs, seems far less ambitious than was once thought. Yet shipheads have the advantage of being able to rely on displacement buoyancy to support their colossal frames, and are unable to operate out of water for long before their joints buckle under their own massive weight. The Gustav, by comparison, required four counter-jointed leg assemblies. This made it very slow when moving under its own power, but did allow for reasonably fast aiming, without the aid of specialized tracks as in the original design.

For long-distance travel, the Schwerer Gustav would fold up into its rail mode, an operation that required a number of hours. The joints in the legs also allowed for adjustments to the exact width of the bogies, meaning that the machine could operate on tracks anywhere in the world, despite the lack of universal standardization. These adjustments also required hours of hard work supported by heavy cranehead engineering vehicles, which traveled on the tracks along with the Gustav.

The Gustav was the largest solid-projectile-firing weapon built during the Neverwar. No naval or even permanently-emplaced weapon comes close to its bore, length, or weight. Once it was able to utilize tankhead mobility, it became practical enough for Kyzeria to initiate the production of multiple Gustavs. At least five models were fully operational with another three or four under construction by the unexpected end of the war. The five that were completed participated in almost all large protracted sieges, both against the Britonica Alliance and the Yamaichi Dominion. The siege of Grigomos saw three Gustav shell the mighty city at the same time, easily the greatest concentration of firepower in history.

SCHWARZNER REAKTOR UND EXPERIMENTELLER STRAHLENKANONEN PANZERKOPF

SPECIFICATIONS

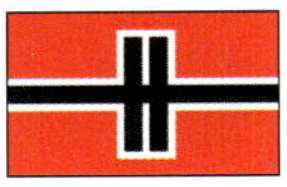

COUNTRY: New Kyzeria
UNITS BUILT: 5
NO. OF CREW: 5
MAIN ARMAMENT: 1 x volton gun
SECONDARY ARMAMENT: none
ARMOR: 15-50 mm
MAX SPEED: 15 km/h

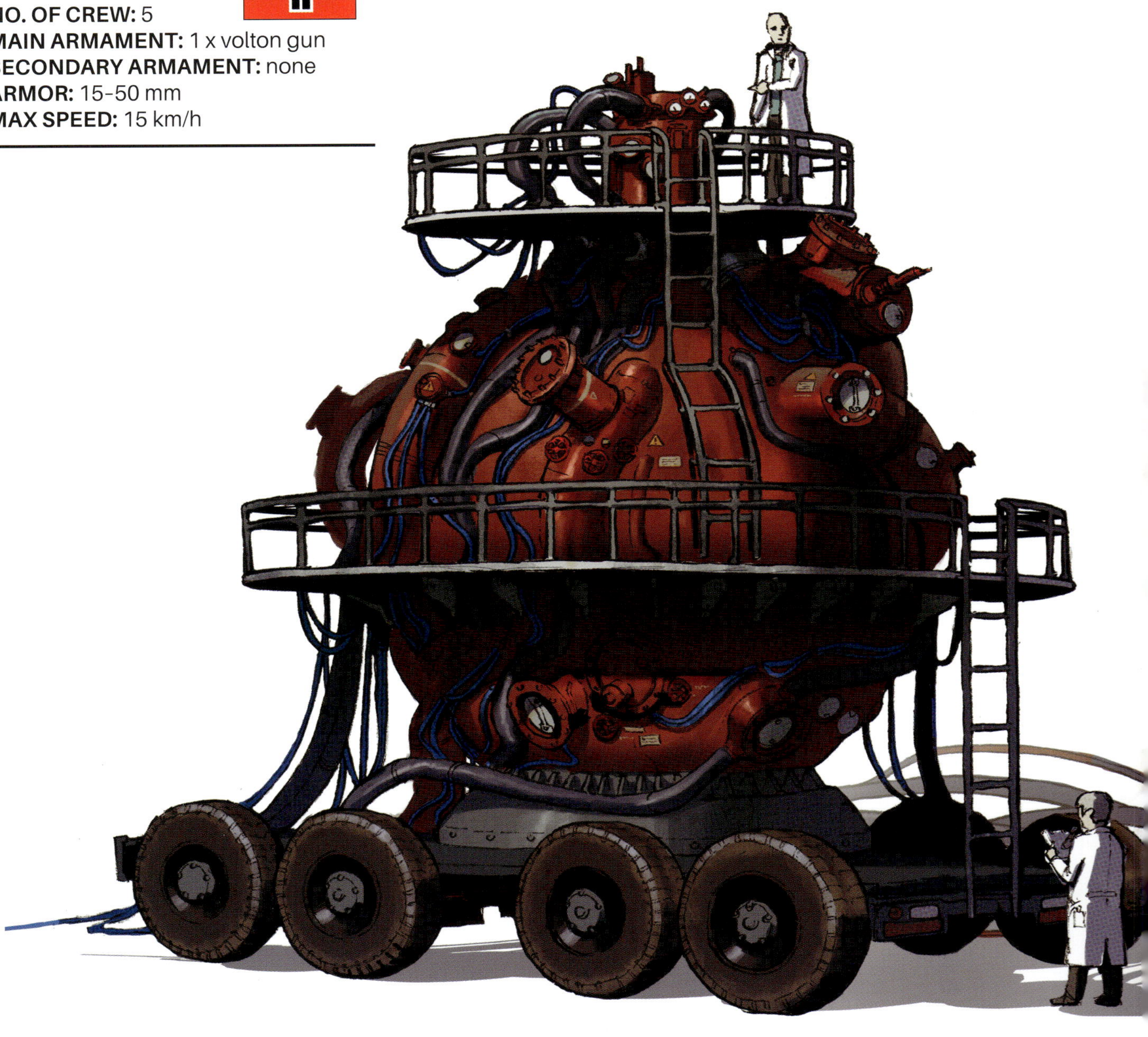

"Having to find a use for this much power is a good problem to have."
-Prof. Karl Otto Schwarzner

By reverse-engineering the mysterious Isle à Von artifact known as the "Ijzerhart," Kyzerian engineers were able to develop a device known as the Schwarzner Reaktor. Named after the head engineer of the Godhead Revival Project, this reactor was a machine that defied known physics at the time—a machine capable of creating vast amounts of energy utilizing some unknown minerals found within the original Ijzerhart artifact. The earlier models were large and cumbersome and only capable of a very short output window. However, that output was so powerful that the Kyzerian engineers had a hard time finding a system that would not be overloaded when connected to the reactor.

As they learned more about how to harness the unprecedented power output, the Kyzerians would pioneer Voltaic Photon Discharge weaponry (aka volton guns) long before other nations. The exact physics of these weapons is somewhat vaguely understood, even to this day, but the observable principles can be described with relative clarity. The voltaic energies are guided along a linear track which creates a kinetic resonance in the particles. The particles rapidly accelerate until they reach a plasma and then semi-photonic state. The resulting beam of light is hugely devastating, concentrating large amounts of energy in a small area with high precision.

During these early Kyzerian experiments, however, the energy could not even be harnessed to power a tankhead, which still required its own combustion engine. As well, the weapons mounted were simply not robust enough to withstand the tremendous energies channeled into them. They would invariably overheat, sometimes even melting entirely after the first firing.

Rather than trying to refine and rein in this power, the Kyzerians went down the opposite path and started development on the volton-beam-equipped Schwerer Gustav, where the sheer mass of the weapons platform allowed for better heat management.

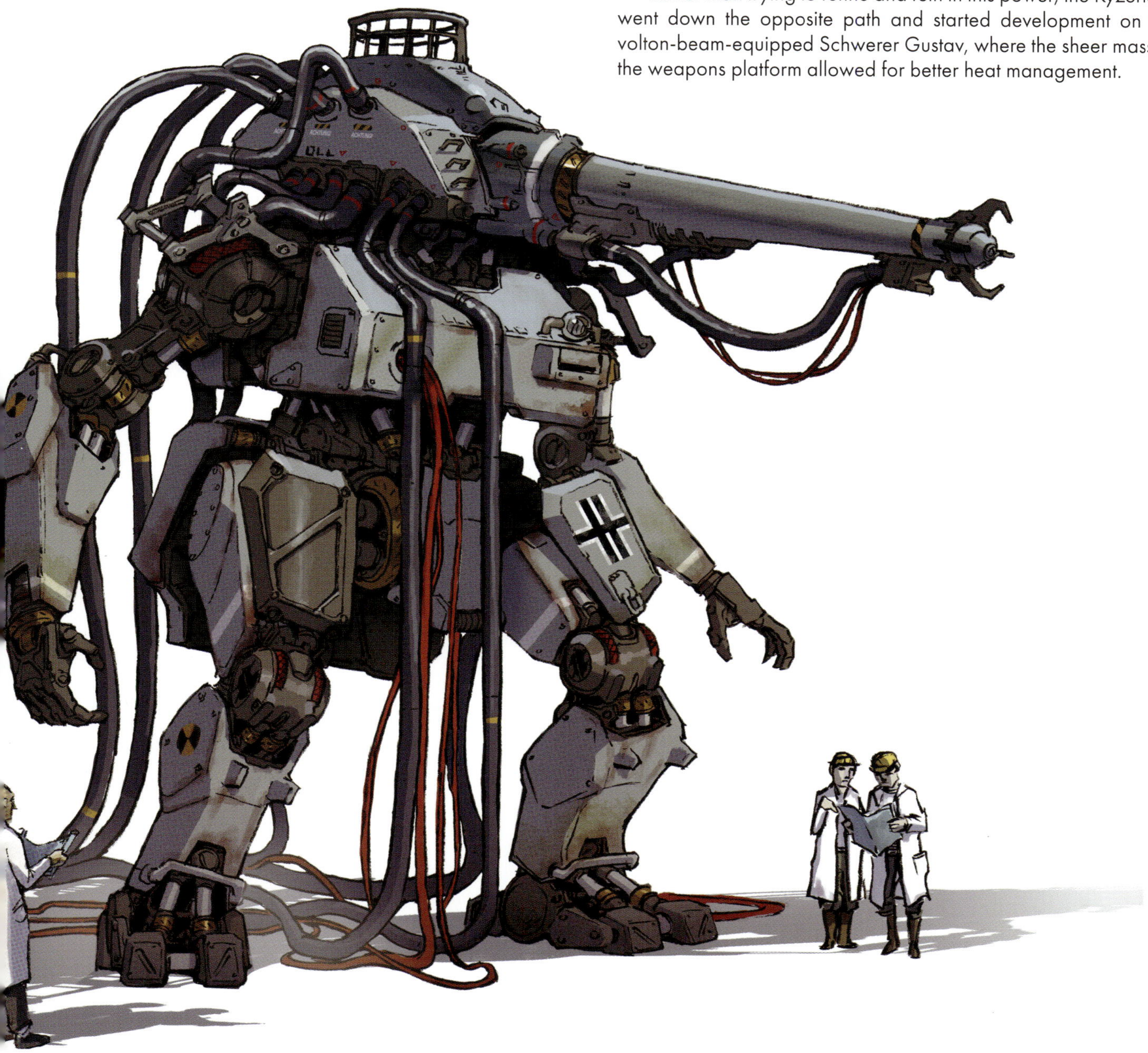

EXPERIMENTELLER STRAHLENKANONE FENRIR

SPECIFICATIONS

COUNTRY: New Kyzeria
UNITS BUILT: 1
NO. OF CREW: 310 gun crew + 2,000 rail crew
MAIN ARMAMENT: 1 x mega-volton cannon
SECONDARY ARMAMENT: 6 x 37 mm guns
ARMOR: 8–30 mm
MAX SPEED: 25 km/h

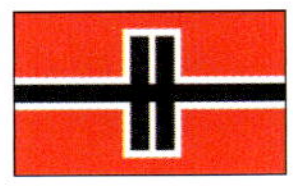

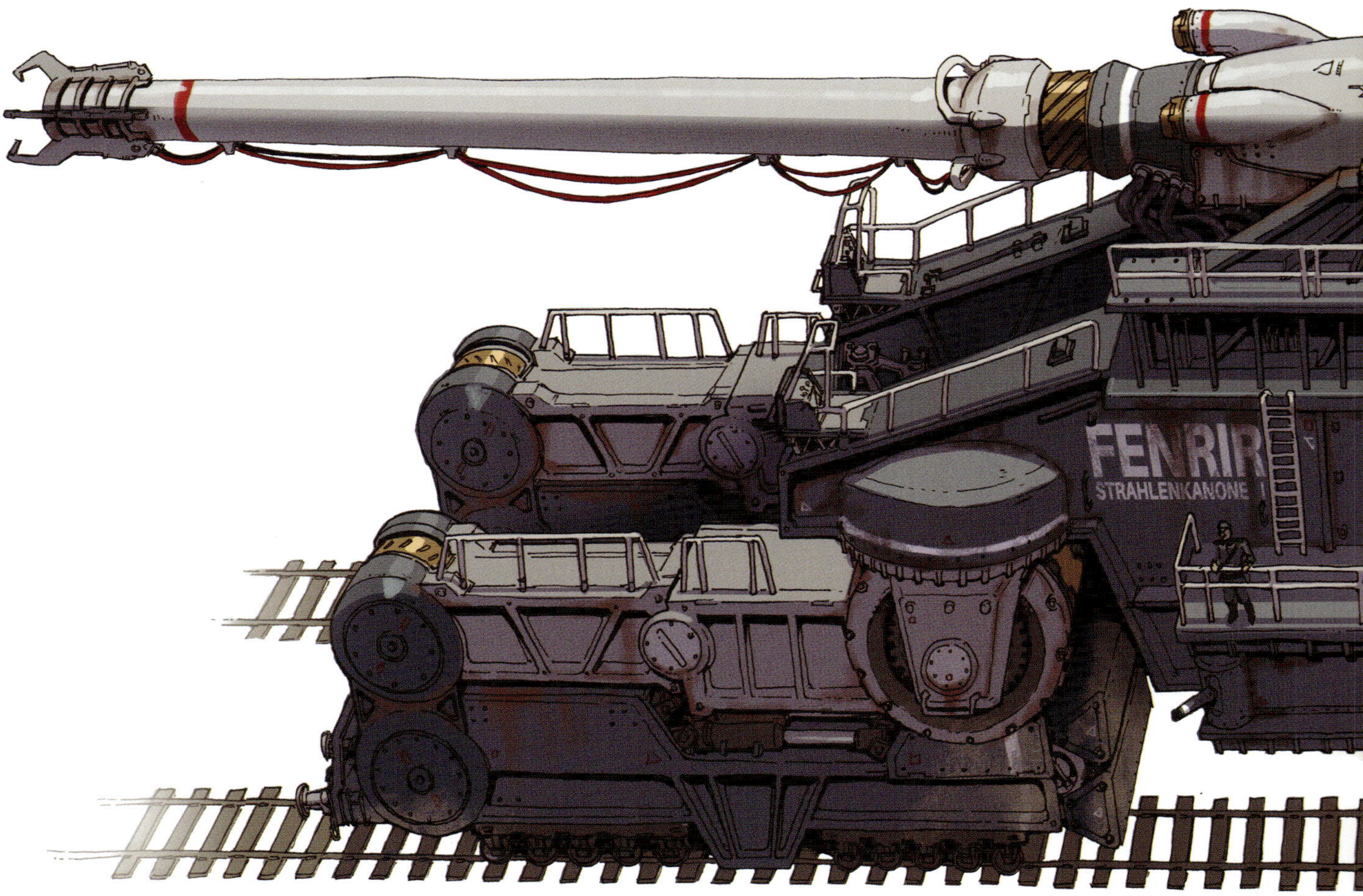

Very little is known about this experimental Kyzerian weapon. Only a handful of photographs survived the war. Historians speculate that it was likely a test platform for the energy weapons that would later be mounted onto their ultimate weapon, the Godhead.

The Alliance only became aware of this monstrosity in WC047, when it was field-tested against three fortified positions on the Rozenian front. The heavy autumn fog had allowed the rail carriage to approach unnoticed to within firing distance of its targets. While its firepower was devastating, this event highlighted a drawback of that type of energy weapon: they required a direct line of fire to the target. The first attack by the beam weapon was a complete surprise to the local Britony forces. By the second, they realized that the direction of the attack could be clearly observed. By the third, an observation team managed to triangulate the position of the weapon. Using this data, the Alliance ordered an artillery barrage in an attempt to destroy the unseen target. While they failed to hit their mark, they managed to obliterate the tracks behind the weapon carriage. The gun's crew then retreated, but not before setting the generator to overload and self-destruct in order to prevent capture by Alliance forces.

"O what terrible light it was. What dreadful, hideous light that turned that hushed night's sky into hellish day."
-Rozenian Shepherd Gulnara Wynmûne

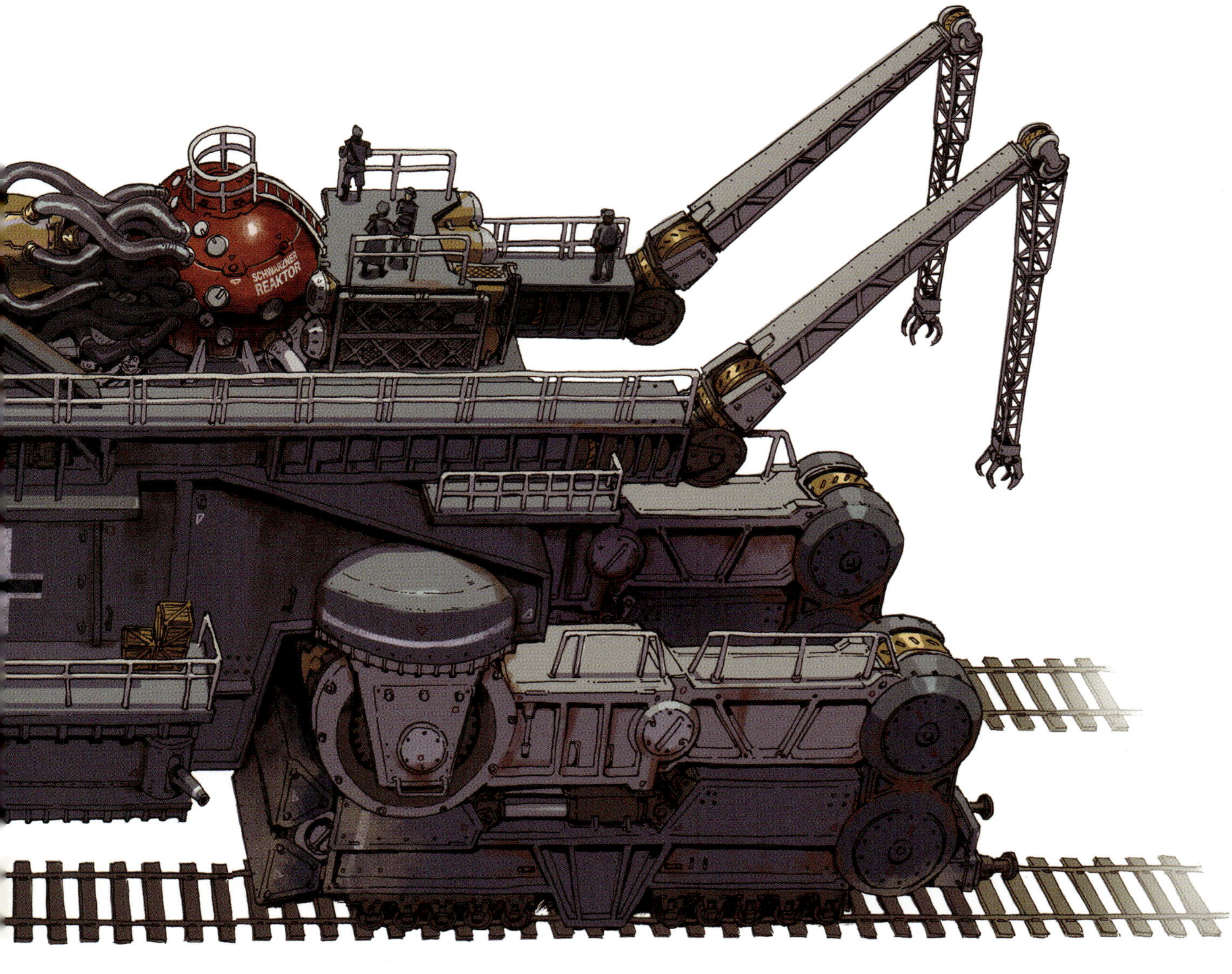

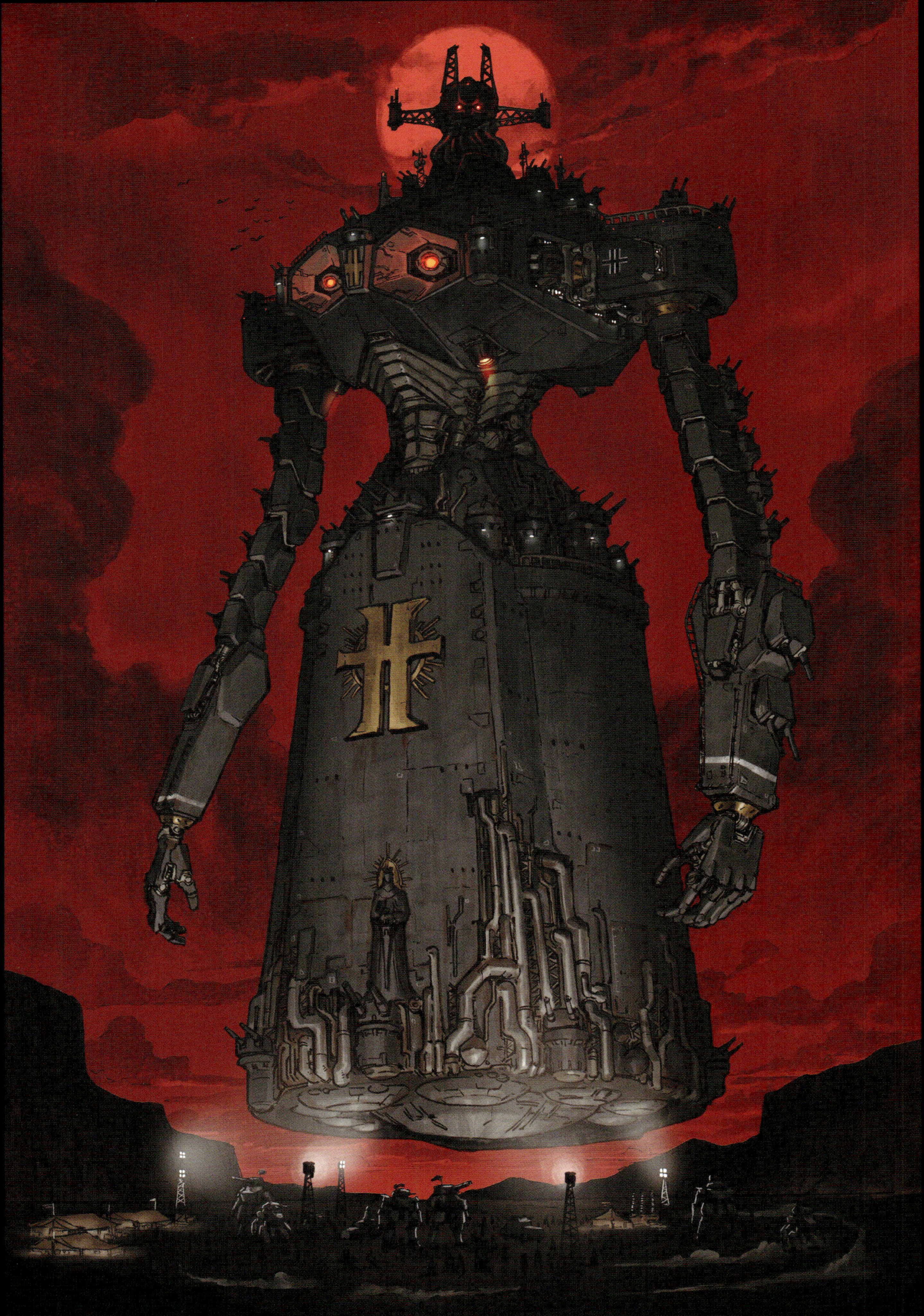

GOTTESHAUPT PROJEKT

The operation to destroy the Godhead resulted in the most important battle in the Neverwar, yet much about that fight and the Godhead remains unknown or classified today. By WC046, the Dominion and Alliance already had some idea that the Kyzerians were building a new superweapon. However, the Kyzerians were known to have a predilection for large-scale projects and were, in fact, engaged in so many that it was hard to tell one from another. Some thought the Ratte or the volton-gun variant of the Schwerer Gustav were this "ultimate weapon" that they had learned of. Things became more clear after the Oberherr Invasion failed in WC047 and New Kyzeria was forced into a hasty and disorganized retreat. Taking advantage of their confusion, in WC048 Yamaichi Intelligence operatives were able to discover the existence of Project Godhead. Even without all the details, the amount of human and material resources dedicated to the project was enough to convince them that halting it had to be considered of utmost strategic importance. This information was intentionally leaked to Alliance Intelligence soon after.

When the Godhead's multiple Schwarzner reactors were fully powered for the first time, it sent out a resonance wave so powerful that it was picked up on radio as far away as the western shores of Yamaichi. In Grand Britony, the radio noise was deafening for a moment, then quickly faded. Confusion reigned in the aftermath. The accepted theory was that the wave had been caused by a meteorological event, such as a high stratosphere thunderburst. Alliance Intelligence was more skeptical and they dedicated their Conundrum Codebreaker Unit, made up of top scientists from many different fields, to analyzing the phenomenon. Turise Drefelsen, a Kyzerian physicist who had fled to Grand Britony after the rise of the Kriegsdiener religion in her home country, realized that the signal was similar to those emitted when the mega-volton gun of the Schwerer Gustav "Fenrir'' was charged.

Using precise logs kept by radio listening posts throughout the Alliance, Drefelsen and her colleagues were also able to triangulate the source of the radio burst to the abandoned Kyzerian town of Mohnstadt. Employing physics theories she had personally developed, Drefelsen even made an estimate of the power of the device that had emitted the wave. Drefelsen calculated and recalculated numerous times, assuming she must have made an error, but the results were the same every time. The amount of power that needed to be generated to create such a wave was of an apocalyptic magnitude. When she informed Alliance Intelligence of her findings, they were quick to connect the dots and rightly assumed the Godhead Project was the source of the signal.

Returning the favor, these reports were deliberately leaked by the Alliance to the Dominion, in hopes that the Yamaichi would somehow respond to the growing threat of the Godhead. Meanwhile, the Alliance worked on plans for a tactical strike on Mohnstadt with the goal of disabling or destroying the weapon before it could become fully operational. The urgency of these plans quickly mounted as more energy bursts were detected. Alliance Command had narrowed down their options to two courses of action. Neither were favorable. On the one hand, they risked being too late if they waited to amass a large enough strike force capable of taking down the Godhead. On the other, if they attacked immediately with available forces they might simply not be strong enough to accomplish the task. In an act of desperation, all known data on the Godhead and their preliminary plans were transferred to the Dominion through trusted intermediaries.

Unbeknownst to the Alliance at the time, a Yamaichi spy jethead had witnessed the test firing of the Godhead's main volton guns and its destructive effect. The pilot had reported that the Godhead, despite its colossal size, was able to levitate at a height sufficient for other tankheads to pass beneath it. Neither the Yamaichi nor the Alliance could conceive even of theoretical technologies which could accomplish such a feat, leading some of those privy to the intelligence to boldly claim it could only be ancient and forgotten sorcery. In face of such stupendous odds, the Yamaichi Dominion and Britonica Alliance, enemy combatants in over thirty years of war, agreed to put their differences aside and coordinate an immediate all-or-nothing strike against the Godhead. The decision to commit to this action bypassed all councils, cabinets, electorates, and planning commissions, with people at the top of their chains using vetos or simply force of will to make this mad idea work. Clearly cognizant of the strategic risks to their respective forces, as well as the personal risks to themselves, the Alliance and Yamaichi commanders gave the operation the codename "Grandmaster Gamble."

The plan outlined three phases of attack. Yamaichi forces would deploy all available assets, sailing as many ships and carriers as they could muster up the Nehalean Channel, straight into the Kyzerian heartland, where they were to launch their airborne forces. The pilots of the Yamaichi jetheads and bomberheads were all volunteers aware that they would be flying to the full extent of their fuel reserves and unable to return to base, no matter the battle's outcome. Meanwhile, West Rozenia and Grand Britony would use their fastests units to break through a recently-observed weak point in the East Rozenian defensive line that bypassed the famous Triesburg Gate and to then outrun the defenders in a sprint for the Kyzerian interior. More than half of the Brit attack force would consist of trucks and bowsers carrying extra fuel, providing the attack units the reserves they needed to penetrate all the way to Mohnstadt. Finally, the Liberterran fleet would approach via the Tramonic Ocean, deploying a combined amphibious landing and airborne force.

When the three forces came face to face with the revived Kyzerian god of war, the Godhead made quick work of the attacking ground forces, utilizing its many guns and volton cannons located in its palms, chest, and head. The Alliance and Yamaichi losses were tragic, all nations sharing equally in the tremendous burden of war. From commanders to tankhead drivers, jethead pilots to maintenance workers, every soldier who took part in the Grandmaster Gamble did so with the knowledge that it would likely be a suicide mission regardless of the outcome. Despite the eventual global peace that resulted from this plan, we can not understate the loss of life that was needed to make it a reality.

There is a second set of reports regarding the Grandmaster Gamble that are almost unbelievable because of how outlandish they were. Officially, these are dismissed as obviously hyperbolic and definitively false. For historians, however, they are not so easily

CONTINUED>>

GOTTESHAUPT PROJEKT (CONTINUED)

dismissed due to the fact that they have been so often corroborated by the accounts of both Alliance and Yamaichi personnel who took part in the operation and by their friends and family, who heard the accounts from survivors. The first repeated account is that multiple air transports landed or crashed on the Godhead's top section, including at least one tankhead transport. Certainly the Godhead was massive enough that a light airborne tankhead could operate on top of it, provided it was able to get past the titan's many anti-aircraft guns. Yamaichi and Alliance boarders, fighting side by side, managed to penetrate to the command bridge where, it is claimed, they discovered the bodies of Vaimund II and Vailinde IV, the twin kaisers, murdered.

The story does not stop there, however. Supposedly—and, again, this has been recorded in both Alliance and Dominion sources—the commander's position was taken by a grizzled, bearded man garbed in the robes of the Kriegsdiener religion. The witnesses have made the seemingly incredible claim that the man was the spitting image of the deceased Razivir, the notorious religious advisor to the old kaiser, and rumored true father of the twin heirs.

Razivir's death in WC011 was well documented, of course. However, stories of a shadowy Kriegsdiener priest advising the twins had been circulating for years, and it is possible an imposter in the sect opportunistically capitalized on the fame of the Kriegsdiener cult leader by pretending to be Razivir. It has been written in the Kriegsdiener texts related to the Godhead prophecy that "the blood of two rivers need flow to breathe life into the god of war made real." Modern Kriegsdiener adherents claim that the twin heirs were killed to fulfill this prophecy to activate the Godhead's full power. What we do know is that by sabotage, accident, or a lucky shot, the bridge went up in a catastrophic explosion, killing everyone present. Without its control systems, the levitating titan faltered and shut down, crashing to the ground.

The defeat of the Godhead led to the most amazing turn of events, a true twist of history. The Alliance forces found themselves deep in enemy territory, fighting side by side with a supposed enemy, the Yamaichi. Without supply lines or established routes for escape, there was no possible retreat back to their own territories that would not result in extreme, likely total, losses. But while they had suffered major casualties, they were still the most advanced weapons and veteran soldiers that two-thirds of the world's nations could muster, a serious threat to the now-leaderless Empire. The Kyzerian forces who had been defending the Godhead stood in the shadow cast by the ruins of their felled god of war and laid down their arms in surrender. Their ultimate weapon had been broken and so had their will to fight.

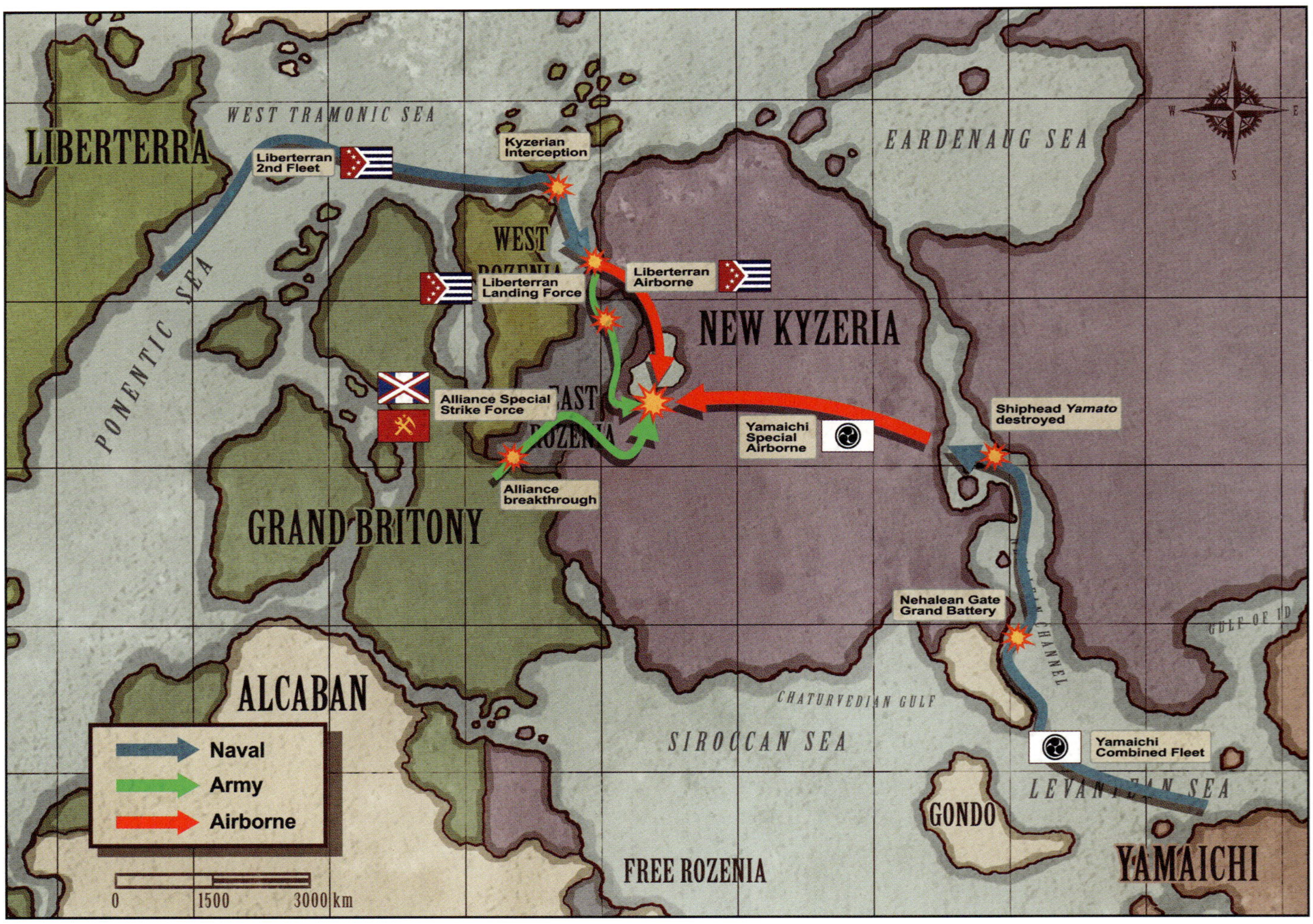

The four-pronged assault of the Grandmaster Gamble plan.

"My brothers and sisters. Today we celebrate our apotheosis! A true God made in the image of Man!"
- Kriegsdiener Priest Friedrich Engel, at the christening of the Godhead

Like a snake with its head cut off, the command structures throughout the Greater Kyzerian Empire collapsed. As reports spread of the combined Alliance and Dominion force now in the heartland of the Empire, many thought they had lost the war. The Kyzerians who had a better overview of the situation realized that they might have a golden opportunity to strike the now-vulnerable other nations, but that if they did, then the estranged strike forces still deep inside New Kyzeria might well unleash devastating acts of retribution with their last remaining fuel and ammo.

A global ceasefire was offered and accepted. The confusion on all sides was so great that having time to regroup was considered advantageous. Hasty agreements were drafted that allowed the stranded elements of the strike force that were lost along the way to return home, and aid was supplied to the strike force that had occupied the Godhead's wreck, fuel and food mostly, as well as medical aid for the many wounded. Some forces from each nation would remain stationed surrounding the Godhead, carefully observing each other to make sure no side could tamper with the titanic machine or abscond with its technologies.

Amazingly, this state of bemusement lasted into the next few days, and then weeks. Though there were many places in which fighting reignited, or rather never truly stopped, the shock seemed to have sapped the will to fight, and thus the momentum, from all the embattled nations. Negotiations were opened with an earnestness that had not been seen for two generations. These negotiations took many weeks and were as hard-fought as the war itself, if not more so. In the end, on the shores of Taubensee Lake, within sight of the wreckage of the Godhead, representatives from all three great powers signed a provisional peace agreement, on the last day of the 49th year of the War Century. "The Neverending War" had finally ended.

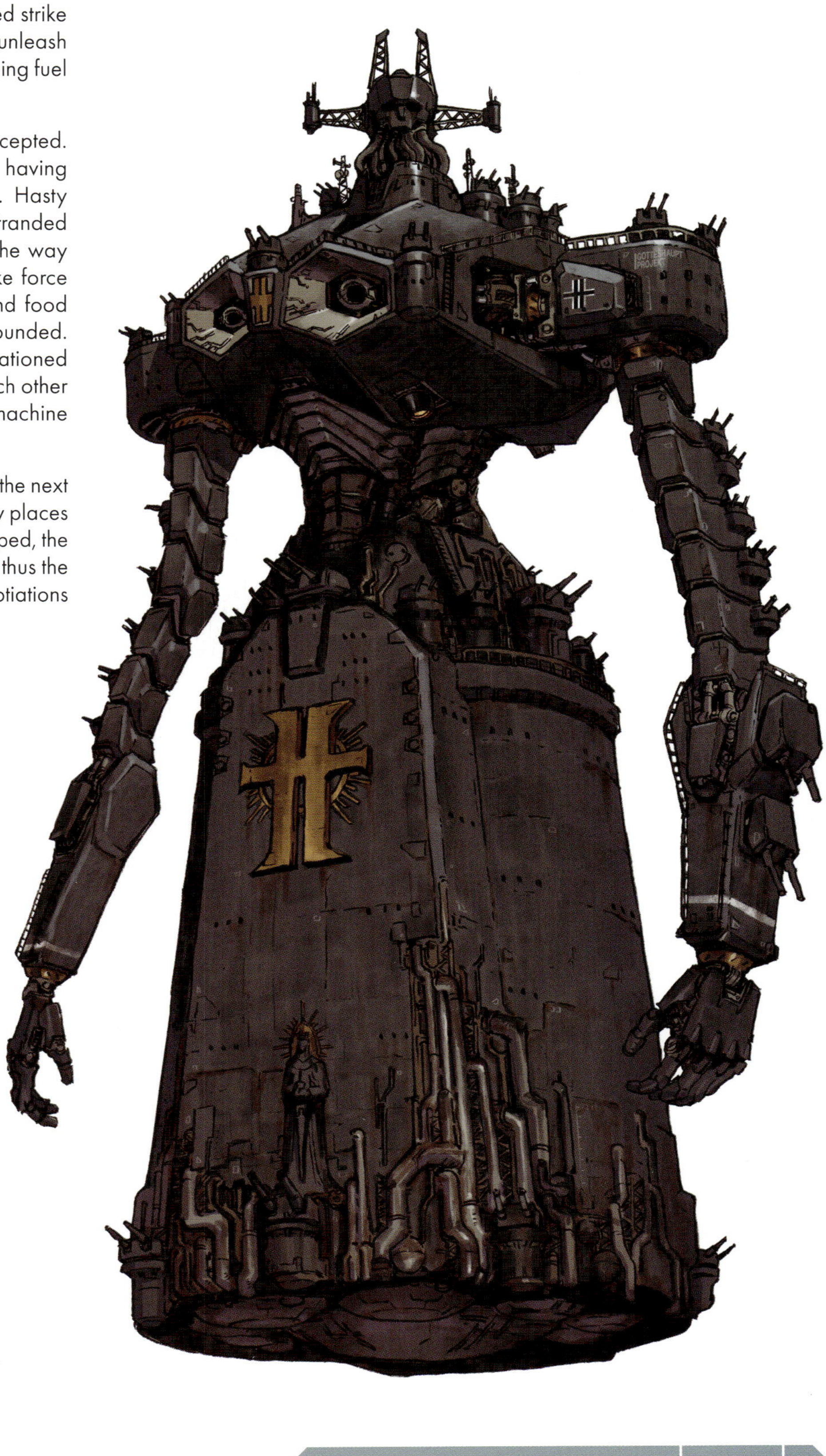

SPECIFICATIONS

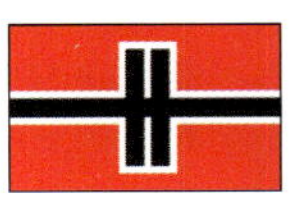

COUNTRY: New Kyzeria
NO. OF CREW: 4,162
ARMAMENT:
- 2 x giga-volton guns
- 11 x volton guns
- 2 x twin 305 mm guns
- 2 x twin 150 mm guns
- 12 x 128 mm anti-tank guns
- 44 x 105 mm anti-aircraft guns
- 12 x 20 mm anti-aircraft guns

ARMOR:
- 320 mm (main hull)
- 360 mm (skirt)
- 120 mm (main deck)
- 50 mm (upper deck)

SPEED: 30 km/h

EPILOGUE

Having come to the end of the Neverwar, we have equally reached the limits of this author's expertise. I would, however, like to share some observations and speculations about our current time, and the shadows the Neverwar still casts on it.

While the defeat of the Godhead and the following impromptu ceasefire heralded an uneasy truce and, eventually, the first real peace in half a century, all was not—and is not—at ease. Many will still remember the paranoia that followed those early days. Rumors of hidden Kyzerian factories secretly producing superweapons for a retaliatory strike have (with one notable exception) proven unfounded. However, governments do not change their thinking as easily as individuals may, especially after fifty years of war, and the peace between the Alliance and the Yamaichi is a fragile one. The Kyzerians were soundly defeated, but how do you just forget and forgive the enemy you have fought across the globe for two generations? Only because they helped you win one fight? Opportunists from all the formerly-embattled nations have found there can be great political value in exploiting such lingering bitterness. Thankfully, their blustering remains just that, and their regrettable warmongering has yet to reignite conflict.

Long-suffering Rozenia was occupied for so long that few Rozenians alive today remember what their own country was like without outside forces directly controlling it. This has led to an identity crisis in the new "United" Rozenia. While their newly-formed government is taking back control of the country, it is also undeniably a puppet whose strings are pulled by two masters, with

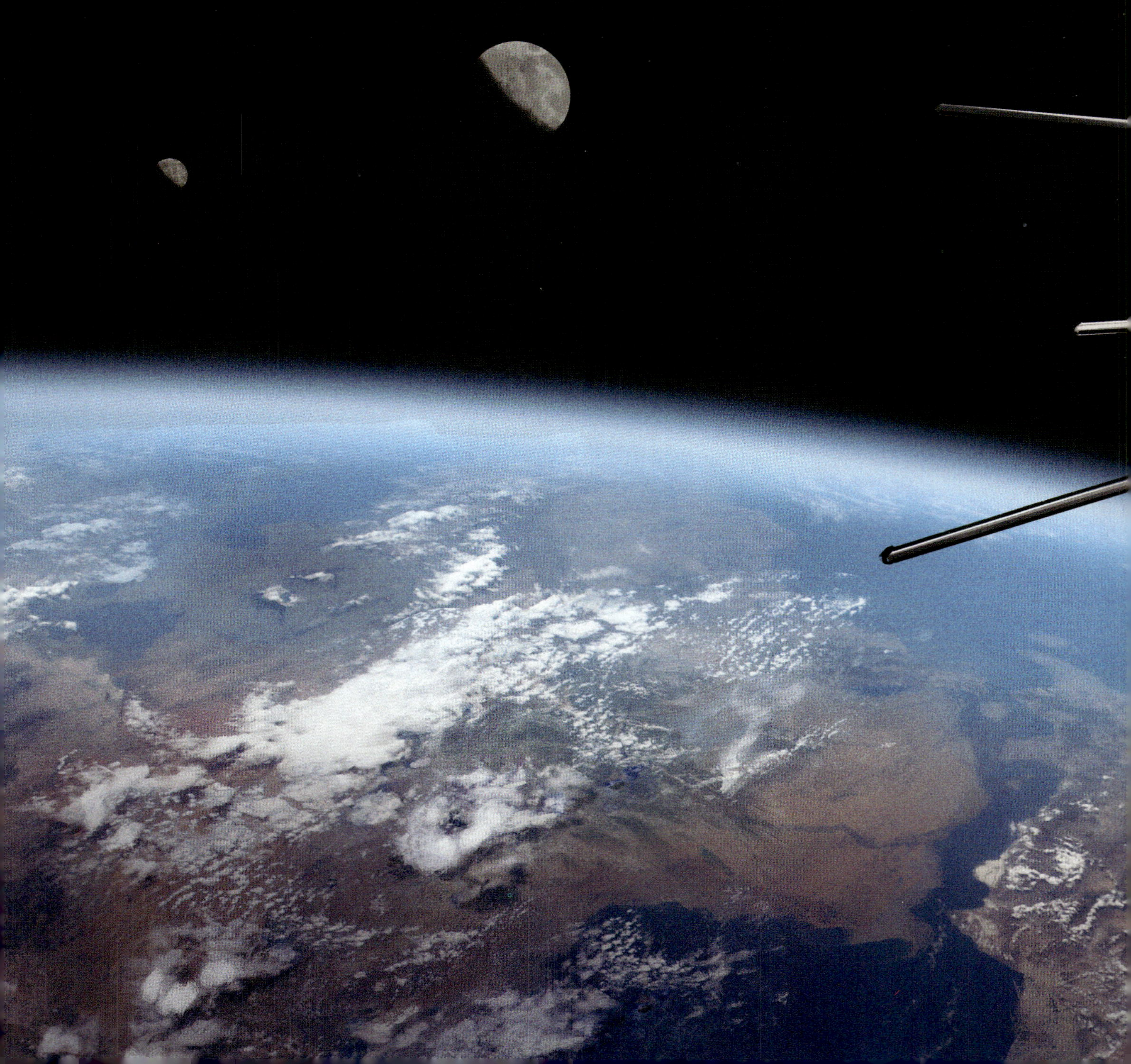

both the Britonicans and Kyzerians evidently unwilling to surrender the control they hold over the region, though they frame their involvement as post-war support for the war-torn country.

In the last throes of war, Kyzerian High Command had the ancient Isle à Von relics, progenitors of all tankheads, destroyed. Many of the nation's lead scientists disappeared soon after. Some say they were killed to keep their secrets out of enemy hands, but unknown numbers of them surely went into hiding. Many, no doubt, defected to or were hunted down and recruited into the Alliance and Yamaichi military machine. Might their knowledge of the Isle à Von artifacts lead to an age of new wonders? Or will they continue to spawn ever more terrible horrors of war? Could the sudden interest in space and our moons be related to the civilization, still so enigmatic, which left behind the titanic war machines discovered by the Kyzerians and the Yamaichi? Is there some unknown treasure to be claimed beyond the reaches of our sky and, if so, what price will it demand?

For now, the Neverwar has finally come to an end and this is undeniably, unequivocally, a good thing. The end of a war does not equate to peace, however. Peace is something we must all work at, constantly, individually and together.

Those of us who study the Neverwar can see, all too clearly, dangerous parallels in the patterns of today. The tinder sits ready, waiting to ignite at the merest spark. As all of Earde turns their attention to space, who knows where the fires of their mighty rockets might ignite the next war?

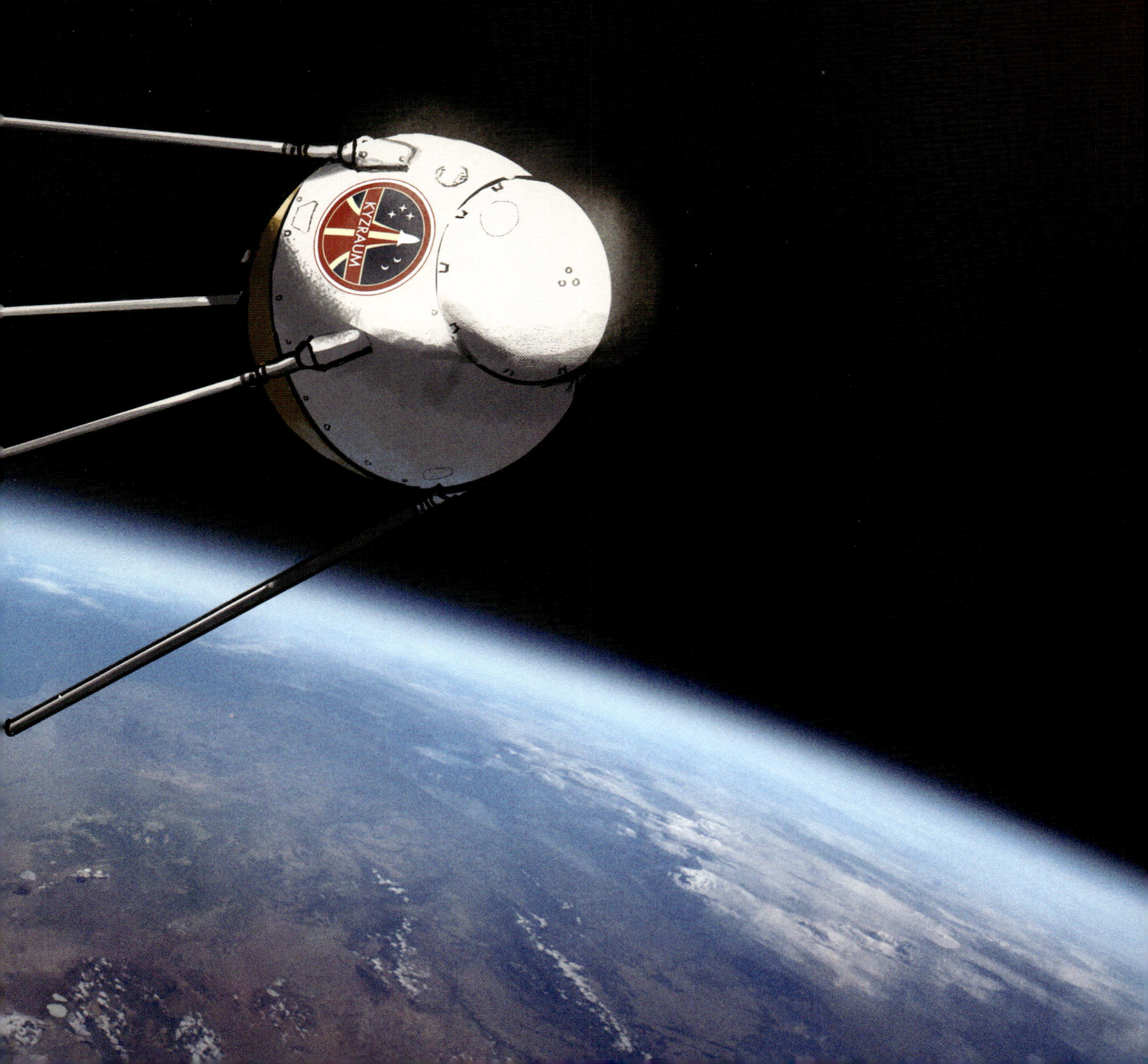

TIMELINE

- **BWC 225–130, Kyzerian Conquest:** During this era, the many small states that make up Kyzeria are united under Kaiser Neapollöwe through complex alliances, brutal warfare, and convoluted plots. This conflict is fought by knights on horseback when it begins and soldiers with guns and cannons when it ends. The newly-formed nation is dubbed Kyzeria.
- **BWC 190–70, Mechanical Evolution:** Revolutionary mechanical innovations cause industrial production and society to advance at an unprecedented rate. Civilization reaches new, previously unthinkable heights—and warfare, new depths of brutality.
- **BWC 98**: The first successful powered Aerostat balloon flight crosses the sea from Northern Britony to Rozenia.
- **BWC 83:** The first armored car ever created, the Simms Battle Car, is used in combat.
- **BWC 10:** Grand Britony Admiral William Trout violates Gondo territorial waters, resulting in a naval conflict lasting multiple weeks before political liaisons manage to resolve the situation. This will later become known as the Trout Incident.
- **BWC 9:** Political power in Kyzeria is concentrated in the Kaiser and their court. After this reformation period, the nation is dubbed New Kyzeria.
- **BWC 8-3, Hilland revolts:** A coalition of Brit and Kyzerian workers in the Hilland region of northern Alcaban declares independence from both nations due to grievances over taxation. The armored cars employed by both countries prove inadequate in the rough terrain and pockets of resistance manage to hold out for nearly half a decade.
- **WC001:** The assassination of Felix Friedrich, the exiled crown prince of Rozenia, and his Britonican wife, Alexandrina, Duchess of Highhill, causes a domino of alliances, threats, and pacts to take effect, rapidly consolidating most countries of Earde into three major factions: the Greater Kyzerian Empire, Britonica Alliance and Yamaichi Dominion. Rozenia forms a neutral zone between the Britonica Alliance and the Kyzerian Empire. The Southern Front, where the Alliance and Empire share a border, quickly becomes a hot zone.
- **WC003:** Advances on the Southern Front have lost all momentum. Both sides dig in for a stagnated trench war.
- **WC004:** The death of the Rozenian Tsar leads to a civil war within Rozenia as multiple scions make a claim for the throne. Both the Britonica Alliance and Kyzerian Empire provide covert and material aid to the scattered factions.
- **WC004, Winter:** The mystic Razivir oversees a meteoric rise to power in the Kriegsdiener cult and becomes personally entangled with Kaiser Walmir VII's household. Razivir's popularity leads the term "War Century" to be adopted in common parlance, as the Kriegsdiener cult predicts that a hundred years of "glorious war" will reshape the world.

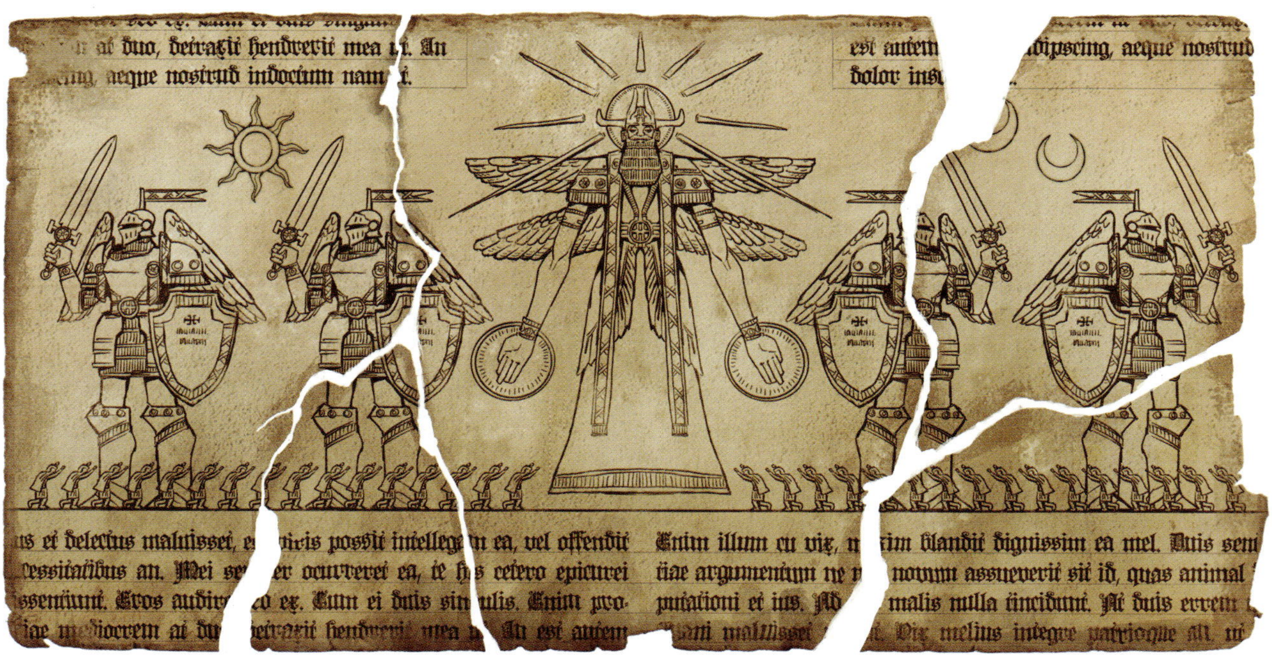

A page from a Kriegsdiener manuscript recovered after the war, supposedly depicting the artifact used as the base for the Godhead project. Uncertain Provenance.

- **WC005, Spring:** Traditionalist reclamation policies in the Yamaichi Dominion lead expeditionary forces into Western territories beyond the mainland province of Hyakkoku, straining relations with the Kyzerian Empire, which fears a war on two fronts. Yamaichi declares they are reclaiming the ruins of monasteries traditionally considered to be sites of origin for one of their national religions.
- **WC005, Fall:** Under the Nolburg Plan, the Kyzerians attempt to circumvent the Southern trenches by going through neutral Rozenia. Britonican "peacekeeping" forces stationed in Rozenia detect the Kyzerian forces and move to intercept, calling for reinforcements on all channels. Bad weather and muddy roads, common in Rozenia at this time of year, slow the Kyzerian forces significantly. Both local resistance and the Britonican vanguard fight a fierce retreating battle in what the Rozenians will later call "Burinoch," the Night of Storms.
- **WC006:** The Night of Storms leads to a populace divided between support for the Britonicans and fear of Kyzeria. Internal conflict in Rozenia solidifies around two branches of nobility. Each aligns with one of the two main combatants based mainly on geographic proximity. Archduchess Adela of Momil, the western pretender to the throne, marries Britonican Viscount Serpenoirewhile. In response, the eastern Court swears a Kriegsdiener blood oath to the Kaiser, effectively becoming a vassal state of the Kyzerian Empire under a religious bond.
- **WC008:** Britonican and Yamaichi commercial operations in the Darumeshia report attacks by war machines never seen before. Both nations initially assume local pirates to be the culprits. However, as sensational reporting on these mysterious machines grows, the Britonicans and Yamaichi begin to suspect one another of the raids. Resentment simmers for the next decade.
- **WC009:** Razivir directs Kriegsdiener cultists, with Kyzerian support, in a covert raid on Britonica's central Isle à Von to excavate and retrieve the remains of a mythological superweapon. The retrieval succeeds, but Britonican naval forces inflict heavy losses on the retreating Kyzerians. Later, all members of the expedition suffer from a terrible disease soon known as "The Curse." Examination of the retrieved artifacts reveals that its power source still radiates energy of an unknown origin, but that its output is too low to be of any practical use. Early attempts to replicate or weaponize the power source fail. In response to the Zeehond Expedition, Grand Britony undertakes new offensives on the Rozenian front, using their superior mobility in a series of lightning strikes.
- **WC010, Early Spring:** Engineering assistant Hans Nacktgeborn starts tinkering with the joints and power couplings of the excavated Isle à Von artifacts, building a tiny model of a walking tank in his spare time. Within three months, New Kyzeria has produced functioning prototypes of the A1 Sturmlaufers. Parts are labeled as "tank turret/head top parts" as misdirection, which leads to the nickname "tankhead."

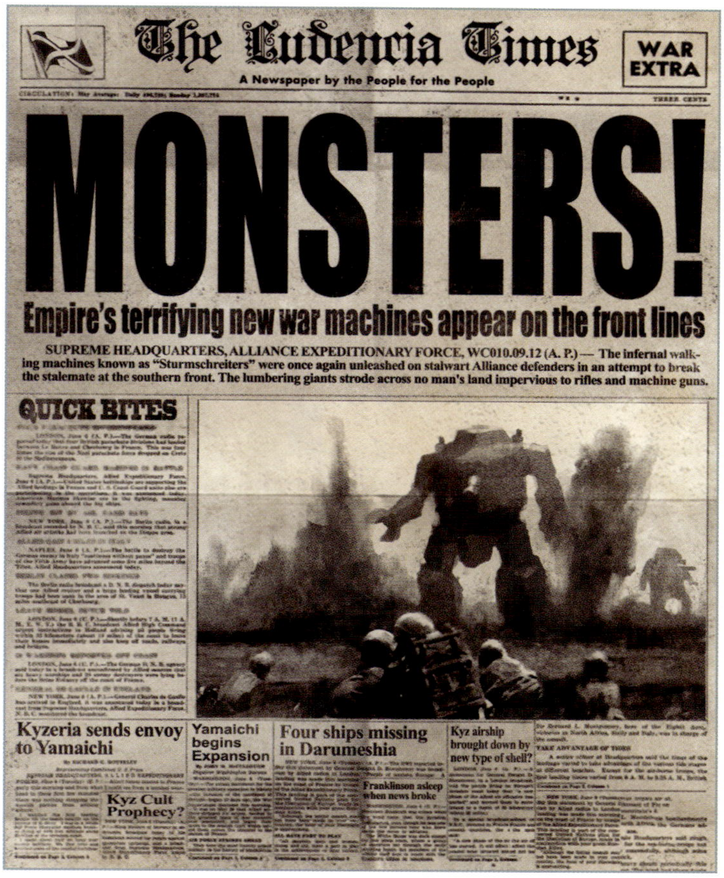
The Ludencia Times

A Newspaper by the People for the People

WAR EXTRA

MONSTERS!

Empire's terrifying new war machines appear on the front lines

SUPREME HEADQUARTERS, ALLIANCE EXPEDITIONARY FORCE, WC010.09.12 (A. P.)— The infernal walking machines known as "Sturmschreiters" were once again unleashed on stalwart Alliance defenders in an attempt to break the stalemate at the southern front. The lumbering giants strode across no man's land impervious to rifles and machine guns.

QUICK BITES

Kyzeria sends envoy to Yamaichi

Yamaichi begins Expansion

Kyz Cult Prophecy?

Four ships missing in Darumeshia

Franklinson asleep when news broke

Kyz airship brought down by new type of shell?

- **WC010, Summer:** The first deployment of an A7V Sturmschreiter breaks the defensive line on the Southern Front between New Kyzeria and Grand Britony, moving the front for the first time in more than seven years. These powerful new weapons of war allow the Kyzerian Empire to push forwards on the Southern Front at a steady pace, never allowing the Alliance to set up adequate defenses to halt their advances. Only during fall does the boggy ground stop them for the rest of the year.
- **WC011, Spring:** Kaiser Walmir VII ratifies a nonaggression pact with the Yamaichi by providing them with tankhead technology in the hopes of creating an impenetrable buffer between Liberterra's western coast and New Kyzeria's eastern holdings.
- **WC011, Summer:** The tankhead Mark I is deployed for the first time by the Alliance. Tankhead-on-tankhead warfare becomes a reality. The first gyrojet cannon is tested soon after as a way to compensate for recoil on tankhead weaponry. It would soon become the standard weapon equipped on tankheads.
- **WC011, Fall:** Razivir is accused of using occult cult rituals and curses to drain the life from the dying Kaiser. When Kaiser Walmir VII eventually passes due to illness, an angry mob murders Razivir. Although Razivir is stabbed, decapitated, and burned, witnesses claim his head is still laughing when it is shot out of a cannon.

TIMELINE (CONTINUED)

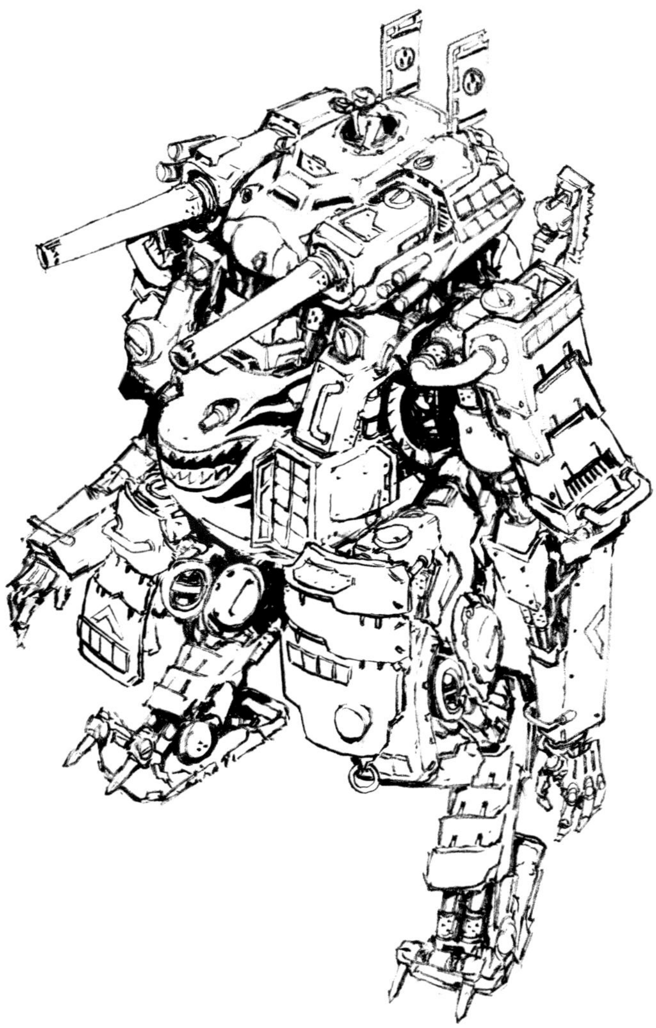

- **WC012, Ceasefire:** With the Kyzerian Empire still mourning the death of its monarch, the Empire's aggressive advance grinds to a halt on all fronts. Political attention turns inward. Vicious competition for guardianship of the twin heirs is waged. Whoever is appointed will be made Regent and de facto leader of the Kyzerian Empire until the twins reach maturity. This eventually results in a strong council of nine guardians who spend more time keeping each other's influence in check than performing their actual duties. With their political leadership in shambles, New Kyzeria's generals fear a coordinated Alliance counterattack and negotiate an ad hoc ceasefire. Both sides agree to temporarily stop production of heavy military vehicles, including large tankheads.
- **WC013:** Vaimund II and Vailinde IV, young twin heirs of the Kaiser, become more and more involved in New Kyzeria's political and military matters despite being only ten years old. Their strong personalities overwhelm their council of stewards. The Kyzerian military begins making plans to not only reclaim lost territories but also to conquer new ones. Those of the Kriegsdiener cult in the court whisper that the twins are Razivir's children in truth, reincarnated prophets of their god of war.
- **WC014, Fall:** A new wave of raids by unknown forces in the Darumeshia Archipelago, reminiscent of those of WC008, brings old resentments to a boil. Localized conflicts erupt on half a dozen islands between workers of Britonican and Yamaichi origins. The Brit government tries to shift blame toward the Kyzerians, claiming they are secretly using unknown amphibious tankheads to bring the Alliance and Dominion into conflict, but the situation continues to escalate.
- **WC014, The Endyear Conflict:** The northeastern Rozenian province of Suoros denies the Kyzerian Empire access to build military checkpoints and defense lines that the Empire considers essential for national security. After the fourth denial of these requests, on the last day of the year WC014, the Kyzerians invade Rozenia. Despite their superior forces and equipment, the harsh winter and grimly determined Rozenian opposition costs the invaders dearly, forcing the Kyzerians to retreat.
- **WC015, Early Spring:** The Kyzerian Empire launches expeditions in force into less defended areas of the stalemated Southern Front, scouting out possible routes of advance. Falsely believing the Alliance to be cowardly and war shy, many instead run into patrols eager to engage. Diplomatic resolutions to these skirmishes are stonewalled by the twin Kaisers and war is officially declared once more in the third month of WC015.
- **WC015, Summer:** A small group of Britonican ex-military officers and navy personnel residing in a Yamaichi-held Royal Bay harbor in the Darumesha Archipelago tries to covertly sink a Kyzerian destroyer anchored in the harbor. They are successful but one of their number is wounded during the raid and is arrested by Dominion law enforcement. While the saboteurs were acting independently, the debacle spirals out of control and escalates into an official war declaration between the Yamaichi and the Alliance.
- **WC015, Winter:** The twin heirs to the Kyzerian Empire (now twelve years old) visit Narakura, the historic Yamaichi capital, to reinforce the bonds between the two allies against the Britonica Alliance. The twins are shown the historic throne room of Yamaichi, where the throne of their mythical emperor has remained empty since his reign, signifying that the prefectures of Yamaichi rule themselves as equals. The twins seat themselves on the oversized throne and declare themselves fit to rule as twin emperors of two nations. This shocks the Yamaichi diplomats beyond speech and the Kyzerians retreat to their quarters. When word of this scandal leaks to the general population, riots erupt in the Yamaichi capital and the Kyzerians flee to their naval transport.
- **WC016, Spring:** Riots continue throughout Yamaichi lands. The Yamaichi government first pleads and then demands that the Kyzerian heirs make formal apologies for their actions. The first two envoys are met with derision; the third is executed on the grounds that he has impugned the dignity and wisdom of the twins. Yamaichi officially declares war on the Kyzerian Empire. The vast majority of the world is now at war. A number of Kyzerian naval ships anchored in Yamaichi harbors are sunk by combined military and civilian forces.
- **WC017:** Deployment of the first walking battleship, the *Dreadnought*.

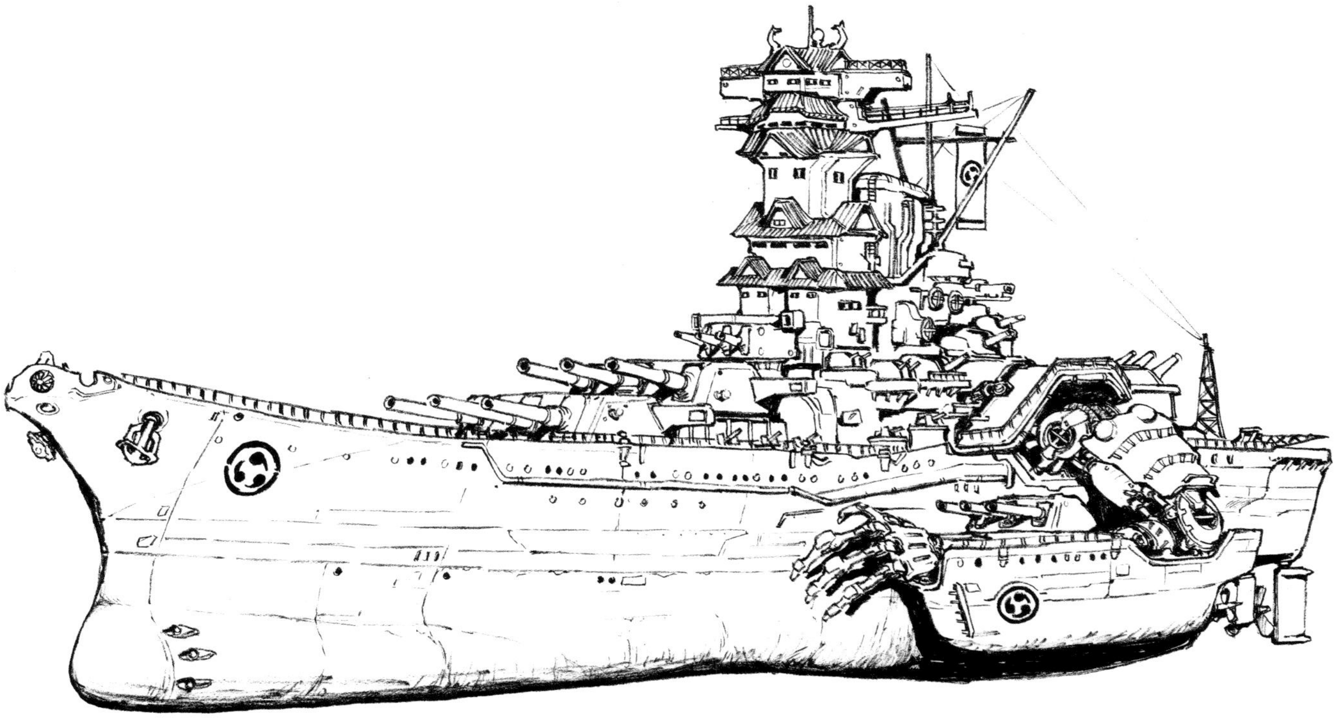

- **WC019, Momil assault:** Utilizing a pair of armored trains modified to run on Rozenian-gauge rails, Kyzerians break through the defenses around the town of Momil in the Honeder province of Rozenia, one of the last truly neutral zones in the region. The assault catches the local defenders off guard, as they have failed to install prepared anti-train "tulip" barricades. The Kyzerians capture and occupy the rest of the region in the same year.
- **WC021–022, Gregale Islands campaign:** Yamaichi forces occupy a chain of remote islands in the northern Gregale Sea, hoping to use them to exert control over both Alliance and Kyzerian naval movement in the northern Libercian Ocean. The extreme weather and terrain make reconnaissance nearly impossible and both the Kyzerians and Liberterrans end up shelling the islands, which had already been abandoned during their counteroffensive.
- **WC022–026, Geannoog Islands occupation:** The island group north of Rozenia, so far seen as being of little strategic value and thus spared heavy fighting, slowly has its defenders assigned to more pressing areas. For the following two years, the islands are occasionally shelled by Kyzerians who still operate under the assumption that there are Alliance forces stationed on the islands. Late in WC025, the Kyzerians finally make landfall on Texsey, the easternmost island, and find it undefended. The rest of the islands are occupied during the first two months of WC026 before Alliance naval troops recapture them later the same year.
- **WC025, The Futile Year:** No progress of real consequence made by any army or force. A year of terrible casualties and destruction, with nothing to show for it.
- **WC028–029, First battle of Grigomos:** Kyzerian forces change the focus of their might and make rapid headway towards the Rozenian capital of Grigomos, but are stopped in their tracks by an Alliance counteroffensive. Kyzerian forces are locked into an entrenched struggle for control for over a year. It will be considered the end of the last effective Kyzerian lightning offensive. The Rozenian capital will be moved away from Grigomos and the front lines after this attack.
- **WC033, Battle of Kerskdyne:** A Kyzerian breakthrough on the Southern Front leaves most of the Alliance forces there surrounded and outnumbered. The fighting retreat and naval evacuation will become the stuff of legends.
- **WC034:** First official Jethead combat deployment. The introduction of these nimble air units radically changes the way aerial combat is fought. When jetheads took over the sky the need for accurate anti-air weaponry became paramount.
- **WC036, Second battle and siege of Grigomos:** In the summer, Kyzerians advance on the old Rozenian capital of Grigomos using new routes through the Novgryov province. The siege will last nearly a thousand days.
- **WC038, Battle of Kraitai Island:** A large Yamaichi airborne assault on the largest island in the Levantean Sea. The Kyzerian defenders, most of them stationed on the island as a form of punishment, put up a better fight than expected. In the end, however, a fatal error results in loss of the island's airport to Yamaichi hands. The Dominion is able to fly in reinforcements and the island is soon captured.
- **WC039, End of the siege of Grigomos:** The besieging Kyzerian forces finally break Grigomos and capture the old Rozenian capital city.

TIMELINE (CONTINUED)

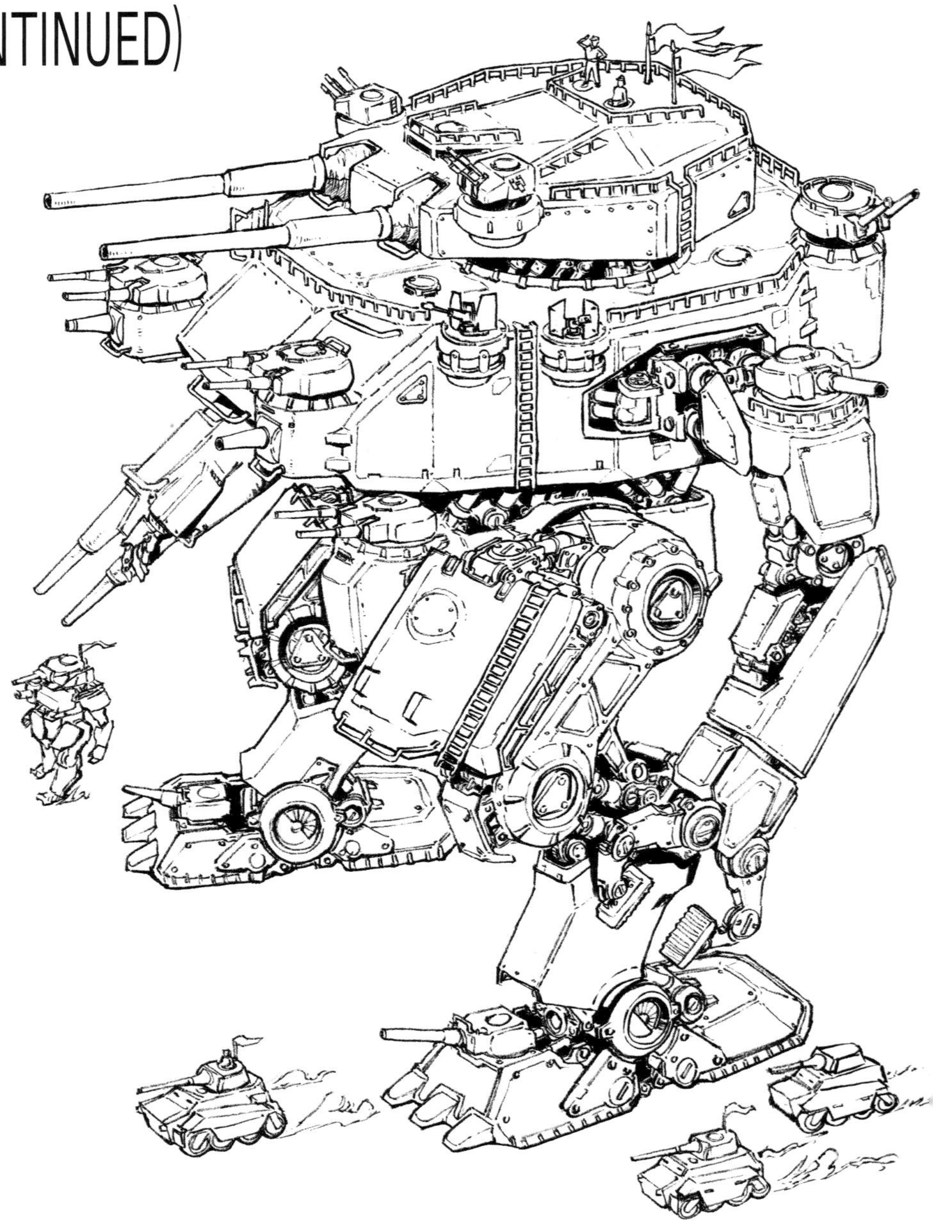

- **WC041, Raid on Asakoro Docks:** A Kyzerian raid catches a large portion of the aging Yamaichi Darumeshian fleet docked and off guard. Over half of its assets are taken out of action or destroyed. The raid marks a turning point in the Darumeshian area, after which all naval forces will engage each other more actively rather than passively defending their respective regions.
- **WC041, First successful reproduction of Ijzerhart Engine:** Using materials discovered beneath Grigomos, the Kyzerians reproduce a stable reverse-engineered Ijzerhart Engine, named the Schwarzner Reaktor. Limited access to the rare fuel it requires hampers the project and the engine can only be operated for a very short time.
- **WC043, Battle for Scala Delta:** An amphibious Alliance assault on the Rozenian Scala river delta attempts to open up passage to the deep water docks of the port city Voetgooi. Delays and miraculous foresight by the twin kaisers allow the Empire to reinforce their defenses on the islands in the delta prior to the attack. A muddy, month-long battle ends with the Alliance victorious.
- **WC044, Hur Initiative:** In one of the most daring naval raids in history, a small Yamaichi force of volunteers, supported by 00/A2 Urashimataro tankheads, destroys important Kyzerian shiphead maintenance facilities on the Gulf of Id. From this point forward, Kyzerian shipheads will have to return to docks on the Nehalean Channel for repairs.
- **WC045, Attack on the Lumpe Line:** Kyzerian forces push back hard on the Western Rozenian front in an attempt to stop Alliance forces from recapturing Grigomos. The Alliance is caught almost entirely off guard and the harsh winter weather cuts off their supply lines. Some of the war's most brutal fighting ensues, with close-quarters fighting in heavily forested areas and extreme cold claiming over 100,000 lives.
- **WC046, First voltaic energy projection weapon test:** Utilizing their Schwarzner Reaktor, the Kyzerian Empire successfully test-fires a Voltaic photon discharge weapon for the first time.
- **WC046, Nuhihi Seal Islands campaign:** As the Alliance and Dominion navies become more entangled in the Libercian Ocean, the Kyzerians try to capitalize on this distraction and launch a large naval offensive to neutralize or capture the Yamaichi forces' Nuhihi Seal Islands. The campaign is marked by heavy fighting and titanic clashes between the mighty Kyzerian and Yamaichi shipheads, including the *Gneisenau*, *Graf Zeppelin*, *Yamato* and *Musashi*.

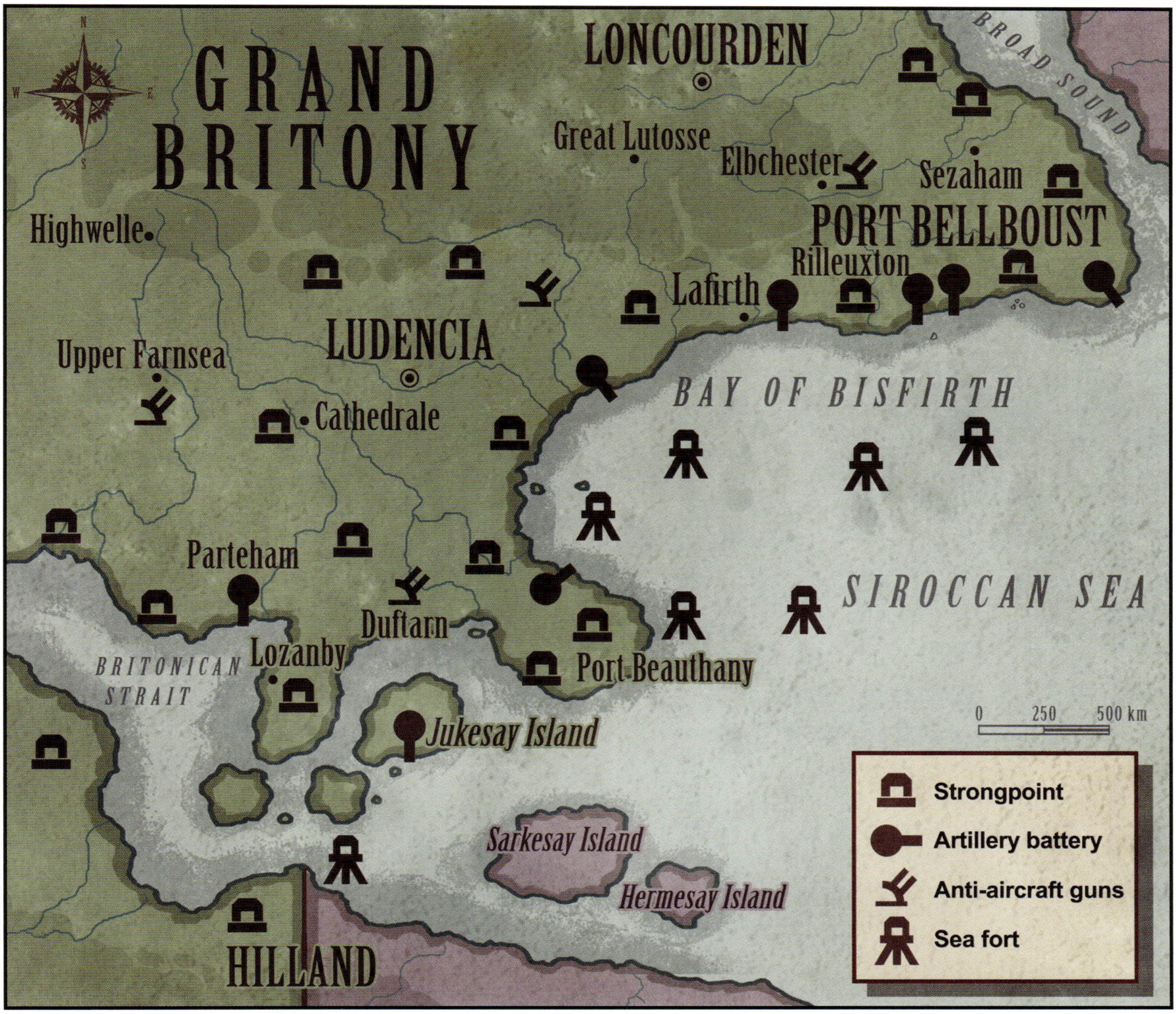

A map illustrating the extent of defenses around Ludencia and the southern coast of Grand Britony by the end of the war in WC049. Note that most sea forts mainly functioned as anti-air emplacements, preventing enemy jetheads from dropping naval mines into essential Alliance shipping lanes.

- **WC047, Yearbreak Day; Ratte unleashed:** On the first day of the year, the Kyzerians unleash the unfinished Ratte prototype on the Southern Front. At this time, it is the heaviest land vehicle in existence. The battle to destroy the rampaging giant takes two full days.
- **WC047, Spring; Operation Oberherr:** The Kyzerians launch a combined naval and airborne invasion on Liberterra in an attempt to disrupt Alliance production capability. The invasion is a short but hard-fought meat grinder for both sides, but the Imperial force is particularly weakened and retreats.
- **WC047, Fall; The Novgryov betrayal:** The West Rozenian province of Novgryov elects Sergey Arurader as their new Chief Magistrate, who immediately shifts allegiance to East Rozenia, marking the loss of the Bol-Obez Tank Kong factory.
- **WC048, The Godhead Project uncovered:** Yamaichi spies, taking advantage of the disarray following the Kyzerian Empire's failed Operation Oberherr, confirm that the Empire is working on a superweapon using relics found below Grigomos. This information is leaked to Alliance Intelligence. It is uncertain if this was a deliberate action or done by a rogue element. Dominion Intelligence generally did not share intel with the other great powers even if it would benefit the Dominion to do so.
- **WC049, Grandmaster Gamble:** The Kyzerian Empire completes its Godhead superweapon much sooner than the other factions expect. Britonica and Yamaichi forge a desperate alliance and launch a joint strike under the codename "Grandmaster Gamble." While eventually successful in deactivating the Godhead, the operation leaves the Alliance and Dominion forces hugely overextended in enemy territory. The Kyzerian heartland is now flooded with desperate enemy forces. With all the combatants vulnerable, a hasty ceasefire is agreed to by all sides, despite continued skirmish. Subsequent negotiations will lead to the Taubensee Agreement, bringing an end to the Neverwar, one day before its 50th year.

GLOSSARY

A

airship
Large, slow-moving aircraft using lighter-than-air gas to fly. Until the latter half of the Neverwar, this was the only mode of air transport and combat in general use.

Alcaban
Large continent south of Grand Britony.

armored car
Wheeled armored vehicle using articulated wheels to steer.

B

Britonica Alliance
[a.k.a: the Alliance]
A democratic union of countries to the west of New Kyzeria that includes Grand Britony, Liberterra and, later, West Rozenia.

Brit
Descriptor referring to the nation of Grand Britony and its people.

BWC (Before War Century)
A common notation to indicate years before the start of the Neverwar. Adopted after WC was popularized globally, replacing many widely-differing methods of accounting for time.

C

Ceasefire period
A ceasefire treaty was signed by all warring parties following the death of Kaiser Walmir VII in WC012. Officially, hostilities were frozen until the end of WC014, but small scale conflicts took place throughout the period.

D

Darumeshia Archipelago
Group of islands to the west of Alcaban and east of the Yamaichi Dominion.

E

Earde
The world in which the Neverwar takes place.

engineering vehicle
Vehicle used for the maintenance and repair of tankheads and other vehicles, as well as for military construction and demolition purposes.

F

Free Rozenia
Rozenian protectorate state on the east coast of Alcaban. It was the only Rozenian territory not directly involved in the war and became a central hub for international refugees.

G

Godhead Project (Gotteshaupt Projekt)
Secret Kyzerian project to revive an ancient weapon found under the old Rozenian capital of Grigomos.

Gondo
Small nation located between the Siroccan and Levanthean Seas that relied on ancient treaties and pacts with both Kyzeria and Yamaichi to maintain a precarious position of neutrality during the Neverwar.

Grand Britony
Country located between Rozenia and Liberterra and initiating figurehead of the Britonica Alliance. Known for its tea and pastries.

Grigomos
Old capital of Rozenia, captured by the Kyzerians during the war. Source of much of the material used to construct the Godhead.

gyrojet
Firearm system using a miniature rocket engine to propel a projectile after firing. Following its invention, most heavy weaponry on Earde has been designed with a gyrojet system of some sort.

H

Hilland

Semi-independent holdings in the north of Alcaban split between Kyzeria and Grand Britony.

howitzer

An artillery weapon that falls between a cannon and a mortar.

Hyakkoku

Northwest province of the Yamaichi Dominion. Shares a large border with Kyzeria.

I

Ijzerhart
Artifact generators seemingly from prehistoric times. Recovered by the Kyzerians along with other Isle à Von relics during the Zeehond Expedition and by the Yamaichi in the Allentii mountain range.

Isle à Von
An island located in the middle of Grand Britony. Site of the Zeehond Expedition.

J

jethead
Flying machine propelled by jet or turbine engines. Utilizes tankhead technology for articulation and agility.

K

Kaiser
The official title for the monarch and emperor of the Kyzerian Empire. During the Neverwar the position was held by Walmir VII until his death in the fall of WC011, when his twin heirs Vaimund II and Vailinde IV inherited the title together.

Kriegsdiener
Religious cult originating in the Kyzerian nation, most famously headed by Razivir.

Kyz
Colloquial adjective for both the nation of New Kyzeria and its people.

Kyzeria
[a.k.a: Greater Kyzerian Empire, Kyzerian Empire, New Kyzeria, The Empire.]
Large nation unified under a ruling dynasty of Kaisers. Creators of the first tankhead.

L

Liberterra
Large nation to the west of Grand Britony. Largest member of the Britonica Alliance in terms of population and area.

Ludencia
Capital of Grand Britony.

Lumpe Line
Extensive line of fortifications built on the border between Rozenia and Britony, intended to prevent Kyzerian forces from reaching Britony by going through Rozenia.

M

mortarhead
Tankhead with a large caliber weapon intended for indirect fire.

N

Neverwar
[a.k.a: Never-ending War]
A commonly used name for the first truly global conflict. The start of the Neverwar marked the beginning of the War Century calendar, WC001. It ended in WC049.

Nolburg Plan
Kyzerian military operation during the fall of WC005 that involved moving troops through Rozenia to circumvent Alliance defenses.

O

Operation Oberherr
Failed amphibious landing on the western Liberterran shores by the Kyzerian Empire.

R

Reginwald (Sir George Reginwald)
Pioneering inventor of the swing arm suspension and gyrating weapon system, better known as turrets today. Credited with the creation of the first modern military tank.

Rozenia
Collection of semi-independent states between Kyzeria and the Britonica mainland. Originally known for its pastoral landscapes and historic cities, it became a focal point during the early years of the Neverwar and was split into East and West Rozenia for most of the war.

Razivir
A famous Kyzerian mystic, leader of the Kriegsdiener cult and personal adviser to Kaiser Walmir VII.

S

Southern Front
The infamous trench lines separating Grand Britony and New Kyzeria. Located between Rozenia and the Siroccan Sea.

shiphead
Walking battleships capable of traversing land for a limited amount of time.

Schwarzner Reaktor
The first fully-operational reverse-engineered Ijzerhart generator. It was named after the head engineer of the Gotteshaupt Projekt, Karl Otto Schwarzner.

subhead
A vehicle intended for underwater use that uses tankhead-style articulation to improve its maneuverability in tight areas as well as its combat flexibility.

T

tank
Wheeled armored vehicle using differential steering.

tankhead
Armored vehicle using two or more legs as its primary mode of mobility.

V

voltaic matter
A transitional state of matter between matter and energy which acts much like slow electricity.

volton weapon
A type of weapon that produces super high density voltaic matter which transitions into light, often described as a beam. Also known as a "voltaic photon discharge" weapon.

W

WC (War Century)
The calendar system commonly in use following the Neverwar. Originally intended to count the hundred years of war prophesied by the Kriegsdiener cult. Even though the war "century" lasted a day under 50 years, the notation has yet to be replaced.

Y

Yamaichi Dominion
Nation consisting of 100 prefectures originally unified under the shogun warlord Hideto Kamijima. The ruling position has since been symbolically left empty and rulership of the Dominion has rested in a rotating council of regents representing the various prefectures.

Z

Zeehond Expedition
Kyzerian covert operation to Britonica's central Isle à Von in WC009, during which the Empire acquired artifacts that would lead to the development of tankheads and the first step towards creation of the Godhead.

SKETCHBOOK

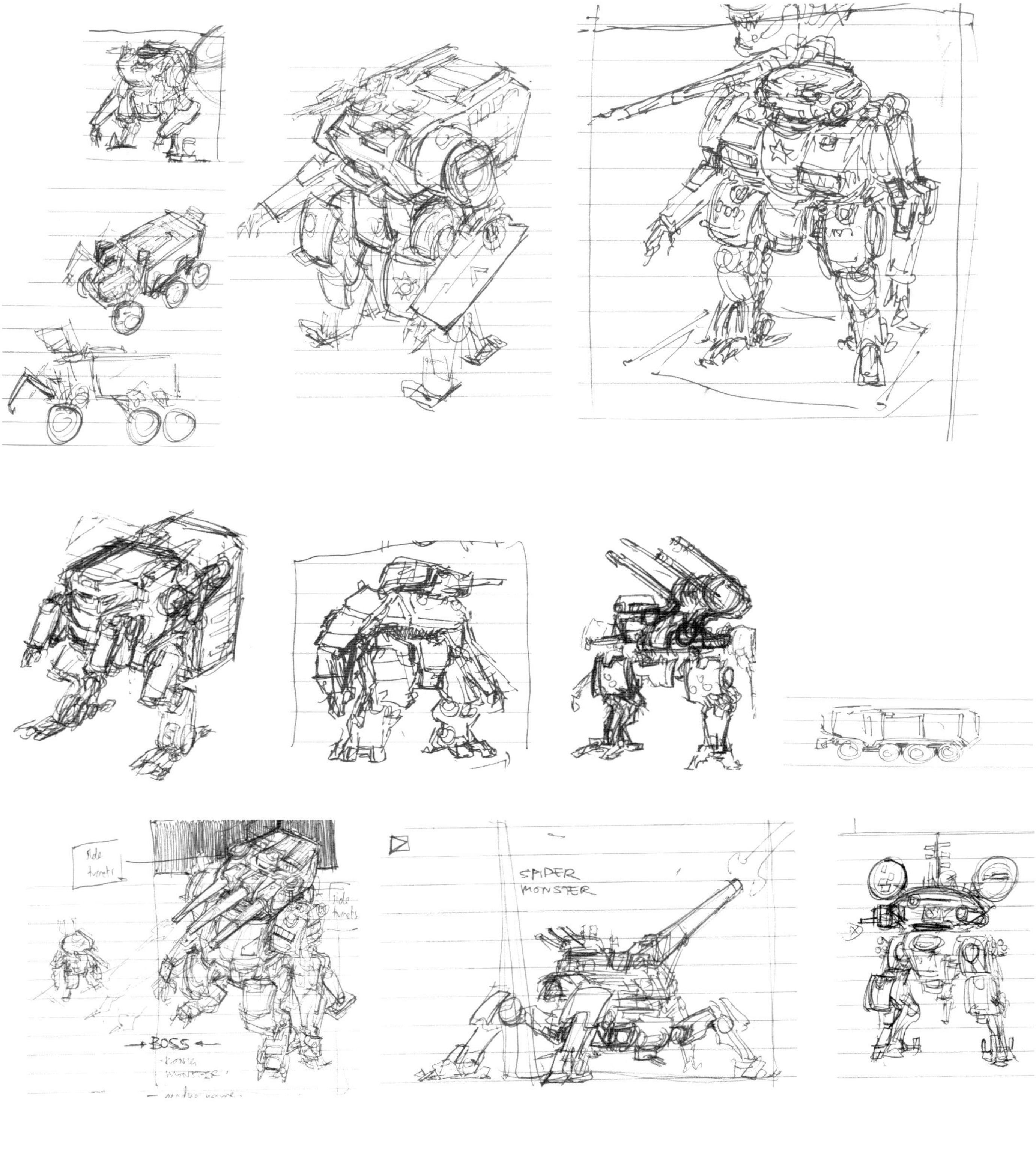

Some of the earliest sketches done for TANKHEAD. The "Boss" sketch at the bottom left would eventually turn into the Ratte many years later.

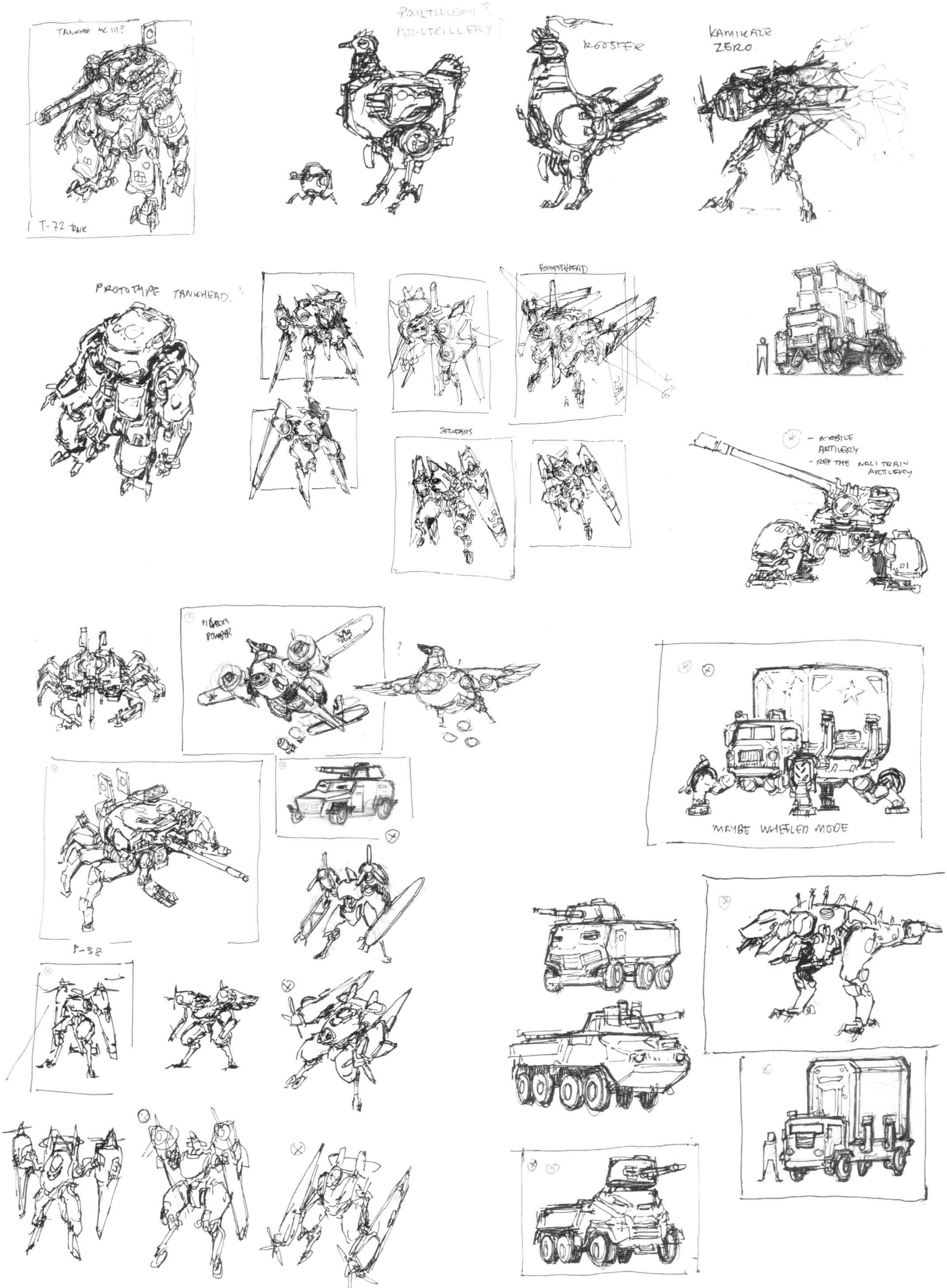

Thumbnail sketches for various units.

SKETCHBOOK

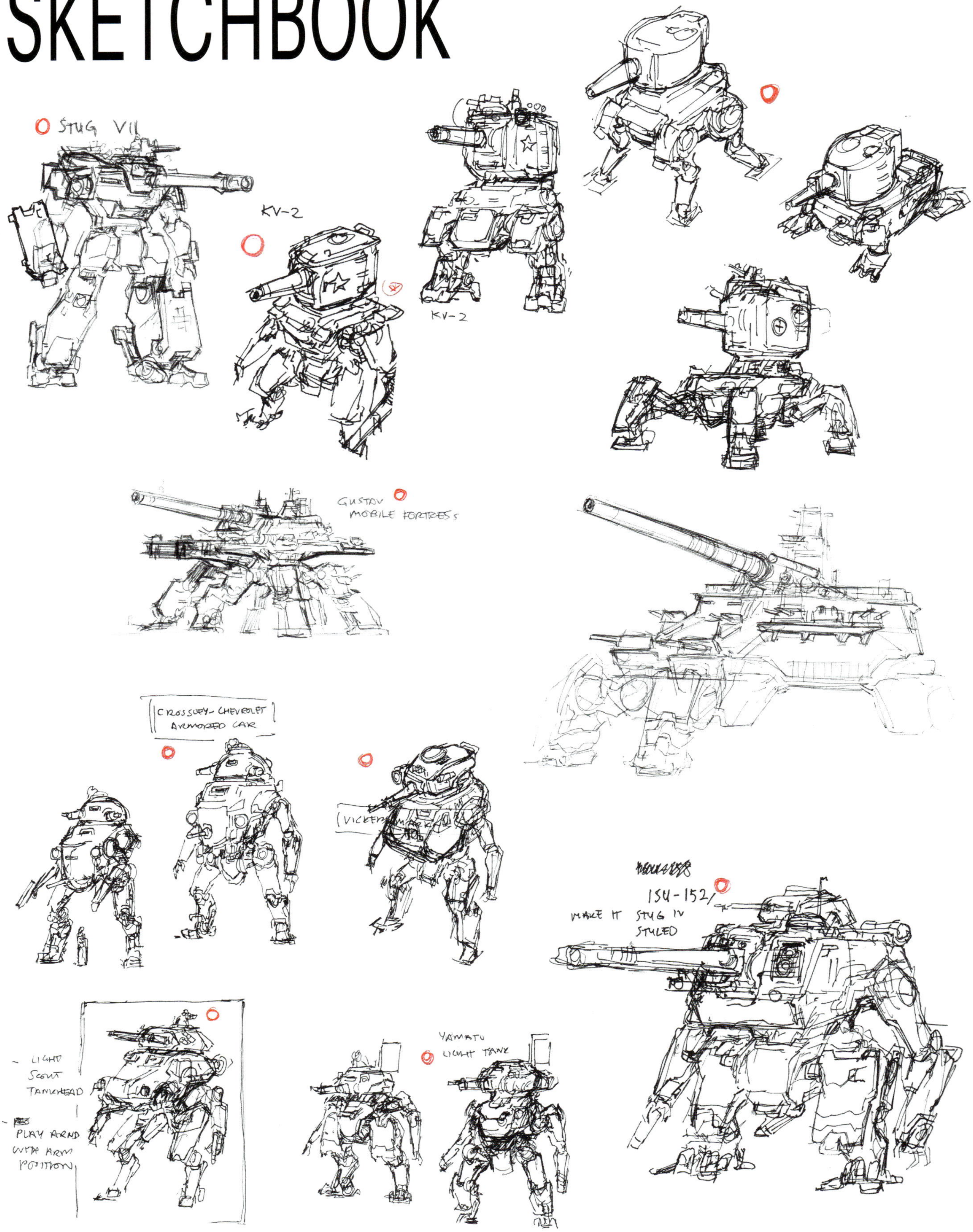

Almost every design in TANKHEAD started out as one of these tiny ink sketches. It's one of my favorite parts of the process where I can stay rough and loose while I try to get my idea on paper.

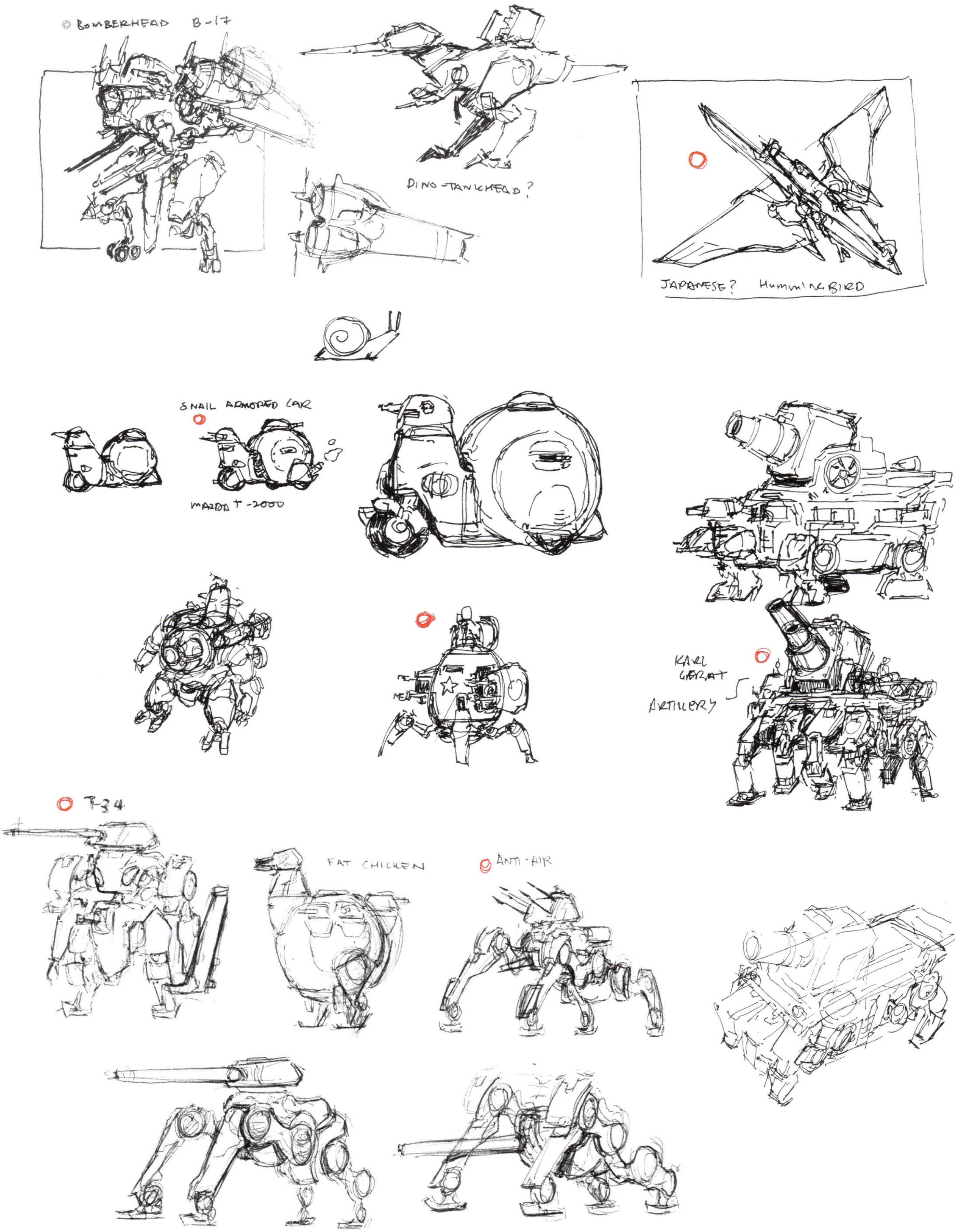
BOMBERHEAD B-17
DINO-TANKHEAD?
JAPANESE? HUMMINGBIRD
SNAIL ARMORED CAR
MAZDA T-2000
KARL GERAT
ARTILLERY
T-34
FAT CHICKEN
ANTI-AIR

SKETCHBOOK

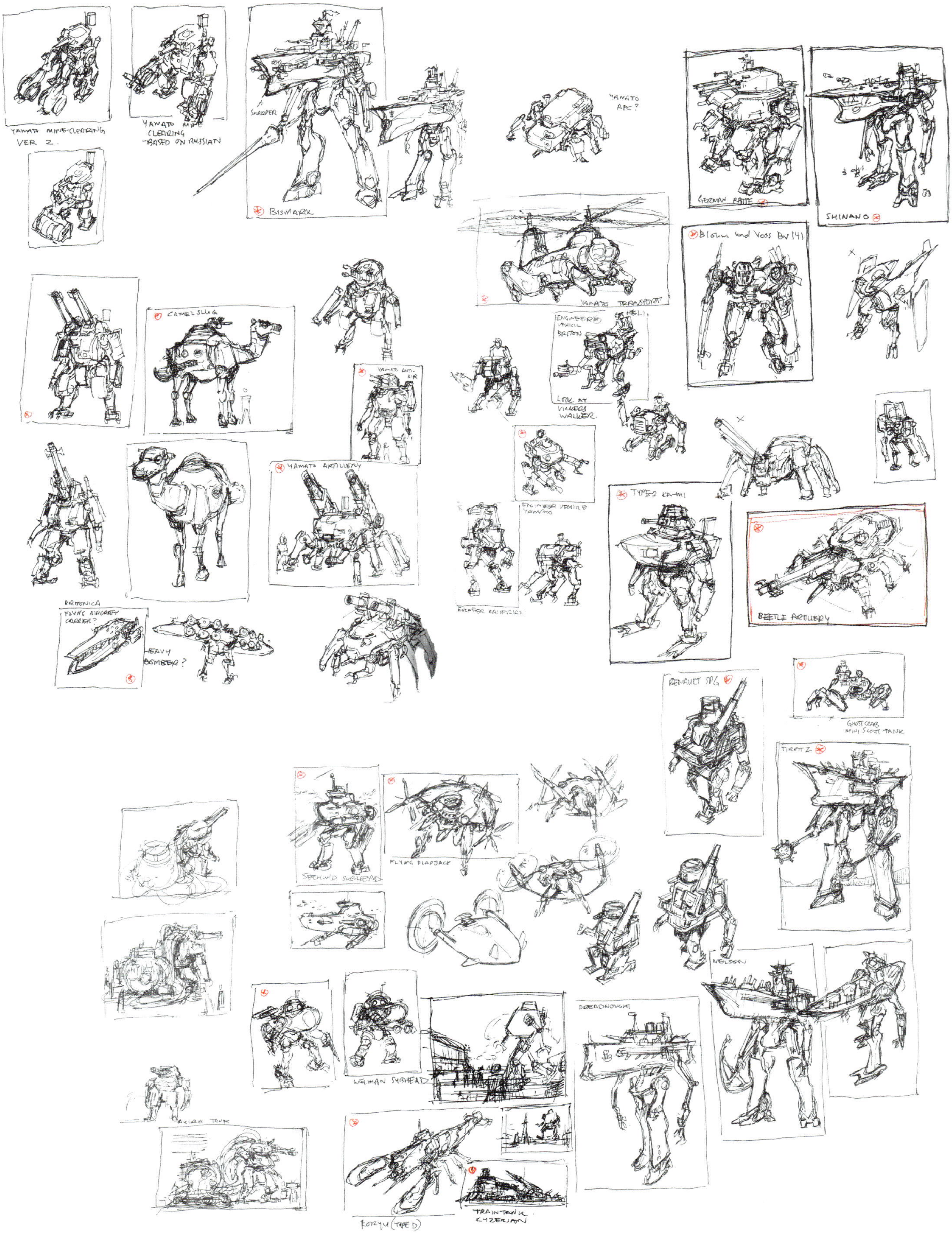

YAMATO MINE-CLEARING VER 2.
YAMATO MINE CLEARING -BASED ON RUSSIAN
SHARPER
BISMARK
YAMATO APC?
GERMAN RATTE
SHINANO
YAMATO TRANSPORT HELI
Blohm und Voss Bv 141
CAMELSLUG
YAMATO ANTI-AIR
LOOK AT VICKERS WALKER
YAMATO ARTILLERY
TYPE 2 KA-MI
BEETLE ARTILLERY
HEAVY BOMBER?
RENAULT SPG
TIRPITZ
FLYING FLAPJACK
NELSON
DREADNOUGHT
AKIRA TANK

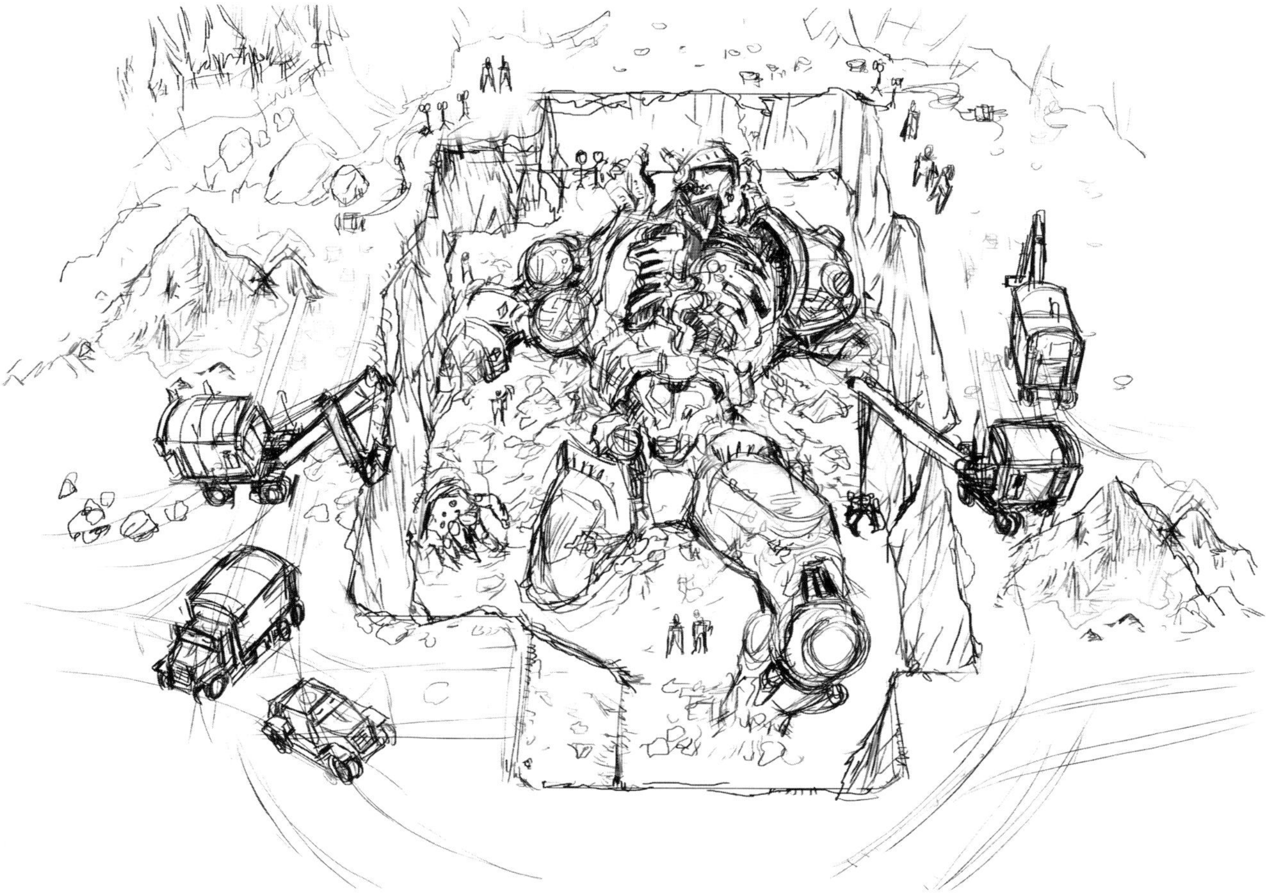

Early ideas for the Zeehond Expedition.

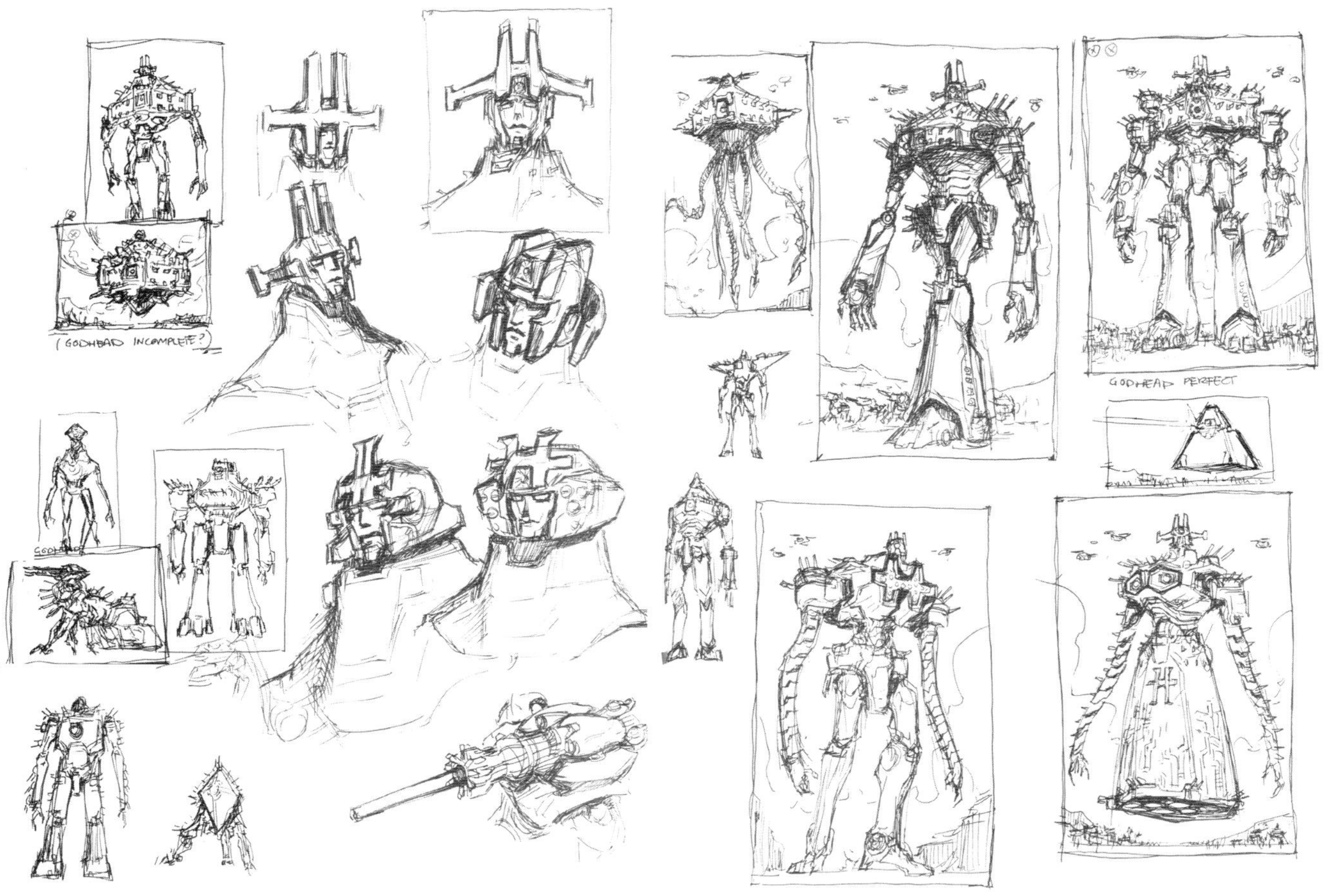

Thumbnail sketches exploring ideas for the Godhead.

SKETCHBOOK

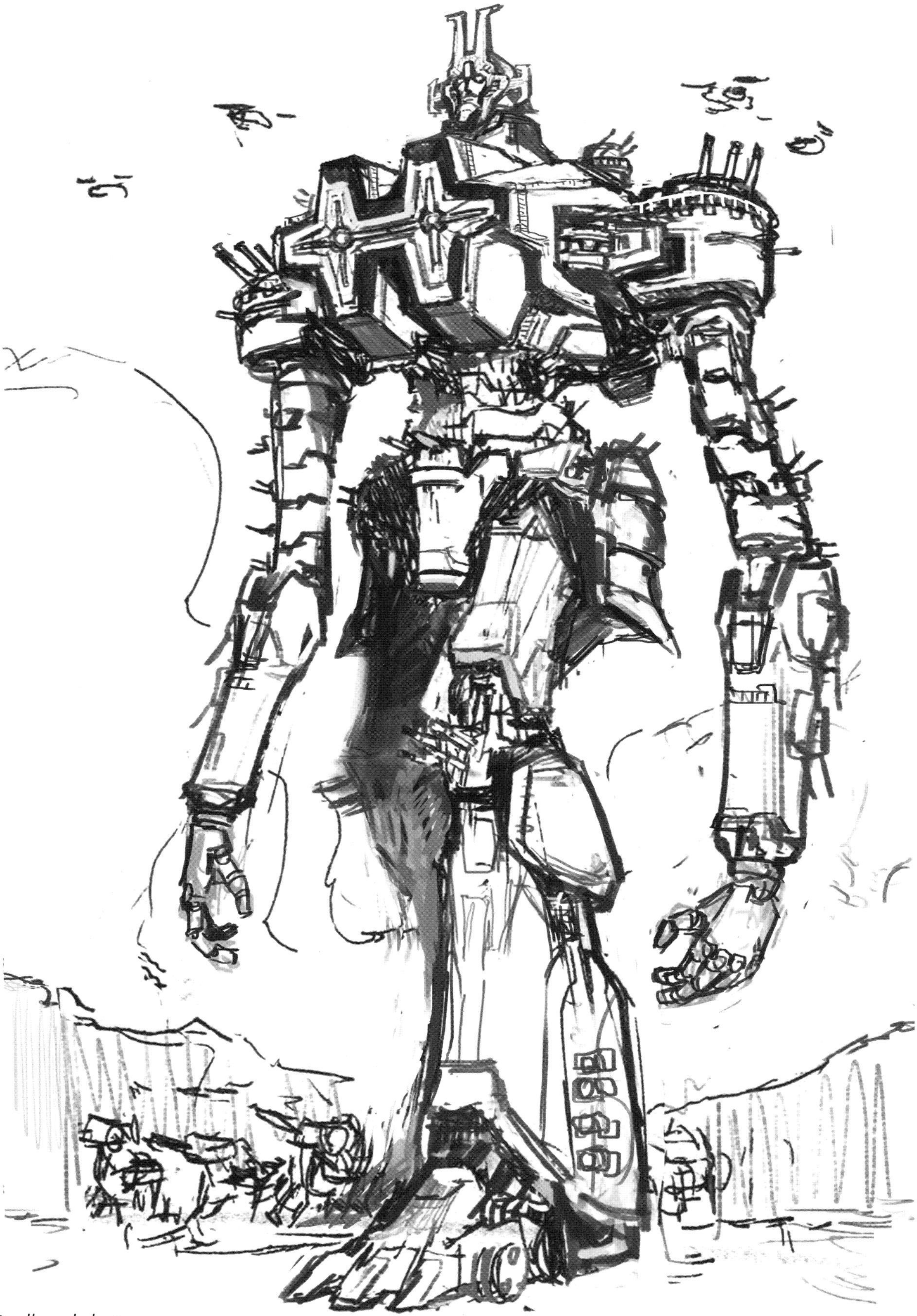

Early Godhead.design.

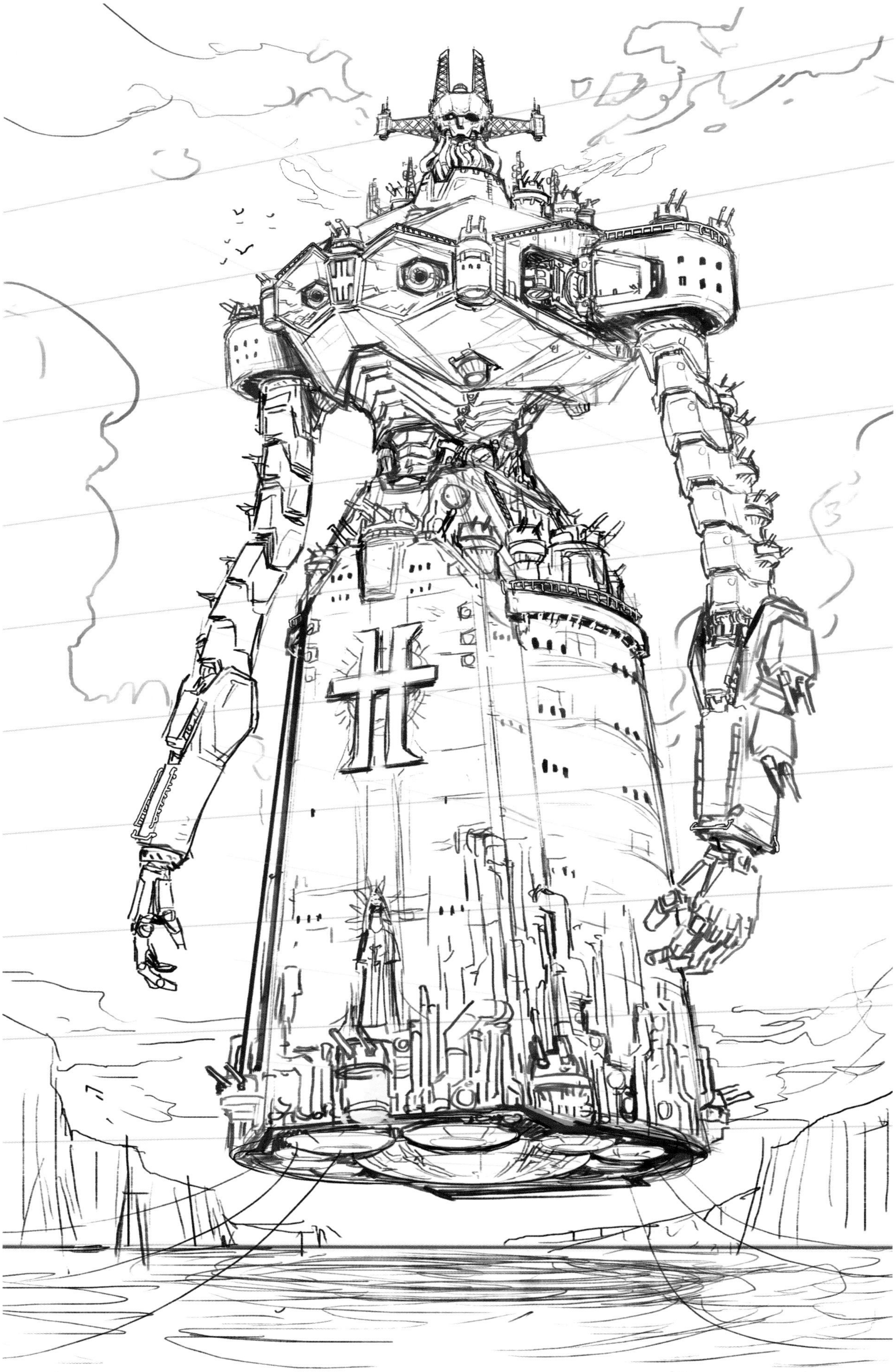

Final design for the Godhead.

タンクヘッド

TANKHEAD®

MECHANICAL ENCYCLOPEDIA ARTBOOK

CREATED BY

EMERSON TUNG

UDON STAFF

CHIEF OF OPERATIONS
ERIK KO

DIRECTOR OF PUBLISHING
MATT MOYLAN

DIRECTOR OF OPERATIONS
MARSHALL DILLON

VP OF BUSINESS DEVELOPMENT
CORY CASONI

DIRECTOR OF MARKETING
MEGAN MAIDEN

ASSOCIATE EDITOR
M. CHANDLER

JAPANESE LIAISONS
MEGUMI CUMMINGS

TANKHEAD: MECHANICAL ENCYCLOPEDIA ARTBOOK HC

www.UDONentertainment.com

Published by UDON Entertainment Corp.
118 Tower Hill Road, C1, PO Box 20008
Richmond Hill, Ontario, L4K 0K0, Canada

Second Printing: July 2025
ISBN: 978-1-77294-354-2

Printed in Malaysia